D1294335

Modern Physics for Engineers

Modern Physics for Engineers

Jasprit Singh

A Wiley-Interscience Publication
JOHN WILEY & SONS, INC.
New York • Chichester • Weinheim • Brisbane • Singapore • Toronto

This text is printed on acid-free paper. ♾

Published simultaneously in Canada.

Library of Congress Cataloging in Publication Data:

Singh, Jasprit.
Modern physics for engineers / by Jasprit Singh.
p. cm.
Includes index.
ISBN 0-471-33044-2 (alk. paper)
1. Physics. 2. Quantum theory. 3. Statistical mechanics.
4. Materials science. I. Title.
QC21.2.S473 1999
530'.024'62—dc21 98-48448
CIP

Printed in the United States of America

10 9 8 7 6 5 4 3 2 1

CONTENTS

8 SCATTERING AND COLLISIONS 266

9 SPECIAL THEORY OF RELATIVITY 319

PREFACE

Over the last few years there have been several important changes in the undergraduate curricula in both Engineering and Physics departments. In engineering schools there is an increased emphasis on design type courses which squeeze the time students have for *fundamental* courses. In physics departments there is an increased need to make a stronger connection between the material studied and modern technological applications. This book is motivated by these changes.

This book deals with important topics from the fields of quantum mechanics, statistical thermodynamics, and material science. It also presents a discussion of the special theory of relativity. Care is taken to discuss these topics not in a disjointed manner but with an intimate coupling. My own experience as a teacher is that applied science students are greatly motivated to learn basic and even esoteric concepts when these concepts are closely connected to applications from real-life technologies. This book strives to establish such connections wherever possible.

The material of the text can be covered in a single semester. The text is designed to be highly applied and each concept developed is followed with discussions of several applications in modern technology.

The emphasis in this text is not on tedious mathematical derivations for the solutions of various quantum problems. Instead we focus on the results and their physical implications. This approach allows us to cover several seemingly complex subjects which are of great importance to applied scientists.

I believe the level of the text is such that it can be taught in the physics, electrical engineering or material science departments of most schools. For physics students, the book may be suitable in the sophomore year, while in the engineering schools it may be used for seniors or even for entering graduate students.

I am extremely grateful to Greg Franklin, my editor, for his support and encouragement. He was able to get valuable input from a number of referees whose comments were most useful. I also wish to extend my gratitude to the following reviewers for their expert and valuable feedback: Professor Arthur Gossard of the Materials Department at the University of California, Santa Barbara; Professor Karl Hess of the Beckman Institute of Advanced Science and Technology at the University of Illinois; and Dr. Michael Stroscio of the Army Research Office.

The figures, typing, cover design, and formatting of this book were done by Teresa Singh, my wife. She also provided the support without which this book would not be possible.

JASPRIT SINGH
Ann Arbor, MI

INTRODUCTION

MODERN PHYSICS AND TECHNOLOGY

Modern technologies are changing our lives at an unprecedented pace. New materials, information technologies of communication and computation, and breakthroughs in medical technologies are bringing new opportunities to us. For students of applied physics and engineering these technologies offer both challenges and opportunities. On one hand, these developments offer solutions to complex problems—problems that seemed unsolvable a decade ago. On the other hand, the amount of knowledge needed to understand and contribute to new developments is also growing tremendously. This places a considerable burden on students.

Modern technologies seem so "magical" that they almost lull us into believing that one simply needs to wave a magic wand and new devices and systems will appear. It is easy to forget that these new inventions are based on well-established *fundamental* principles of physics derived through the application of the scientific method. Most of these principles have been known to humanity for only about a hundred years or less. Quantum mechanics, quantum statistics, theory of relativity, and structure of materials provide us the basis for modern technologies. Yet all of this understanding was developed in the twentieth century.

This book deals with the basic principles of modern physics that have led to revolutionary new technologies. Our focus is on the basic principles of quantum mechanics and material science and how these principles have been exploited to generate modern technologies. The emphasis in this text is not on tedious mathematical derivations for the solutions of various quantum problems. Instead we focus on the results and their physical implications. In Chapter 9 we present a treatment of the special theory of relativity along with some important applications.

In Figs. 1 and 2 we show an outline of some of the applications we link to the basic principles discussed in this text. As one can see, a number of different fields have benefited from modern physics. The connection between basic principles and applications is a very important one—one that is used by applied scientists as they go about inventing new technologies. It is important for students of applied physics and engineering to appreciate this connection.

I.1 Guidelines for the Instructor

This book can be used for engineering students as well as for physics majors. For engineering students there are two possible roles of this book, depending on the undergraduate curricula at a particular university. In some schools a modern physics course is offered in the physics department for undergraduate engineering students. This book should be particularly attractive for such a course, since there is a close

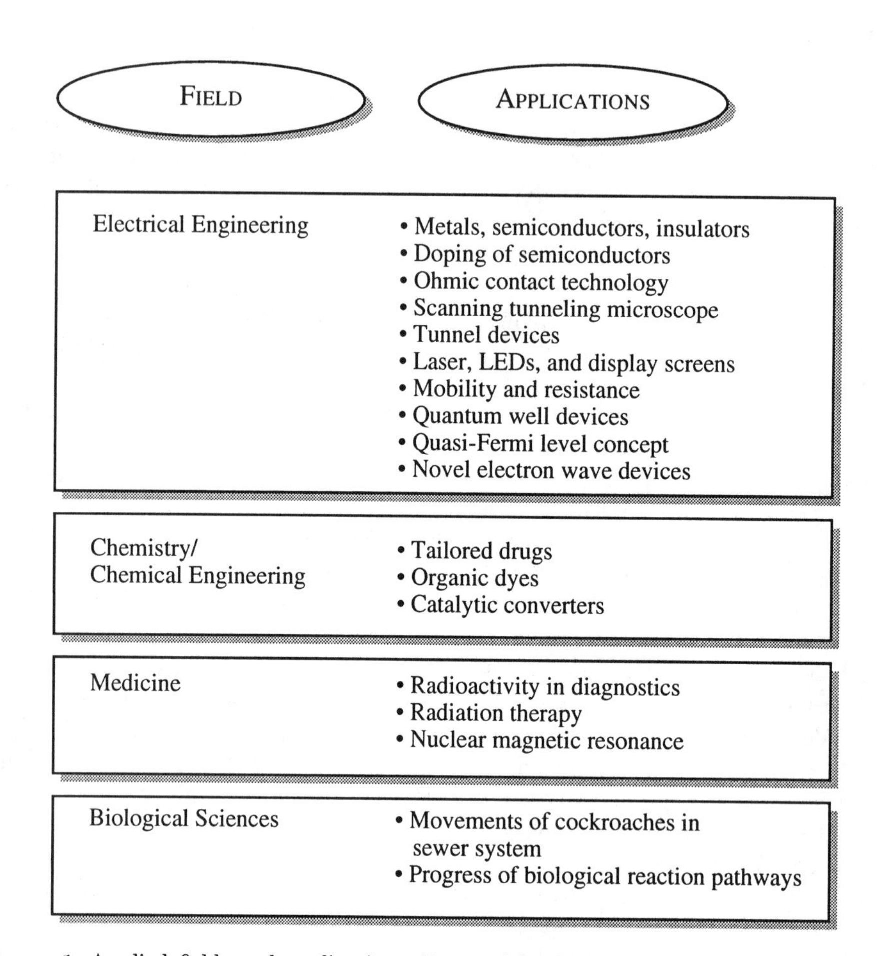

Figure 1: Applied fields and applications discussed in this textbook.

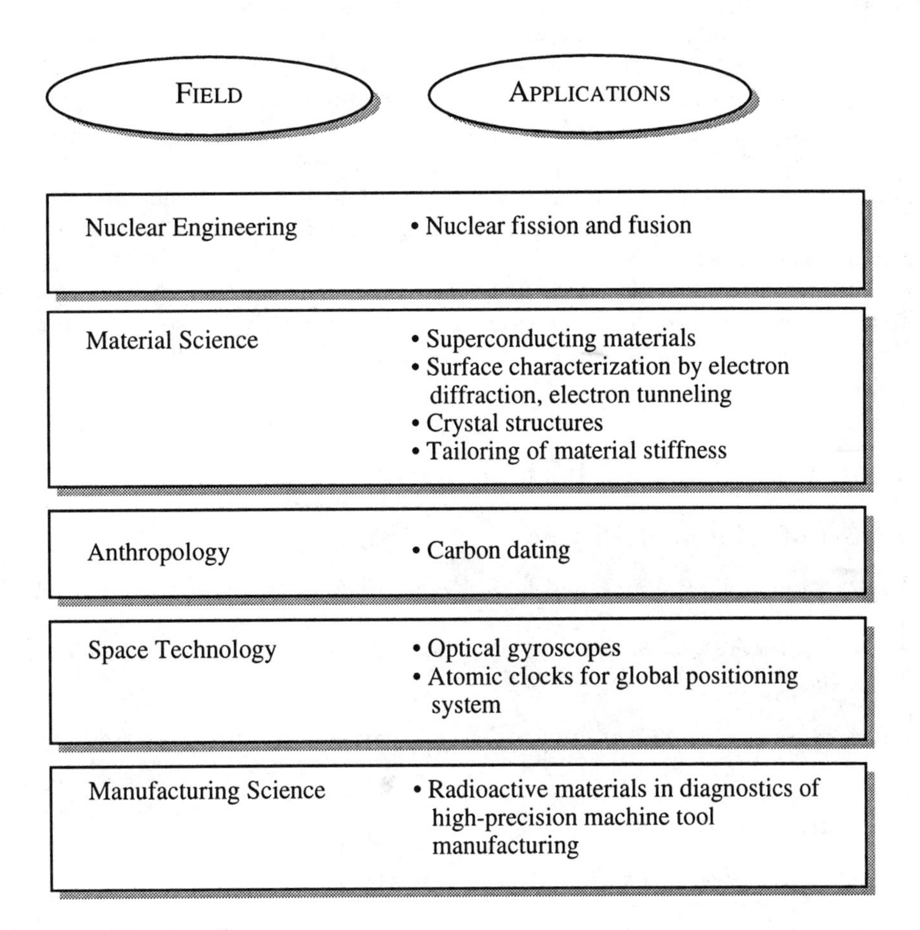

Figure 1: (Continued).

connection between basic principles and applications. For such students the instructor should cover the first six chapters and Chapter 9 in some detail and pick selected topics from Chapters 7 and 8. For example, the coupled well problem could be discussed from Chapter 7. A physical discussion of optical processes could be discussed from Chapter 8. Derivations of Fermi golden rule (Appendix A) or Boltzmann transport theory (Appendix B) should be avoided for students at this level.

The book would be very useful for students who are seniors or entering graduate students in the electrical engineering, material science, or chemical engineering departments. For such students the first couple of chapters could be covered rather quickly (some sections could be given as reading assignments). Topics in Chapters 7 and 8 as well as Appendices A and B could be covered in detail.

Finally, for physics majors the book could be used for a modern physics course. The book covers most topics of relevance to such a course and provides an important link to technical applications. Such a link is normally not provided to physics majors, but given the changes occurring in the physics profession, will most likely be appreciated by them.

The system of units used throughout the text is the SI system—a system that is widely used by technologists. The text contains nearly 100 solved examples, most of which are numerical in nature. It also contains about 200 end-of-chapter problems.

SOME IMPORTANT REFERENCES

Review of Classical Physics

- R. Resnick, D. Halliday, and K. S. Krane, *Physics*, Wiley, New York, 1992.
- R. A. Serway, *Physics for Scientists and Engineers*, Saunders, Philadelphia, 1990.
- H. D. Young, *University Physics*, Addison-Wesley, Reading, MA, 1992.

Quantum Mechanics

- K. Krane, *Modern Physics*, Wiley, New York, 1996.
- P. A. Lindsay, *Introduction to Quantum Mechanics for Electrical Engineers*, 3rd edition, McGraw-Hill, New York, 1967.

Special Relativity

- R. Resnick, *Introduction to Special Relativity*, Wiley, New York, 1996.
- K. Krane, *Modern Physics*, Wiley, New York, 1996.
- E. F. Taylor and J. A. Wheeler, *Spacetime Physics*, Freeman, New York, 1992.

General

- R. P. Feynman, R. B. Leighton and M. Sands, *The Feynman Lectures of Physics*, Vol. III, Addison-Wesley Publishing Company, Reading, MA, 1964. This and the accompanying volumes are a *must read.*

CHAPTER 1

CLASSICAL VIEW OF THE UNIVERSE

CHAPTER AT A GLANCE

1.1 INTRODUCTION

The term classical physics sounds old, but it is the term used for the physics in existence until the beginning of the twentieth century. This is the physics that was developed by scientists such as Archimedes, Newton, Maxwell, Boltzmann, ... and even though the developements due to quantum physics and relativistic physics have ushered in the more accurate *modern* physics, classical physics continues to be used in a vast array of applications quite successfully.

Classical physics, it turns out, gives us an approximate description of how our world behaves. This approximation is usually adequate for a very large class of problems. However, on some levels, the inaccuracies of classical physics are manifested quite dramatically. Modern physics can be classified as a much better approximation and is suitable for cases where classical concepts fail.

In this chapter we will take a quick tour of important classical concepts. It is important to keep in mind that modern physics reduces to classical physics when certain conditions are met. These conditions may be stated as, "mass of the particle is large," or "potential energy is slowly varying in space," etc. Thus it is important to understand the classical concepts.

We start this chapter with an overview of the approach known as the scientific method. This approach must be followed in the understanding and development of any physics related concept. This is the method that has brought physics to the present level—a level where enormously complex physical phenomena are well understood.

1.2 SCIENTIFIC METHOD

Physics is the branch of science that attempts to answer the questions related to the *'hows'* and *'whys'* of our physical universe. In this great attempt, some very rigorous paths are to be taken. There must be a logical flow between one assertion and the next. It may be fine to explain the twinkling of stars to a child by claiming there are angels lowering and raising their lanterns. But this is not the approach we can follow in physics.

While there are still phenomena we cannot explain by simply invoking our understanding of physics, it is quite remarkable that a great range of observations can now be explained through modern physics. Physics has developed from explaining how cannon balls fly and pulleys work to what happens inside a white dwarf star or what happens to protons when they are smashed into each other at close to the speed of light.

It is important to understand the approach known as the *scientific method*, which has allowed us to understand our universe so well. This approach has been used by intellectual giants like Galileo, Newton, Maxwell, Einstein, Feynman, etc. It is also an approach used by thousands of basic and applied scientists around the world. The scientific method is used by the electrical engineer when he or she is trying to build a gigabit memory chip. It is used by the biophysicist who is assembling a special molecule to attack cancer cells. Or by the mechanical engineer developing a more crash-safe car. The scientific method allowed Galileo to do his famous experiment to show that objects with different masses drop to the earth with the same velocity. It also allowed Copernicus to show that the earth went around the sun. It allowed Newton to describe how gravity worked and Einstein to come up with the theory of relativity. It ushered in the quantum age and the information age based on computers.

Let us examine the components of the scientific process. Fig. 1.1 gives an overview of this approach. Here are the building blocks:

• **Observation**: This is the core of the scientific method. Careful measurements of a physical phenomenon set the stage for most scientific endeavors. Sometimes the experiment may be motivated by a theoretical prediction based on existing knowledge. *In the scientific method no one can argue against a fundamentally sound experiment.* The experiment can never be subverted to fit an existing line of thought.

• **Postulates**: The scientific method now starts, with the ultimate aim to explain experiments and to predict phenomena. The method sets up some postulates which seem reasonable to the scientists. These postulates are the starting points from which the theory describing the phenomena springs forth. At the most fundamental level the postulates do not have any proofs, in the sense that they cannot be derived from other laws.

• **Physical Laws**: Next, physical laws are set up, which describe how physical quantities evolve from one value to another in space and time. These *equations of motion* along with the powerful language of mathematics allow the scientist to relate experiments and the scientific formalism. These laws not only should be able to explain existing observations, but also must predict phenomena that can then be observed by properly designed experiments. It is also very important to keep in mind what a good scientific method tries to do:

• *Make assumptions that are absolutely necessary.* If we are building a formalism for how an airplane flies we must not assume anything about whether a cockroach evolved 10 million years ago or 100 million years ago!

• *Accept the guidance of experiments.* A good scientist should not retain a formalism which explains everything except one lonely experiment. It is this lonely experiment that may ultimately advance our knowledge.

• *Consider our present understanding as an approximation.* It is important to realize that when a scientist writes down certain laws or equations of motion, he or she is only describing how the real world behaves in an approximate way. This is because the laws may describe nature quite well in some regimes but may not be able to describe things well in a different regime. And even though we know a great deal of the rules of nature, we don't know all the rules. And some rules may manifest themselves under conditions that are almost impossible to realize in a controlled manner in a laboratory.

It sometimes happens that a certain set of laws or rules work quite well for a long period in our history. Then something is observed which does not fit into the scheme we have developed. One then looks for a broader set of rules. *These new rules must explain all the previous observations, along with the new ones.* One of the most dramatic instances of such a development is the subject of this book. Toward the end of the nineteenth century, a series of remarkable experiments were conducted that completely baffled existing physics. This created a tremendous excitement among

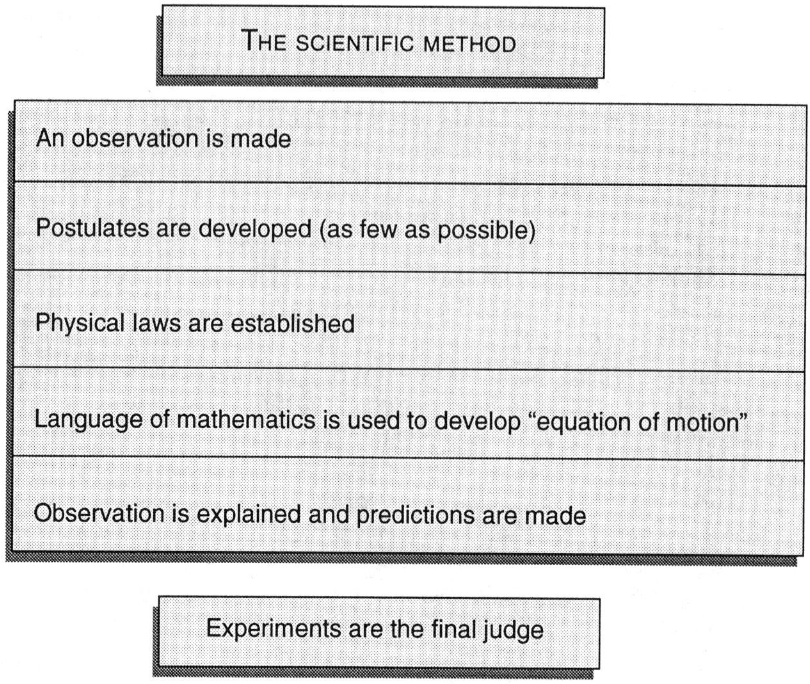

THE SCIENTIFIC METHOD

An observation is made

Postulates are developed (as few as possible)

Physical laws are established

Language of mathematics is used to develop "equation of motion"

Observation is explained and predictions are made

Experiments are the final judge

Figure 1.1: An overview of the scientific method.

the physicists who then gradually built what we now call *modern physics*. Before we start our discussions of modern physics we need to review *classical physics*, which is the name often given to physics existing up to the beginning of the twentieth century.

1.3 OVERVIEW OF CLASSICAL PHYSICS: BASIC INTERACTIONS

In classical physics our physical universe is distinctly divided into two categories—particles and waves. When we talk about particles, we do not necessarily refer to point particles or very tiny particles. The earth is a particle; so are a car, a refrigerator, and a tennis ball. The laws which describe particles and those which describe waves are very different, and we have no problem distinguishing a particle from a wave. In Fig. 1.2 we show an overview of classical concepts. We have particles which are represented by masses, momentum, position, etc., and waves which are described by amplitude, wavelength, phase, etc. We also have an important branch of physics which deals with systems of large numbers of particles. This field is known as thermodynamics.

In classical physics particles interact with each other via two kinds of interactions. As shown in Fig. 1.2, the first kind of interaction is the gravitational interaction. If we have two particles of masses m_1 and m_2 placed at a separation of

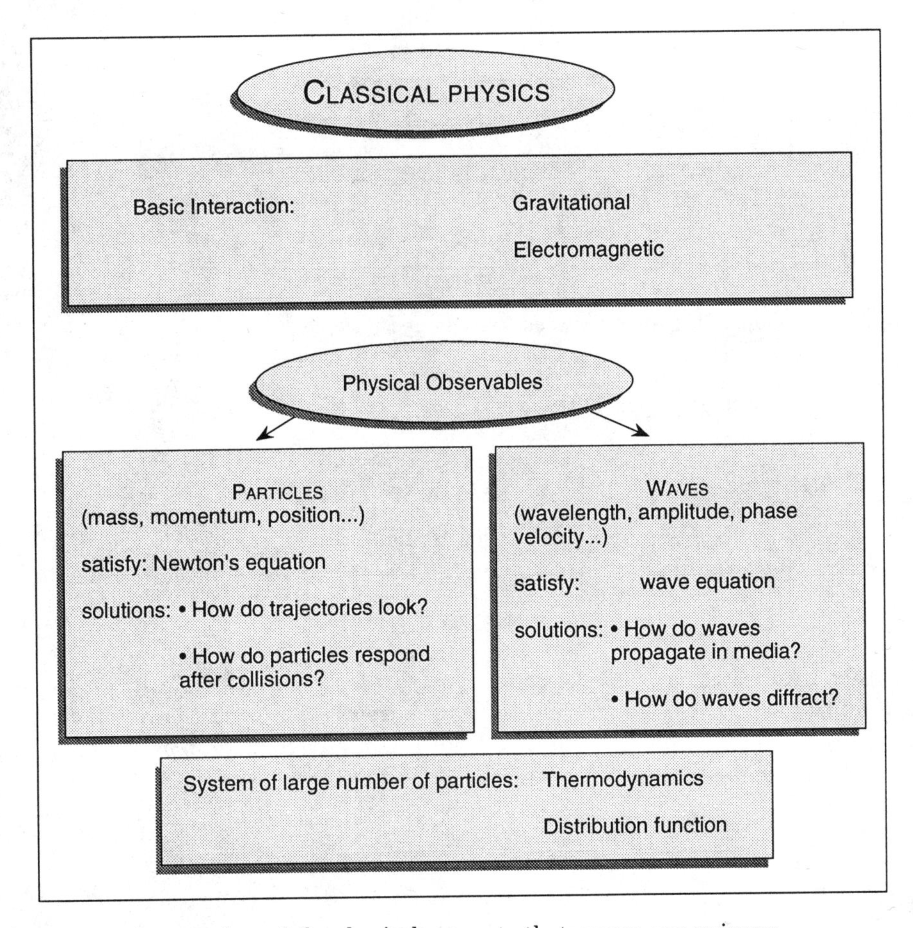

Figure 1.2: An overview of the classical concepts that govern our universe.

r, they have an attractive force given by

$$\mathbf{F} = G\frac{m_1 m_2}{r^2}\hat{r} \tag{1.1}$$

where G is a coefficient with a value 6.67×10^{-11} N.m^2/kg^2.

The second form of interaction between particles is the electromagnetic interaction. A manifestation of this interaction is that if we have two particles with charges q_1 and q_2 separated by a distance r, there is a force felt by each particle given by (see Fig. 1.3)

$$\mathbf{F} = \frac{1}{4\pi\epsilon_0}\frac{q_1 q_2}{r^2}\hat{r} \tag{1.2}$$

where

$$\frac{1}{4\pi\epsilon_0} = 8.988 \times 10^9 \text{ N.m}^2/\text{C}^2$$

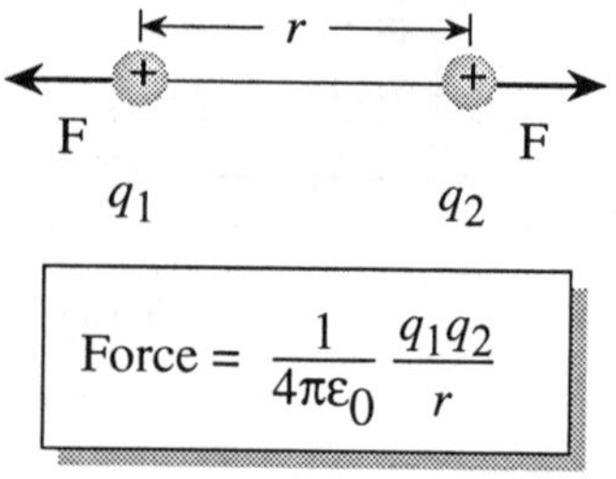

r

F F

q_1 q_2

$$\text{Force} = \frac{1}{4\pi\varepsilon_0}\frac{q_1 q_2}{r}$$

Figure 1.3: Coulombic force between two charged particles.

The potential energy of the system is

$$U = \frac{1}{4\pi\epsilon_0}\frac{q_1 q_2}{r} \tag{1.3}$$

If a charged particle is moving in an electric field $\mathbf{E}$ and a magnetic field $\mathbf{B}$ with a velocity $\mathbf{v}$, it sees a force given by (see Fig. 1.4a)

$$\mathbf{F} = q\mathbf{v} \times \mathbf{B} \tag{1.4}$$

Another manifestation of the electromagnetic interaction is that a magnetic field is produced by an electric current I. The current is just the motion of charged particles. If, for example, the current is flowing in a circular loop as shown in Fig. 1.4b, the magnetic field at the center is

$$\mathbf{B} = \frac{\mu_0 I}{2r}\hat{n} \tag{1.5}$$

with

$$\mu_0 = 4\pi \times 10^{-7}\ \text{Ns}^2/\text{C}^2$$

The direction of the field is given by the right-hand rule, i.e., if you hold the wire in the right hand with the thumb pointing along the current direction, the fingers point in the field direction.

A number of interesting interactions arise from the equations described above. These include the attraction or repulsion between current-carrying wires, torque on a current-carrying wire in a magnetic field, etc. All these disparate looking phenomena were placed into a single coherent picture by James Clark Maxwell. The classical electromagnetic phenomena is described completely by the Maxwell equations.

The properties of electromagnetic fields in a medium are described by the four Maxwell equations. Apart from the electric ($\mathbf{E}$) and magnetic ($\mathbf{B}$) fields and velocity of light, the effects of the material are represented by the dielectric constant, permeability, electrical conductivity, etc. We start with the four Maxwell equations

$$\nabla \times \mathbf{E} + \frac{\partial \mathbf{B}}{\partial t} = 0$$
$$\nabla \times \mathbf{H} - \frac{\partial \mathbf{D}}{\partial t} = \mathbf{J}$$

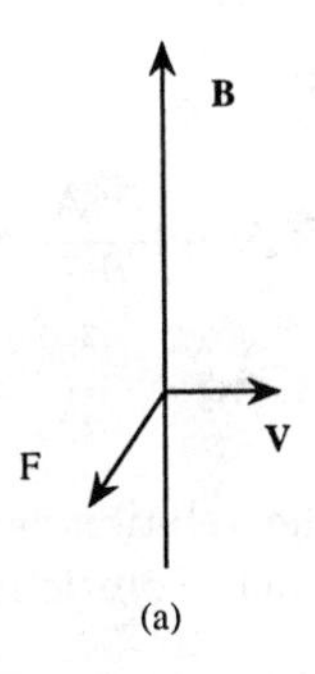

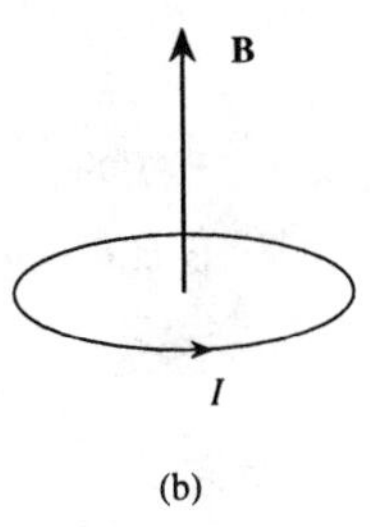

Figure 1.4: (a) Force on a charged particle moving with velocity v in a magnetic field. (b) Magnetic field produced by a current.

$$\nabla \cdot \mathbf{D} = \rho$$
$$\nabla \cdot \mathbf{B} = 0 \tag{1.6}$$

where **E** and **H** are the electric and magnetic fields, $\mathbf{D} = \epsilon\mathbf{E}$, $\mathbf{B} = \mu\mathbf{H}$, and **J** and ρ are the current and charge densities. In electromagnetic theory it is often convenient to work with the vector and scalar potentials **A** and ϕ, respectively, which are defined through the equations

$$\mathbf{E} = -\frac{\partial \mathbf{A}}{\partial t} - \nabla\phi$$
$$\mathbf{B} = \nabla \times \mathbf{A} \tag{1.7}$$

The first and fourth Maxwell equations are automatically satisfied by these definitions. The potentials **A** and ϕ are not unique but can be replaced by a new set of potentials $\mathbf{A}^{'}$ and $\phi^{'}$ given by

$$\mathbf{A}^{'} = \mathbf{A} + \nabla\chi$$
$$\phi^{'} = \phi - \frac{\partial\chi}{\partial t} \tag{1.8}$$

The new choice of potentials does not have any effect on the physical fields **E** and **B**.

It can be shown that the equations for the vector and scalar potential have the form (for $\mu = \mu_0$)

$$\begin{aligned}
\frac{1}{\mu_0}\nabla^2\mathbf{A} - \epsilon\frac{\partial^2\mathbf{A}}{\partial t^2} &= \mathbf{J} \\
\nabla^2\phi - \frac{\partial^2\phi}{\partial t^2} &= -\frac{\rho}{\epsilon}
\end{aligned} \tag{1.9}$$

It is useful to establish the relation between the vector potential $\mathbf{A}$ and the optical power. The time-dependent solution for the vector potential solution of Eqn. 1.9 with $\mathbf{J} = 0$ is

$$\mathbf{A}(\mathbf{r}, t) = \mathbf{A}_0 \left\{\exp\left[i(\mathbf{k}\cdot\mathbf{r} - \omega t)\right] + \text{ c.c.}\right\} \tag{1.10}$$

with

$$k^2 = \epsilon\mu_0\omega^2$$

Note that in the MKS units $(\epsilon_0\mu_0)^{-1/2}$ is the velocity of light c (3×10^8 ms^{-1}). The electric and magnetic fields are

$$\begin{aligned}
\mathbf{E} &= -\frac{\partial\mathbf{A}}{\partial t} \\
&= 2\omega\mathbf{A}_0\sin(\mathbf{k}\cdot\mathbf{r} - \omega t) \\
\mathbf{B} &= \nabla \times \mathbf{A} \\
&= 2\mathbf{k} \times \mathbf{A}_0\sin(\mathbf{k}\cdot\mathbf{r} - \omega t)
\end{aligned} \tag{1.11}$$

The Poynting vector $\mathbf{S}$ representing the optical power is

$$\begin{aligned}
\mathbf{S} &= \mathbf{E} \times \mathbf{H} \\
&= \frac{4}{\mu_0}vk^2\left|\mathbf{A}_0\right|^2\ \sin^2(\mathbf{k}\cdot\mathbf{r} - \omega t)\hat{k}
\end{aligned} \tag{1.12}$$

where v is the velocity of light in the medium ($= c/\sqrt{\tilde{\epsilon}}$) and $\hat{k}$ is a unit vector in the direction of $\mathbf{k}$. Here $\tilde{\epsilon}$ is the relative dielectric constant. The time-averaged value of the power is

$$\begin{aligned}
\langle\mathbf{S}\rangle_{\text{time}} &= \hat{k}\frac{2vk^2\left|\mathbf{A}_0\right|^2}{\mu_0} \\
&= 2v\epsilon\omega^2\left|\mathbf{A}_0\right|^2\hat{k}
\end{aligned} \tag{1.13}$$

since

$$|\mathbf{k}| = \omega/v \tag{1.14}$$

The energy density is then

$$\left|\frac{\mathbf{S}}{v}\right| = 2\epsilon\omega^2\left|\mathbf{A}_0\right|^2 \tag{1.15}$$

The electromagnetic spectrum spans a vast array of wavelengths. In Fig. 1.5 we identify the important regimes of this spectrum.

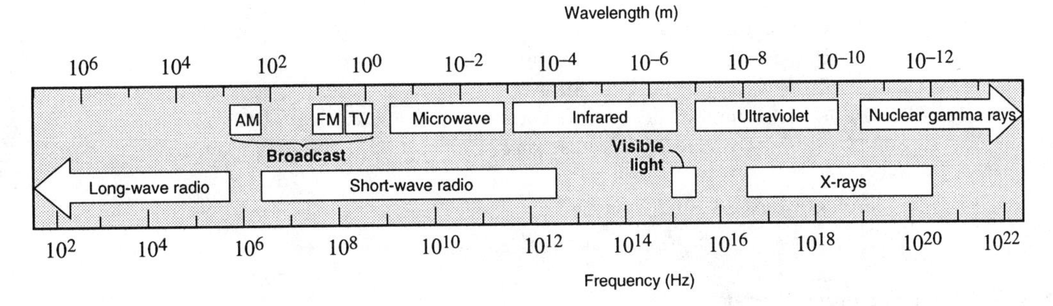

Figure 1.5: The electromagnetic spectrum along with the terms used for different frequency regimes.

EXAMPLE 1.1 A diver jumps from a 10 m diving board and takes 1.43 seconds to hit the water. If the earth's radius is 6400 km, calculate the mass of the earth.

The distance d traveled in time t under acceleration a is

$$\begin{aligned} d &= \frac{1}{2}at^2 \\ &= \frac{1}{2}\frac{Gm_e}{r_e^2}t^2 \end{aligned}$$

This gives

$$\begin{aligned} m_e &= \frac{2dr_e^2}{Gt^2} \\ &= \frac{2 \times (10 \text{ m})(6400 \times 10^3 \text{ m})^2}{(6.67 \times 10^{-11} \text{ N.m}^2/\text{kg}^2)(1.43 \text{ s})^2} \\ &= 6 \times 10^{24} \text{ kg} \end{aligned}$$

EXAMPLE 1.2 A ball weighing 1 kg is moving with a velocity of 10^4 cm/s. What would the change in velocity be if the kinetic energy of the ball increased by 100 eV?

The kinetic energy of the ball, initially, is

$$\begin{aligned} K \cdot E_1 = \frac{(1.0 \text{ kg})(10^2 \text{ms})^2}{2} &= 5 \times 10^3 \text{ J} \\ &= 3.125 \times 10^{22} \text{ eV} \end{aligned}$$

The change in velocity is related to the change in energy by the relation

$$\begin{aligned} \Delta v &= \frac{\Delta E}{mv} \\ &= \frac{(100 \times 1.6 \times 10^{-19} \text{ J})}{(1.0 \text{ kg})(10^2 \text{ ms})} \\ &= 1.6 \times 10^{-19} \text{ ms} \end{aligned}$$

This is a negligible amount! We will see later that a few eV are important energies in quantum systems. The example here shows that in most systems we can observe with our senses, a few eV are impossible to detect.

EXAMPLE 1.3 In quantum mechanics, an important unit of angular momentum is denoted by $\hbar$ (Planck's constant). The value of $\hbar = h/2\pi$ is 1.05×10^{-34} J.s. Quantum mechanics usually becomes important for systems when the angular momentum is in the range of a few $\hbar$. Consider a child swinging a ball weighing 200 gm at the end of a string 1 m long at a rate of 2 revolutions per second. Calculate the angular momentum of the ball in units of $\hbar$.

The magnitude of the velocity of the ball is

$$\mid v \mid = 4\pi \text{ ms}$$

The angular momentum is

$$\begin{aligned} \mid L \mid &= (1 \text{ m})(0.2 \text{ kg})(4\pi \text{ ms}) \\ &= 2.513 \text{ J.s} \\ &= 2.39 \times 10^{34} \ \hbar \end{aligned}$$

We can see that the angular momentum is enormous in units of $\hbar$.

EXAMPLE 1.4 The solar constant gives the power density incident on the earth from the sun. Assuming this to be 2.0 calories/minute/cm^2, calculate the following:
i) The magnitude of the Poynting vector in sunlight in units of W/m^2.
ii) The rms electric field in sunlight.
iii) The rms magnetic field in sunlight.

The Poynting vector magnitude is simply (1 calorie = 4.186 J)

$$S = \frac{2 \times 4.186 \text{ J}}{(60 \text{ s})(10^{-4} \text{ m}^2)} = 1.395 \ kW/m^2$$

To calculate the electric field, we note that

$$B = \frac{kE}{\omega}$$

or

$$H = \frac{E}{c\mu_0}$$

Using $\mu_0 = 4\pi \times 10^{-7}$ (kg.m A^{-2} s^{-2}) and $c = 3 \times 10^8$ ms, we get

$$\mid E_{rms} \mid^2 = c\mu_0 \ S = 5.25 \times 10^5 (\text{V/m})^2$$

or

$$E_{rms} = 725 \text{ V/m}$$

In a similar manner, the magnetic field has a value

$$H_{rms} = 1.93 \text{ A/m}$$

EXAMPLE 1.5 Calculate the relative magnitude of the force felt by a charged particle in the electric and magnetic field in electromagnetic radiation. Calculate this ratio for a particle moving at 10^7 cm/s.

The relationship between the electric field E and magnetic induction B in electromagnetic radiation is

$$B = \frac{k}{\omega} E$$

where k is the wavevector and ω is the angular frequency. The ratio of the force felt by a charged particle is

$$\frac{F(E)}{F(B)} = \frac{qE}{qvB} = \frac{\omega}{kv} = \frac{c}{v}$$

For a particle moving at 10^7 cm/s, the ratio is

$$\frac{F(E)}{F(B)} = 3000$$

EXAMPLE 1.6 Compare the forces felt by an electron in a field of 1.0 kV/cm and in the earth's gravitational field.

In the electric field, the force is

$$\begin{aligned} F(E) &= (1.6 \times 10^{-19}\ \text{C})(10^5\ \text{V/m}) \\ &= 1.6 \times 10^{-14}\ \text{N} \end{aligned}$$

In the gravitational field, the force is

$$\begin{aligned} F(\text{earth}) &= (9.1 \times 10^{-31}\ \text{kg})(9.81\ \text{N/kg}) \\ &= 8.93 \times 10^{-30}\ \text{N} \end{aligned}$$

We see that the gravitational force is much weaker.

1.4 CLASSICAL PARTICLES

When classical physics attempts to explain the behavior of particles, it is answering questions such as:

- What is the trajectory of a particular satellite around the earth?
- If a missile is fired from a ship, where will it land?
- If a billiard ball is hit at a certain angle, what will the outcome be?
- If an automobile crashes into a barrier, what will be the benefit of the specially designed *crumple zone*?

To answer some of these questions one may need powerful computers. However, the procedure used to find the answers is quite simple and is given by Newtonian mechanics.

1.4.1 Newtonian Mechanics

Newtonian mechanics describes everything we may wish to know about particles. The key properties of a classical particle are

$$\begin{array}{rl} \text{mass} & m \\ \text{charge} & q \\ \text{position} & x \\ \text{momentum} & p \\ \text{kinetic energy} & \frac{1}{2}mv^2 = \frac{p^2}{2m} \end{array}$$

$$\text{potential energy} \quad V(x)$$

$$\text{equation of motion} \quad m\frac{d^2\mathrm{x}}{dt^2} = \mathrm{F} \tag{1.16}$$

where F is an applied force. If the particle is charged, the force can arise from an electric field and/or a magnetic field. It could also be a mechanical force. Once the force is known, one can completely describe the particle in the physical world. There are several interesting points that can be made about the particle in classical physics:

- *There is no quantization of particle energy.* What we mean by this is that the total energy of the particle is a continuous variable; i.e., the energy can change in infinitesimal steps (as opposed to finite steps). For example, if we have a pendulum hanging from a point, the energy of the pendulum can start from zero (amplitude of vibration is zero) and increase continuously as the amplitude is increased.
- *Physical observables such as momentum, position, energy, etc., can all be defined or measured with complete certainty.* Of course, it is possible that the measurement system may have some error which may not allow an accurate measurement. But in principle, we can make completely precise measurements on any or all physical measurements.

These points seem rather trivial and obvious, but, as we now know, they are not satisfied by particles under all conditions.

1.5 CLASSICAL WAVE PHENOMENA

In classical physics particles and waves are quite distinct. In some kinds of waves particle motion may be involved. Examples are waves on the ocean surface, where water molecules move up and down to create the wave. In other waves *fields* may be vibrating in space and time. Light waves are ones where electric and magnetic waves vibrate to produce the wave.

Properties of the wave are described not by any mass associated with the wave, but by the amplitude and wavelength (or frequency) of the wave. An appropriate differential equation represents a wave. A simple form of a wave equation is:

$$\frac{1}{v^2}\frac{d^2\psi}{dt^2} - \frac{d^2\psi}{dx^2} = 0 \tag{1.17}$$

This equation has general solutions

$$\psi(x,t) = A\exp[i(kx - \omega t)] \tag{1.18}$$

and

$$\psi(x,t) = B\exp[i(kx + \omega t)] \tag{1.19}$$

Here A and B are constants which determine the amplitude of the wave. Their values are determined by the energy density of the wave and by boundary conditions the wave has to satisfy. For the simple problem outlined above we have

$$\frac{\omega^2}{k^2} = \frac{1}{v^2}; \quad v = \frac{k}{\omega} \tag{1.20}$$

The simple equation we have considered describes a one-dimensional problem and the solutions are simple plane waves travelling along $+x$ and $-x$ directions. Something very interesting happens if boundary conditions are imposed on the wave solutions.

1.5.1 Boundary Conditions and Quantization

The wave equation we have discussed above is a second-order differential equation. In order to solve it we need to impose two conditions on the solutions. In the case of waves, the solutions for the allowed wavelengths (frequencies) come from solving the wave equation with appropriate boundary condtions. *Not all possible frequencies may be allowed for a given problem.* This feature is exploited in musical instruments to produce certain frequencies. Boundary conditions determined by the nature of the problem have a profound effect on the wavelength or wavenumber of the soulutions that are allowed. For example, the system of interest may be such that the wave amplitude goes to zero at the boundaries. An example is that of a vibrating string which is clamped at two ends as shown in Fig. 1.6a. In the case of the string shown the wave amplitude must go to zero at $x = 0$ and $x = L$. As a result, the solutions are of the form

$$\psi(x) = A \sin kx \tag{1.21}$$

where, due to the boundary conditions we have

$$k = \frac{n\pi}{L} \tag{1.22}$$

We note a very interesting outcome of the simple boundary conditions—only certain discrete solutions are allowed. Other solutions are *forbidden*. This is a generic outcome of enforcing stationary (i.e., amplitude goes to zero) boundary conditions.

Another boundary condition which is often relevant to waves is one where the solutions are periodic. In this case the wave can propagate in all space but the solutions are chosen so that the solution repeats itself. A two-dimensional space in which periodic boundary conditions are applied is shown in Fig. 1.6b. We divide the infinite space into cells each of volume V and dimension $L \times L \times L$. We impose the boundary conditions

$$\begin{aligned} \psi(x, y, z + L) &= \psi(x, y, z) \\ \psi(x, y + L, z) &= \psi(x, y, z) \\ \psi(x + L, y, z) &= \psi(x, y, z) \end{aligned} \tag{1.23}$$

The solutions are now of the form

$$\psi(x, y, z) = A \exp[i(k_x x + k_y y + k_z z)] \tag{1.24}$$

where, as a result of the periodic conditions, we have

$$\begin{aligned} k_x &= \frac{2\pi n}{L}; \ \ n = 0, \pm 1, \pm 2, \ldots \\ k_y &= \frac{2\pi m}{L}; \ \ m = 0, \pm 1, \pm 2, \ldots \\ k_z &= \frac{2\pi l}{L}; \ \ l = 0, \pm 1, \pm 2, \ldots \end{aligned} \tag{1.25}$$

The wave solutions fill up the entire space but once again the k-vectors are quantized due to the boundary conditions.

The quantization aspect discussed above is a feature quite unique to waves. As we noted in the previous section, it does not occur when we discuss the classical behavior of particles. Another unique feature of waves is discussed in the next section.

1.6 WAVES, WAVEPACKETS, AND UNCERTAINTY

In classical physics when we deal with wave phenomena we are aware of a little "fuzziness" in the description of certain features of the wave. For example, let us imagine a child creating a wave on the surface of a pond by throwing a stone into the pond. We know from experience that a "wavepacket" or wave "pulse" intially localized around the point where the stone hit the water surface is produced. This wave then propagates toward the edges of the pond. Can we, at any time, *precisely* define the location of the wave and its wavelength? We know from experience and from classical physics that this is not possible. If we try to create a wavepacket highly localized in space we lose the knowledge of the wave's wavelength. On the other hand, if we try to create a "plane wave" with a well-defined wavelength, we lose the knowledge regarding the spatial position of the wave.

The uncertainty described above is not of concern in classical physics when we deal with particles. For example, we have no problem defining *precisely* any combinations of physical observables of a particle. However, a wave description will inevitably bring in an uncertainty in the precision with which we can simultaneously define certain physical observables. To see how this occurs, we examine the uncertainty arising in the wavelength or, for convenience, the wavevector k ($k = 2\pi/\lambda$), and the position of waves.

To describe the wavepacket, let us begin from a plane wave given by

$$\psi_k(x) = e^{ikx} \tag{1.26}$$

The position of the wave is completely undefined (as shown in Fig. 1.7a). To create a wavepacket localized at some point x_0 in space, we have to combine several plane waves. One example is to use an equal amplitude combination of waves centered around k_0 with a spread $\pm\Delta k$. The resulting function, say k_0, but from a spread $\pm\Delta k$, then the function

$$F(x, x_0) = \int_{k_0-\Delta k}^{k_0+\Delta k} dk \; e^{ik(x-x_0)}$$

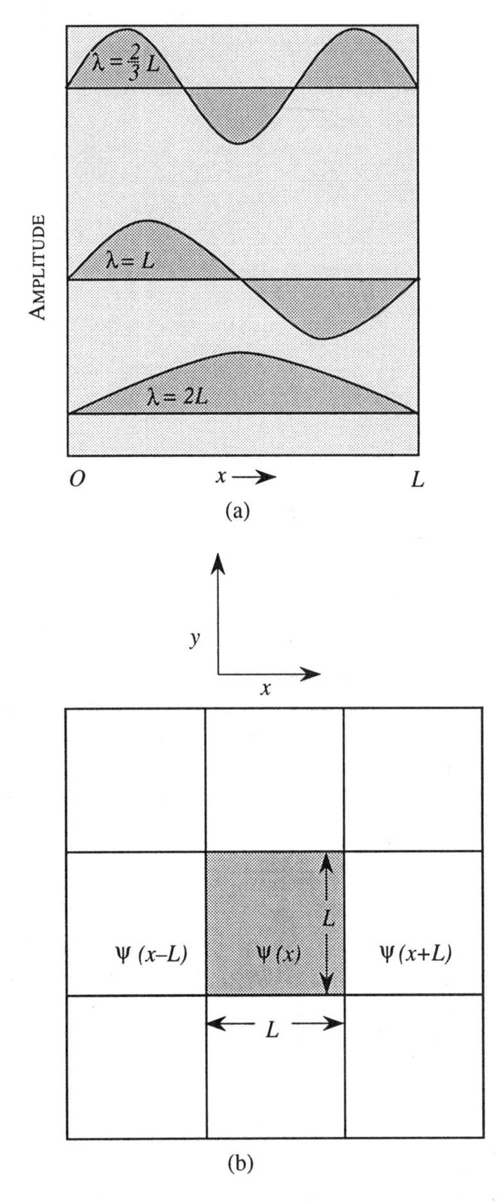

Figure 1.6: A schematic showing (a) the stationary boundary conditions applicable to a string clamped at $x = 0$ and $x = L$; and (b) periodic boundary conditions leading to plane wave solutions.

$$= \frac{2\sin(\Delta k\ (x - x_0))}{x - x_0} e^{ik_0(x-x_0)} \tag{1.27}$$

is centered around the point x_0 and the probability ($|F|^2$) decays from its maximum value at x_0 to a very small value within a distance $\pi/\Delta k$, as shown in Fig. 1.7b.

A more useful wavepacket is constructed by multiplying the integrand in the wavepacket by a Gaussian weighting factor:

$$f(k - k_0) = \exp\left[-\frac{(k - k_0)^2}{2(\Delta k)^2}\right] \tag{1.28}$$

and extending the range of integration from $-\infty$ to $+\infty$. This wave packet has the form

$$\begin{aligned}
\psi(x, x_0) &= \int_{-\infty}^{\infty} \exp\left[-\frac{(k - k_0)^2}{2(\Delta k)^2}\right] \exp\left[ik(x - x_0)\right]\, dk \\
&= \exp\left[ik_0(x - x_0) - \frac{(x - x_0)^2}{2}(\Delta k)^2\right] \\
&\times \int_{-\infty}^{\infty} \exp\left[-\frac{(k - k_0)^2}{2(\Delta k)^2} + i(k - k_0)(x - x_0) + \frac{(x - x_0)^2}{2}(\Delta k)^2\right] \\
&= \sqrt{2\pi}\Delta k\ \exp\left[ik_0(x - x_0) - \frac{1}{2}(x - x_0)^2(\Delta k)^2\right]
\end{aligned} \tag{1.29}$$

$\psi(x, x_0)$ represents a Gaussian wavepacket in space which decays rapidly away from x_0. We note that when we considered the original state $\exp(ik_0x)$, the wave was spread infinitely in space, but has a precise k-value. By constructing a wavepacket, we sacrificed its precision in k-space by Δk and gained a precision Δx in real space. In general, the width of the wavepacket in real and k-space can be seen to have the relation

$$\Delta k\ \Delta x \approx 1 \tag{1.30}$$

This "uncertainty relation" which exists for waves is quite important. In the next chapter we will see that under some conditions, particles behave as waves. Similar uncertainty relations then exist in their properties. These relations are then called *Heisenberg uncertainty relations.*

1.6.1 Propagation of a Wavepacket

In classical wave propagation we are often interested in the question: How does a wave signal or pulse propagate in a medium? For example, we may create an optical pulse by switching a laser and this pulse may move down an optical fiber. We may be interested in the velocity at which the pulse moves and whether it distorts as it moves.

The simplest solution to wave equations have the form

$$\psi_{\mathbf{k}}(\mathbf{r}) \sim \mathbf{A}e^{i\mathbf{k}\cdot\mathbf{r}} \tag{1.31}$$

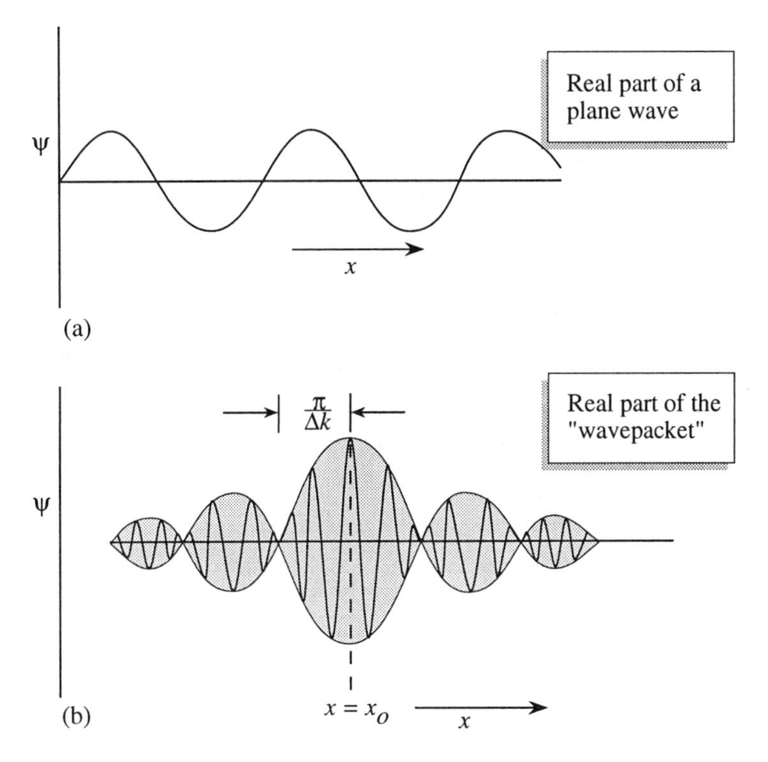

Figure 1.7: (a) A schematic description of a one-dimensional wave $e^{ik\cdot x}$ which is extended over all space; (b) a wavepacket produced by combining several waves produces a packet that is localized in space with a finite spread. The wavepacket is shown centered at x_o and having a spread Δx. The spread is such that $\Delta k \cdot \Delta x \sim 1$. This is an "uncertainty relation" in classical physics for waves. No such uncertainty exists in classical physics for particles.

In such a plane wave the probability density of the the wave, $\psi\psi^*$, is the same in all regions of space. Such a description is not useful if one wants to discuss transport of an optical pulse or of a particle from one point to another. For example, in describing electron transport we wish to describe an electron which moves from one point to another. Thus the wavefunction must be peaked at a particular place in space for such a description. This physical picture is realized by constructing a wavepacket picture.

Construction of a Wavepacket

Let us examine a one-dimensional plane wave state with a wave vector k_0

$$\psi_{k_0}(x) = e^{ik_0x} \tag{1.32}$$

We note that if a state was constructed not from a single k_0 component, but from a spread $\pm\Delta k$, then the function

$$\begin{aligned} F(x, x_0) &= \int_{k_0-\Delta k}^{k_0+\Delta k} dk\ e^{ik(x-x_0)} \\ &= \frac{2\sin(\Delta k\ (x-x_0))}{(x-x_0)}\ e^{ik_0(x-x_0)} \end{aligned} \tag{1.33}$$

is centered around the point x_0 and the probability ($|F|^2$) decays from its maximum value at x_0 to a very small value within a distance $\pi/\Delta k$.

If Δk is small, this new "wavepacket" has essentially the same properties as ψ at k_0, but is localized in space and is thus very useful to describe motion of the particle. A more useful wavepacket is constructed by multiplying the integrand in the wavepacket by a Gaussian weighting factor

$$f(k-k_0) = \exp\left[-\frac{(k-k_0)^2}{2(\Delta k)^2}\right] \tag{1.34}$$

$$\begin{aligned} \psi(x, x_0) &= \int_{-\infty}^{\infty} \exp\left[-\frac{(k-k_0)^2}{2(\Delta k)^2} + ik(x-x_0)\right] dk \\ &= \exp\left[ik_0(x-x_0) - \frac{(x-x_0)^2}{2}(\Delta k)^2\right] \\ &\times \int_{-\infty}^{\infty} \exp\left[-\frac{(k-k_0)^2}{2(\Delta k)^2} + i(k-k_0)(x-x_0) + \frac{(x-x_0)^2}{2}(\Delta k)^2\right] \\ &= \sqrt{2\pi\Delta k}\ \exp\left[ik_0(x-x_0) - \frac{1}{2}(x-x_0)^2(\Delta k)^2\right] \end{aligned} \tag{1.35}$$

$\psi(x, x_0)$ represents a Gaussian wavepacket in space which decays rapidly away from x_0. We note that when we considered the original state $\exp(ik_0x)$, the wave was spread infinitely in space, but has a precise k-value. By constructing a wavepacket, we sacrificed its precision in k-space by Δk and gained a precision Δx in real space. In general, the width of the wavepacket in real and k-space can be seen to have the relation

$$\Delta k\ \Delta x \approx 1 \tag{1.36}$$

We can repeat this procedure for a wave of the form

$$\psi \sim e^{i\omega t} \tag{1.37}$$

and also obtain a wavepacket which is localized in time and frequency, the widths again being related by

$$\Delta\omega\ \Delta t \approx 1 \tag{1.38}$$

Let us now consider how a wavepacket moves through space and time. For this we need to bring in the time dependence of the wavefunction, i.e., the term $\exp(-iEt/\hbar)$ or $\exp(-i\omega t)$.

$$\psi(x, t) = \int_{-\infty}^{\infty} f(k-k_0)\ \exp\{i[k(x-x_0) - \omega t]\}\ dk \tag{1.39}$$

If ω has a simple dependence on k

$$\omega = ck \tag{1.40}$$

we can write

$$\psi(x,t) = \int_{-\infty}^{\infty} f(k-k_0)\ \exp\left[ik(x-x_0-ct)\right]\ dk \tag{1.41}$$

which means that the wavepacket simply moves with its center at

$$x - x_0 = ct \tag{1.42}$$

and its shape is unchanged with time. If, however, we have a dispersive media and the ω vs. k relation is more complex, we can, in general, write

$$\omega(k) = \omega(k_0) + \left.\frac{\partial\omega}{\partial k}\right|_{k=k_0} \cdot (k-k_0) + \frac{1}{2}\left.\frac{\partial^2\omega}{\partial k^2}\right|_{k=k_0} (k-k_0)^2 + \cdots \tag{1.43}$$

Setting

$$\begin{aligned} \omega(k_0) &= \omega_0 \\ \left.\frac{\partial\omega}{\partial k}\right|_{k=k_0} &= v_g \\ \left.\frac{\partial^2\omega}{\partial k^2}\right|_{k=k_0} &= \alpha \end{aligned} \tag{1.44}$$

we get

$$\begin{aligned} \psi(x,t) &= \exp\left[i(k_0(x-x_0)-\omega_0 t)\right] \int_{-\infty}^{\infty} f(k-k_0) \\ &\times \exp\left[i(k-k_0)(x-x_0-v_g t) - \frac{i\alpha}{2}(k-k_0)^2 t\right]\ dk \end{aligned} \tag{1.45}$$

If α were zero, the wavepacket would move with its peak centered at

$$x - x_0 = v_g t \tag{1.46}$$

i.e., with a velocity

$$v_g = \left.\frac{\partial\omega}{\partial k}\right|_{k=k_0} \tag{1.47}$$

However, for nonzero α, we show that the shape of the wavepacket also changes. To see this, let us again assume that

$$\begin{aligned} f(k-k_0) &= f(k') \\ &= \exp\left(\frac{-k'^2}{2\Delta k^2}\right) \end{aligned} \tag{1.48}$$

Then

$$
\begin{aligned}
\psi(x,t) &= \exp\left\{i\left[k_0(x-x_0)-\omega_0 t\right]\right\} \\
&\times \int_{-\infty}^{\infty} \exp\left[ik^{'}(x-x_0-v_g t)\right. \\
&\left. -\frac{k^{'2}}{2}\left(i\alpha t+\frac{1}{(\Delta k)^2}\right)\right] dk^{'}
\end{aligned} \tag{1.49}
$$

To evaluate this integral we complete the square in the integrand by adding and subtracting terms

$$
\begin{aligned}
\psi(x,t) &= \exp\left\{i\left[k_0(x-x_0)-\omega_0 t\right]-\frac{(x-x_0-v_g t)^2(\Delta k)^2}{2\left[1+i\alpha t(\Delta k)^2\right]}\right\} \\
&\times \int_{-\infty}^{\infty} \exp\left\{\frac{-1}{2}\left[\frac{1+i\alpha t\,(\Delta k)^2}{(\Delta k)^2}\right]\right. \\
&\times \left.\left[k^{'}-i\frac{(x-x_0-v_g t)\,(\Delta k)^2}{1+i\alpha t\,(\Delta k)^2}\right]^2\right\} dk^{'}
\end{aligned} \tag{1.50}
$$

The integral has a value

$$
\sqrt{\frac{2\pi\,(\Delta k)^2}{1+i\alpha t\,(\Delta k)^2}}
$$

Further multiplying and dividing the right-hand side exponent by $(1-i\alpha t\,(\Delta k)^2)$ we get

$$
\begin{aligned}
\psi(x,t) &= \exp\left\{i\left[k_0(x-x_0)-\omega_0 t\right]\right\}\sqrt{\frac{2\pi\,(\Delta k)^2}{1+i\alpha t\,(\Delta k)^2}} \\
&\times \exp\left[-\frac{(\Delta k)^2}{2}\,\frac{(x-x_0-v_g t)^2}{1+t^2\,(\Delta k)^4\alpha^2}\right] \\
&\times \exp\left[\frac{i\alpha t\,(\Delta k)^4}{2}\,\frac{(x-x_0-v_g t)^2}{1+t^2\alpha^2\,(\Delta k)^4}\right]
\end{aligned} \tag{1.51}
$$

The probability $|\psi|^2$ has the dependence on space and time given by

$$
|\psi(x,t)|^2 = \exp\left[-\frac{(\Delta k)^2\,(x-x_0-v_g t)^2}{1+\alpha^2 t^2\,(\Delta k)^4}\right] \tag{1.52}
$$

This is a Gaussian distribution centered around $x = x_0 + v_g t$ and the mean width in real space is given by

$$
\begin{aligned}
\delta x &= \frac{1}{\Delta k}\sqrt{1+\alpha^2 t^2\,(\Delta k)^4} \\
&= \delta x(t=0)\sqrt{1+\frac{\alpha^2 t^2}{[\delta x(t=0)]^4}}
\end{aligned} \tag{1.53}
$$

For short times such that

$$\alpha^2 t^2 (\Delta k)^4 \ll 1 \tag{1.54}$$

the width does not change appreciably from its starting value, but as time passes, if $\alpha \neq 0$, the wavepacket will start spreading.

1.7 SYSTEMS WITH LARGE NUMBER OF PARTICLES

We have seen in Section 1.4 that the principles of mechanics given by Newton's equations are capable of describing the behavior of particles. What happens when the number of particles, all interacting with each other via collisions and mutual interactions, starts to increase? If the particle number is small, say less than a hundred or so, it is possible to use a powerful computer to find how each particle will behave in time and space. However, as the particle number increases, it becomes impossible to use Newtonian mechanics to describe how the system will behave. For example, if we were to examine the air in a room, we would find a mixture of oxygen, nitrogen, and carbon dioxide molecules. These molecules are bouncing off the wall and interacting with each other. The number of the particles and their densities are so large that it is simply not possible to apply the principles of mechanics to follow their trajectories. Fortunately, to describe the *measurable properties* of this and other such systems, we don't need to know the precise trajectories of the individual molecules. Such systems containing large number of particles are described by *statistical averages*. The measurable properties we are referring to are pressure, temperature, volume, etc.

Systems containing a large number of particles include: gases (air in a room, gases in a combustion engine, etc.); liquids (particles suspended in a liquid, chemical reagents, etc.); and solids (atoms and molecules in a piece of solid). Properties of these systems are described by the field of thermodynamics.

Let us consider the molecules in the air inside a room. While we cannot describe the individual trajectories of the molecules, we can ask and answer the following questions:

(i) What is the relation between the pressure of the gas and its volume? How is this relation dependent on the density of molecules?

(ii) What is the average kinetic energy of the molecules?

(iii) What is the probability that the molecules have an energy E?

(iv) Is there a difference between the average energy of oxygen molecules and nitrogen molecules?

The field of thermodynamics gives us answers to such questions. We will now state the important concepts of thermodynamics. The essence of thermodynamics is contained in the laws of thermodynamics outlined in Fig. 1.8. We will briefly review some important issues.

Thermal Equilibrium

Let us examine a system consisting of a large number of particles. If we were to observe the system on a microscopic level, we will see a lot of activity going on. Molecules are moving helter-skelter, sometimes suffering a collision and changing their directions and speeds. However, under thermodynamic equilibrium the macroscopic properties of the system will stay constant in time. There will be no net transfer of energy between the system under observation and the rest of the universe.

An important outcome of thermodynamics is that if two systems are in thermal equilibrium with a third system, then they are also in equilibrium with each other. This is known as the zeroeth law of thermodynamics.

If we examine a system on a microscopic level, each particle can be assigned certain *degrees of freedom.* For example, an electron moving in space has three degrees of freedom since it can move in the x, y, and z directions. If we have a molecule with r atoms, there are $3r$ degrees of freedom of which 3 correspond to the motion of the molecule (center of mass motion) and $3(r-1)$ correspond to the internal motion (vibration and rotation).

Consider a system at thermal equilibrium containing a collection of different species of masses m_1, m_2, m_3, and so on. Under equilibrium we have the following equality:

$$\left\langle \frac{m_1 v_1^2}{2} \right\rangle = \left\langle \frac{m_2 v_2^2}{2} \right\rangle = \left\langle \frac{m_3 v_3^2}{2} \right\rangle \tag{1.55}$$

i.e., the average kinetic energy of the species is equal. Thus the mean kinetic energy is not a function of the particle masses but is a property of the system. *This allows us to use the mean kinetic energy of a particle in a system with a large number of particles to define the temperature of the material.*

The mean kinetic energy per degree of freedom k has a value (see Example 1.7)

$$\langle E_k \rangle = \frac{1}{2} k_B T$$

In 3-dimensional space we have

$$\langle E \rangle = \frac{3}{2} k_B T \tag{1.56}$$

The definition of temperature given here creates what is known as the absolute temperature and T is measured in the units of Kelvin (K).

Internal Energy, Free Energy, and Entropy

An important question that thermodynamics answers for us is the following: If we have a system in thermodynamic equilibrium at a temperature T, what is the probability that a particle has an energy E? The equilibrium state is given by the state in which the free energy of the system is minimum. The free energy is different from the internal energy which is simply given by

$$U = \Sigma_i E_i \tag{1.57}$$

where E_i is the energy of the ith particle. The difference between the free energy and the internal energy arises from the quantity known as entropy of the system. A thermodynamic system is described not only by its internal energy but also by another property related to the heat contained in the system and the temperature.

A macroscopic definition of the entropy S is

$$S = \frac{Q}{T} \tag{1.58}$$

where Q is the heat contained in the system and T is the temperature. The entropy of the system is zero at $T = 0$. (This is the third law of thermodynamics.)

There is another definition of entropy that is very useful in defining the distribution function, i.e., the function that gives us the probability of finding a particle at an energy E. The entropy is defined as

$$S = k_B lnW \tag{1.59}$$

where W is the degeneracy of the system, i.e., the number of different ways in which particles can be arranged in the system to create the same total energy.

The free energy F of a system is given by

$$F = U - TS \tag{1.60}$$

As noted above, the equilibrium state is that where the free energy of the system is a minimum. Note that if we simply minimize the internal energy of the system, all the particles would occupy the lowest energy of the system. In this case, the degeneracy of the system is just unity and $S = 0$. At finite temperature there is a competition between the internal energy and entropy. The entropy term $-TS$, decreases as the particles are arranged in higher energy states where the degeneracy is large, but the internal energy term increases. At any given temperature one has to find the minima of F to find the actual distribution of particles. We will carry out this exercise in Chapter 3.

An important manifestation of the classical thermodynamics is the distribution function, which is called the Maxwell-Boltzmann distribution. Consider an ideal gas consisting of non-interacting particles of density N, mass m at equilibrium at a temperature T. The distribution of these particles in energy E or speed w is given by the following expressions:

$$\frac{dN}{dw} = \frac{4N}{\sqrt{\pi}}\left(\frac{m}{2k_BT}\right)^{3/2} w^2 \exp\left(-\frac{1}{2}\frac{mw^2}{k_BT}\right) \tag{1.61}$$

$$\frac{dN}{dE} = \frac{2N}{\sqrt{\pi}} E^{1/2}(k_BT)^{-3/2} \exp\left(-\frac{E}{k_BT}\right) \tag{1.62}$$

Fig. 1.9 shows the distribution of particles as a function of speed at various temperatures. As the temperature increases, the average particle speed increases, as one expects intuitively.

EXAMPLE 1.7 Show that the average of the square of the molecular speeds in an ideal gas is $3k_BT/m$.

The average of the square of the speeds is given by

$$\langle \omega^2 \rangle = \frac{4}{\sqrt{\pi}}\left(\frac{m}{2k_BT}\right)^{3/2} \int_0^\infty \omega^4 \exp\left(-\frac{1}{2}\frac{m\omega^2}{k_BT}\right) d\omega$$

LAWS OF THERMODYNAMICS

ZEROTH LAW: Two systems in thermal equilibrium with a third system are in thermal equilibrium with each other.

FIRST LAW: Represents the law of conservation of energy $\Longrightarrow$ Heat put into a system, dQ + Work done on a system, dW = Increase in internal energy of the system, dU:

$$dQ + dW = dU$$

SECOND LAW: A process whose only effect on all systems is to take heat from a reservoir and convert it to work is impossible $\Longrightarrow$ This law in its mathematical form places limits on the most efficient heat engine, i.e., it says that no heat engine taking heat θ_1 from a system at temperature T_1 and delivering heat θ_2 to a system at T_2 can do more work than a reversible engine for which

$$W = \theta_1 - \theta_2 = \theta_1 \left(\frac{T_1 - T_2}{T_1}\right)$$

The second law is also stated in terms of entropy: The entropy of the universe is always increasing. In a reversible process the entropy is unchanged.

THIRD LAW: The entropy of a system is zero at $T = 0$

Figure 1.8: Laws of thermodynamics.

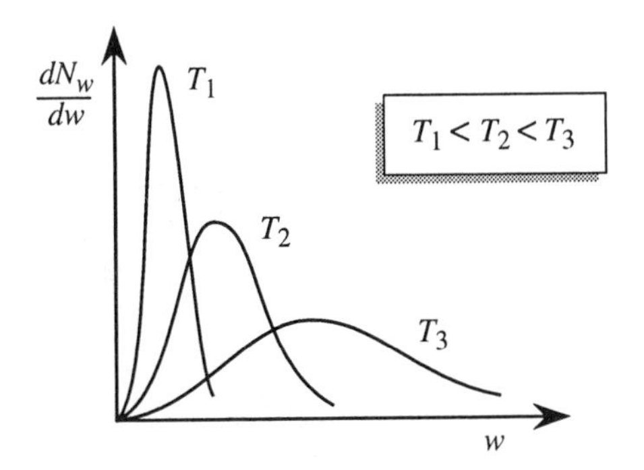

Figure 1.9: Distribution of particle speeds for different temperatures.

The integral has a value $\frac{3}{8}\sqrt{\pi}\,(2k_BT/m)^{5/2}$. Thus

$$< \omega^2 >= \frac{3k_BT}{m}$$

This is, of course, consistent with the observations that the average energy is

$$< E >= \frac{1}{2}m < \omega^2 >= \frac{3}{2}k_BT$$

1.8 CHAPTER SUMMARY

Summary table 1.1 covers key findings and topics studied in this chapter.

1.9 PROBLEMS

Problem 1.1 A child is swinging a 100 gm mass attached to a 1 m string at a rate of two swings per second. Express the angular momentum of the mass in units of $\hbar = 1.05 \times 10^{-34}$ J.s. How small does the string have to be for the angular momentum to be 100 $\hbar$?

Problem 1.2 Consider a pendulum made from a string of length 1 m and mass 100 gm. What is the amplitude of the pendulum (displacement from equilibrium) if the energy of the pendulum is to be 10 $\hbar\omega$ where $\hbar = 1.05 \times 10^{-34}$ J.s and ω is the angular frequency of the pendulum.

Problem 1.3 Calculate the speed of a satellite so that it orbits the earth. How does this speed compare to the speed at which a rocket must be fired in order to leave the earth completely?

Problem 1.4 The separation of the nuclei in a silicon crystal is 2.35 Å. Calculate the gravitational potential energy due to the attraction between the nuclei. Assume that a Si atom is surrounded by four neighbors.

Problem 1.5 In Bohr's model of the hydrogen atom an electron moves in a circular orbit of radius 0.53 Å with an angular momentum of 1.05 $\times 10^{-34}$ J.s. Calculate the

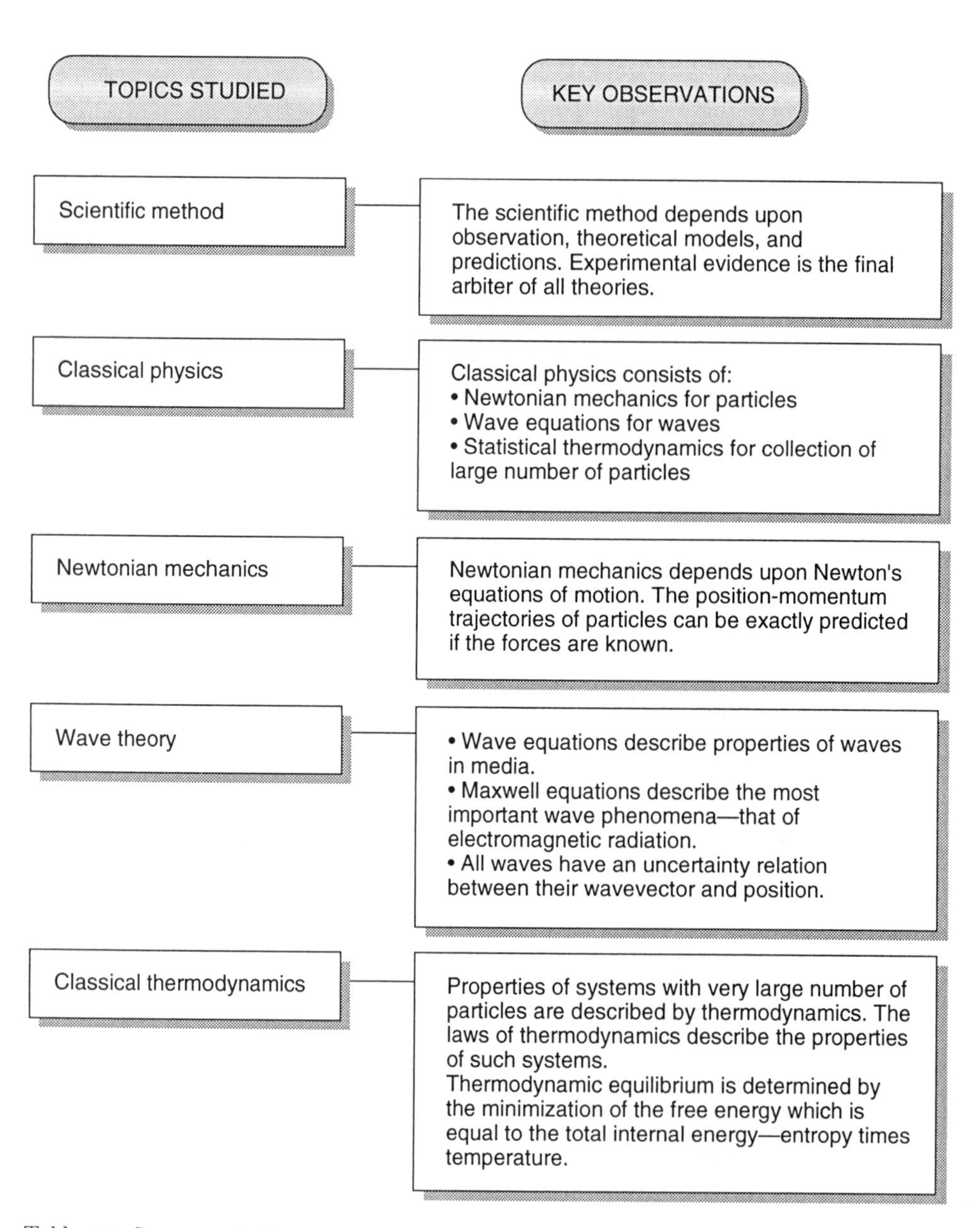

Table 1.1: Summary table

speed of the electron.

Problem 1.6 The optical power density impinging upon a detector has a value of 10^{-6} mW/cm^2. Calculate the electric field amplitude associated with this power.

Problem 1.7 A detector is designed to detect a minimum electric field of 2.0 mV/cm. Calculate the minimum power density level this detector can detect.

Problem 1.8 A microwave oven is designed to produce a maximum electric field (rms) of 1 kV/cm. Calculate the maximum electromagnetic power density produced by this oven.

Problem 1.9 A typical silicon MOSFET "breaks down" when the electric field reaches 2×10^5 V/cm. Calculate the optical power density needed to cause breakdown. Assume that the field needed for breakdown is the rms field of the radiation.

Problem 1.10 A calibrated tuning fork is used to determine the frequency of an instrument by observing the beat frequencies. How long will the observation time have to be to have an accuracy of 0.01 Hz?

Problem 1.11 Consider a diffraction grating with n lines. Show that the resolving power for an mth order beam is

$$\frac{\Delta\lambda}{\lambda} = \frac{1}{mn}$$

Problem 1.12 A gas cylinder contains n molecules of mass 4.7×10^{-26} kg. The temperature is changed from 273 K to 200 K. Calculate the change in the average kinetic energy of the molecules. Also calculate how much the height of the cylinder will have to be altered with respect to the earth's surface to produce the same change in potential energy.

Problem 1.13 Estimate the average speeds of oxygen, nitrogen, and carbon dioxide molecules in air at room temperature.

Problem 1.14 Calculate the probability of the density of atoms of a species in air as a function of height from earth's surface.

Problem 1.15 Calculate the ratio of the density of oxygen molecules at earth's surface to their density at 40 km from the surface. Repeat this problem for hydrogen molecules.

CHAPTER 2

QUANTUM MECHANICS AND THE UNIVERSE

Chapter at a Glance

2.1 INTRODUCTION

2.1.1 Need for Quantum Mechanics

In the previous chapter, while discussing the scientific method, we have argued that theoretical formalisms developed to explain that experiments should be as simple as possible and should be free of unnecessary assumptions. The scientific method also demands that experiments be the final judge on the validity of a model. Classical physics was found to be adequate for explaining experiments, at least at a superficial level, until the end of nineteenth century. Consider an experiment like Ohm's law. At a fundamental level, we now know that we can only explain this law on the basis of quantum mechanics. At this fundamental level we can calculate the resistance of a material. On a superficial level we can explain Ohm's law using classical physics. However, experiments which could not be explained by classical physics started to appear toward the end of the nineteenth century. Scientists were unable to explain these experiments *even at the most superficial level.*

Some of the experiments that challenged classical physics involved very detailed measurements of microscopic particles like electrons, protons, etc. Other experiments were less sophisticated and involved measurements of macroscopic quantities like current flow, heat radiated by a hot object, etc. Fig. 1.1 lists some of the key experiments that ushered in the quantum physics age. We will briefly review some of these critical experiments in this chapter and see how they forced physicists to radically alter their thinking.

2.2 SOME EXPERIMENTS THAT DEFIED CLASSICAL PHYSICS

As we have already pointed out, the classical description of the 1900s was quite adequate for describing complex phenomena, such as planetary motion, machine tools of various kinds, reflection, refraction and interference of light, etc. In fact, almost everything one could observe with the senses could be (at least superficially) examined by classical physics.

Then, at the beginning of the twentieth century, some observations were made which started shaking the foundations of classical thinking. In this section we will examine some of the critical experiments that eventually led to quantum mechanics. In this text we will focus only on non-relativistic quantum mechanics. This means that we will consider situations where the speed of particles (other than photons) is much slower than the speed of light. It is important to note that quantum mechanics does not end with the Schrödinger equation. Beyond this equation we have the Dirac equation, which describes particles and their anti-particles. Beyond that we have a unified field theory linking weak interactions and electromagnetic interactions, proposed by Steven Weinberg and Abdus Salam. Then there is the quark model for the basic building blocks of *elementary particles* like electron, protons, neutrons, etc. The story is still not finished.

1895	Wilhelm Conrad Röntgen discovers X-rays: Mysterious radiation that can penetrate cardboard and human flesh $\Longrightarrow$ What is the origin?
1896	Henri Becquerel discovers radioactivity
1890s	It was discovered that specific heat of metals was not explained by classical thermodynamics
1890s	Detailed measurements of thermal radiation from blackbodies could not be understood on the basis of classical thermodynamics
1887	Heinrich Hertz observes photoelectric effect $\Longrightarrow$ could not be understood on the basis of classical electromagnetic theory which treats light as waves
1905	Albert Einstein introduces the concept of a photon to explain the photoelectric effect
1911	Heike Kamerlingh-Onnes discovers superconductivity: Resistance of some materials goes to zero at low temperatures $\Longrightarrow$ completely stumped classical physics
1913	William H. Bragg and Willaim L. Bragg study X-ray diffraction from crystals $\Longrightarrow$ showed that X-rays have wave-like properties
1914	James Franck and Gustav Hertz show evidence for quantized energy levels in atoms: Electrons in atoms do not behave as classical particles
1921	Otto Stern and Walter Gerlach show the need to introduce an intrinsic magnetic moment of electron

Figure 2.1: Some of the key experiments that ushered in the quantum physics age.

2.2.1 Waves Behaving as Particles: Blackbody Radiation

An extremely important experimental discovery which played a central role in the development of quantum mechanics was the problem of the spectral density of a blackbody radiation. If we take a body with a surface that absorbs any radiation (a blackbody) we find that it emits radiation at different wavelengths. The intensity (power per unit area) of the radiation emitted between wavelengths λ and $\lambda + d\lambda$ is defined as

$$dI = R(\lambda)d\lambda \tag{2.1}$$

where $R(\lambda)$ is called the radiancy. The spectral dependence of $R(\lambda)$ has a well-defined dependence on the wavelength and temperature of the blackbody. In Fig. 2.2 we show how the experimentally observed $R(\lambda)$ behaves at different temperatures. Several interesting experimental observations are made in regard to the emitted radiation:

• The total intensity has the the behavior

$$I = \int_0^\infty R(\lambda)d\lambda \propto T^4$$

or

$$I = \sigma T^4 \tag{2.2}$$

This is known as Stefan's law. The constant σ is called the Stefan-Boltzmann constant. It is found to have a value

$$\sigma = 5.67 \times 10^{-8} \text{ Wm}^{-2}\text{K}^4$$

• The radiancy maximizes at a certain wavelength λ_{max} as can be seen from Fig. 2.2. The temperature dependence of this wavelength is given by

$$\lambda_{max} \propto \frac{1}{T} \tag{2.3}$$

The proportionality constant is given by the relation

$$\lambda_{max}T = 2.898 \times 10^{-3} \text{ mK}$$

This relation is known as Wien's displacement law.

According to classical physics, radiancy is given by the Rayleigh-Jeans law,

$$R(\lambda) = \frac{8\pi}{\lambda^4}k_BT\frac{c}{4} \tag{2.4}$$

However, when careful experiments were carried out and the spectral density tabulated, it was found that the Rayleigh-Jeans law was applicable only in a small frequency range. In fact, as can be seen from the equation, the classical law predicts an infinite energy density at very short wavelengths—an obviously unphysical result. In Fig. 2.2 we show the spectral output for blackbody radiation at different blackbody temperatures. It can be seen that while the classical law gives a reasonable fit

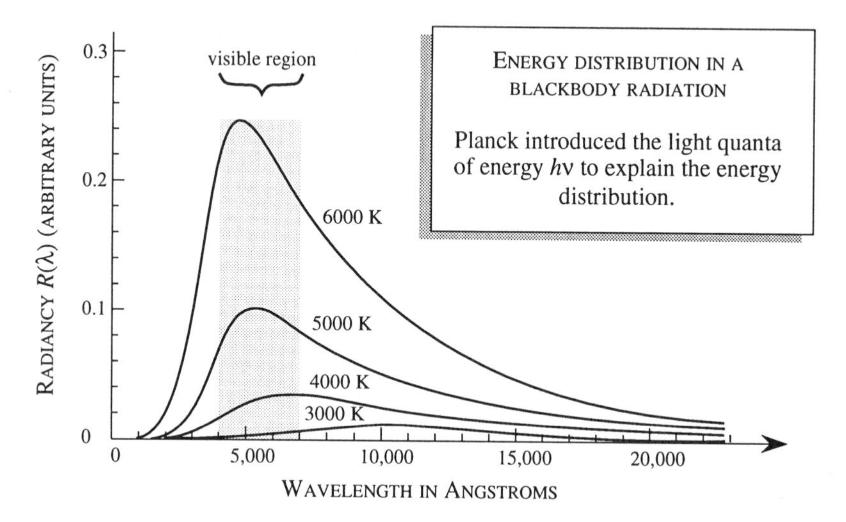

Figure 2.2: Measured spectral energy distribution of a blackbody radiation. The explanation of such experimental observations forced Planck to introduce a constant h (now called Planck's constant).

to experiments for long wavelengths, it completely fails at short wavelengths. The entire spectrum was only understood when Planck suggested that an electromagnetic wave with frequency ω exchanges energy with matter in a "quantum" given by

$$E = h\nu = \hbar\omega \tag{2.5}$$

Here h is a universal constant called Planck's constant. The symbol $\hbar$ stands for $h/2\pi$. This assumption seemed to suggest that light waves have a well-defined energy just as particles do. In Chapter 3 we will see how quantum mechanics allows us to understand the blackbody radiation measurement.

The quantity h or $\hbar$ that has been introduced is called Planck's constant and has a value

$$\begin{aligned} h &= 6.6261 \times 10^{-34} \text{ J.s} \\ \hbar &= \frac{h}{2\pi} = 1.05 \times 10^{-34} \text{ J.s} \end{aligned} \tag{2.6}$$

2.2.2 Waves Behaving as Particles: Photoelectric Effect

If one follows the history of the development of the field of optics one finds that Newton believed that light was made up of particles. He proposed the corpuscular theory in which light particles or corpuscles traveled from one point to another in straight lines. When they encountered a boundary they were reflected or refracted according to Snell's laws.

Huygens, on the other hand, proposed that light was made up of waves. Both the corpuscular theory and the wave theory explained reflection and refraction laws. However, when the phenomena of interference and diffraction were discovered, the wave theory was established and the particle picture was discarded. Eventually the wave picture was understood through Maxwell equations.

An important outcome of the wave theory is that the energy of a light beam can change continuously. As the intensity of the wave increases, the energy carried by a light beam increases. In most experiments this expectation is indeed verified. However, in 1887 Heinrich Hertz carried out an experiment which the wave theory of light was unable to explain. The experiment is known as the photoelectric effect.

In the photoelectric experiment, optical radiation impinges upon a material system and electrons are knocked out due to the interaction of the light with electrons. A typical experimental arrangement used is shown in Fig. 2.3. A potential is applied between the emitter and the collector. If the impinging light cannot knock an electron out of the metal emitter, there will be no photocurrent. If electrons are knocked out these electrons can make it to the collector *if their energy is larger than the potential energy* eV_{ext} *between the emitter and the collector.*

Let us assume that the electrons in the metal need to overcome an energy $e\phi$ in order to escape from the metal. The quantity $e\phi$ is called the workfunction of the metal and arises from the binding of the electrons to the metal ion. If the impinging light beam gives the emitted electrons an energy E_{em}, the electrons will emerge from the metal with an energy $E_{em} - e\phi$. An opposing bias is applied to the emitter-collector and the value of this bias is adjusted so that the electrons emitted are just unable to make it to the collector. This value of the applied voltage V_s is called the stopping voltage. The current will go to zero when

$$E_{em} - e\phi = eV_s \tag{2.7}$$

Experimentally one can measure V_s as a function of the intensity and frequency of light.

Based on the classical understanding of electromagnetic waves, we expect the following to occur:

• The energy with which the electrons should emerge from the metal should be proportional to the intensity of the light beam. Thus as the intensity increases, the stopping voltage should also increase.

• The electron emission should occur at any frequency provided the intensity of the light beam is sufficiently high.

• There should be a time interval Δt between the switching on of the light beam and the emission of electrons. If A is the area over which the electron is confined (roughly equal to the area of an atom or 10^{-19} m^2), the time it should take the electron to gain an energy ΔE is

$$\Delta t = \frac{\Delta E}{IA}$$

where I is the light intensity. If we use $\Delta E \sim 1.0$ eV and $I \sim 1$ W/cm^2, we find that $\Delta T \sim 10^{-3}$ s.

The three observations expected from classical physics all seem consistent with our physical intuition. However, actual experiments show the following to be true:

• If the frequency of light is below a certain value, there is no emission of electrons, regardless of intensity as shown in Fig. 2.3b.

• *The stopping potential is completely independent of the intensity of light.* A typical result is shown in Fig. 2.3c. As can be seen from this figure, the stopping voltage is unaffected by intensity although the photocurrent scales with intensity.

• The initial electrons are emitted within a nanosecond or so of the light being turned on. There is essentially no delay between the impingement of light and electron emission.

The experimental observations were thus completely opposed to what was expected on the basis of the wave theory for electromagnetic radiation. It was clear that a radical new interpretation of light was needed. As noted in the previous subsection, Max Planck had developed his formalism to explain the spectral density of blackbody radiation. Based on Planck's ideas, Einstein saw that the photoelectric effect could be explained if light was regarded as made up of particles with energy

$$E = \hbar\omega = h\nu \tag{2.8}$$

Thus light was to be regarded not as waves but discrete bundles or *quanta* of energy. These quanta were later called photons.

In Einstein's theory electrons are emitted by a single photon knocking the electron out. Thus the kinetic energy of the emitted electron is

$$E_e(K.E) = h\nu - e\phi = eV_s \tag{2.9}$$

There is no dependence of the electron energy on the intensity of light. A beam with higher intensity has more photons, but each photon has the same energy. The cutoff frequency for electron emission is given by the relation

$$h\nu = e\phi \tag{2.10}$$

In Fig. 2.4 we show the work function values for several metals. Also shown are the dependence of stopping voltage on the frequency of light. The slope of this curve is h/e, as can be seen from Eqn. 2.9.

2.2.3 Particles Behaving as Waves: Specific Heat of Metals

Some of the most careful experiments being carried out in the late 1800s involved heat capacities or specific heats of materials. To develop an understanding of the heat capacities of materials, a model was developed in which materials were characterized as insulators and metals. The difference between the two kinds of materials was assumed to arise from the metals having electrons which were "free" to carry current. It was therefore assumed that these extra electrons would also contribute to the heat capacity of metals. It was thus expected that the electron contribution would make metals have a higher heat capacity than that of an insulator. In the insulator, the heat capacity is due to the thermal energy associated with the vibration of the atoms. From classical physics models it can be shown that this gives a maximum heat capacity at high temperatures of $3R$ ($R = Nk_B$ when N is the Avogadro number).

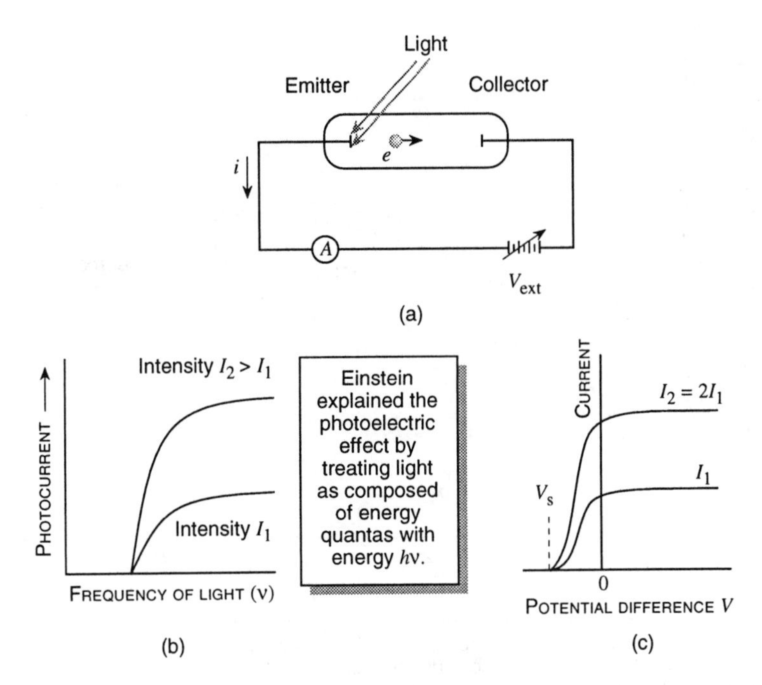

Figure 2.3: (a) A schematic of the experimental setup used for studying the photoelectric effect; (b) photocurrent as a function of the frequency of the impinging light for a fixed applied bias; and (c) photocurrent versus applied bias for when the optical signal frequency is above the threshold frequency.

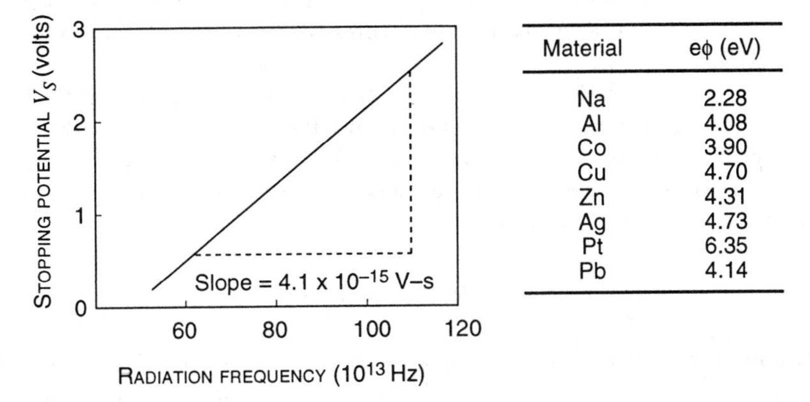

Material	eϕ (eV)
Na	2.28
Al	4.08
Co	3.90
Cu	4.70
Zn	4.31
Ag	4.73
Pt	6.35
Pb	4.14

Figure 2.4: Stopping voltage versus frequency results for sodium. The slope of the curve is h/e. Also shown are the work functions of several metals.

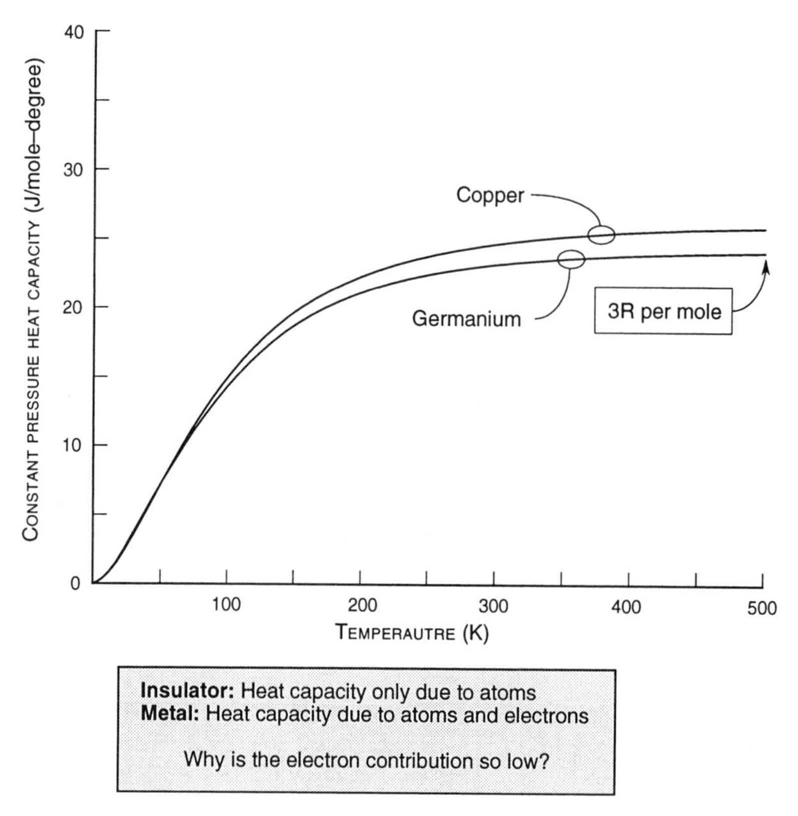

Figure 2.5: Heat capacity of germanium and copper as a function of temperature. The small difference between the metal and insulator heat capacity presented a dilemma to classical physics.

In Fig. 2.5 we show the heat capacity of a metal (copper) and an insulator (germanium). We see that at high temperatures the heat capacity per mole reaches a value of $3R$.

If we assume that electrons in metals can be treated as forming a gas (like molecules in the air in a room) we can use thermodynamics to calculate their average energy. According to classical thermodynamics, the average energy per mole of a metal coming from electrons is (n is the valence of the metal so that each atom has n electrons that form the electron gas)

$$\langle U(\text{electron}) \rangle = \frac{3}{2} n N k_B T \tag{2.11}$$

This would give a heat capacity (which is just the derivative of the total energy with respect to temperature) of $\frac{3}{2}nR$. If the valence is one we would then see an increased heat capacity of $\frac{3}{2}R$ for the metal in comparison to insulators.

In Fig. 2.5 we show the heat capacity of copper which has a valence of 1. The figure shows that the heat capacity of the metal is in fact almost the same as that of an insulator. This observation created a dilemma which could not be understood on the basis of any classical concept. It was only resolved when Sommerfeld applied quantum mechanics and quantum statistics to the electron problem in metals. This involved treating the electrons as waves and applying what is known as Pauli exclusion principle to them. In Chapter 4 we will see how the results of Fig. 2.5 are explained by quantum mechanics.

2.2.4 Particles Behaving as Waves: Atomic Spectra

We now know that a major area where classical physics broke down was the area of atomic physics. One of the earliest models for the atom was proposed by J. J. Thomson, who was the first to identify the electron and measure the ratio of its charge to mass. Thomson built the atomic model on the basis of classical physics. He assumed that the atom was made up of uniform sphere of positive charge Ze and radius R in which negatively charged electrons were embedded, as shown in Fig. 2.6. The size of the positively charged sphere is assumed to be of the order of an Angstrom or so.

The electrons are assumed to be embedded in the uniform charge at various distances. If r is the distance of an electron from the center, the force on it is (from classical electrostatics)

$$\mathbf{F} = \frac{Ze^2\mathbf{r}}{4\pi\epsilon_0 R^3} \tag{2.12}$$

At equilibrium there would be a balancing force from the other negatively charged electrons. Away from the equilibrium position there is a linear restoring force on the electron and it is assumed the electron will oscillate about its mean position just like a pendulum does.

The frequency of the oscillation is

$$\nu = \frac{1}{2\pi}\sqrt{\frac{Ze^2}{4\pi\epsilon_0 R^3 m_0}} \tag{2.13}$$

where m_0 is the electron mass. The oscillating electron would radiate electromagnetic radiation of frequency ν.

Thomson's model for an atom did not survive for long. Several problems besieged it. The frequencies of radiation emitted from atoms were not in agreement with what the model predicted. Also, scattering experiments in which scattering of alpha particles from atoms was studied showed that most of the atom was *empty*. It was found that the positive charge was not distributed over an Angstrom but over a much smaller region. Rutherford, in whose laboratory the scattering experiments were done, proposed an atomic model in which the positive charge Ze was confined in an extremely small nucleus along with some other neutral particles (neutrons).

The most serious challenge for any of the pre-quantum mechanics model arose from the detailed measurements of the frequencies of the emitted and absorbed radiation by atoms. When atoms are excited (say, by electromagnetic radiation), they can absorb the radiation. Once they absorb radiation, they can emit radiation as well. It was found experimentally that emitted and absorbed spectra from a species of atoms consisted of several series of sharp lines i.e., discrete frequencies.

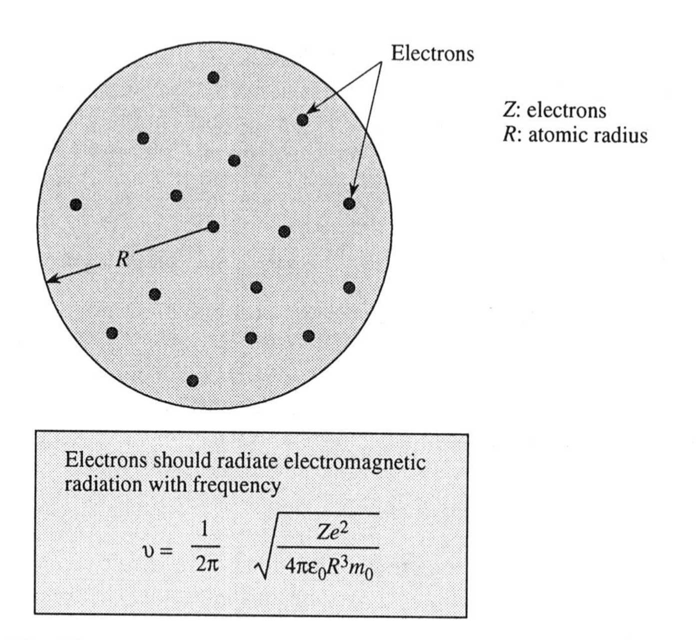

Figure 2.6: The Thomson model for an atom. The Z electron with charge $-e$ are point particles embedded in a uniformly charged positive sphere of radius R.

It was possible to fit simple relations to the positions of these lines. For example, Johannes Balmer found that the emission wavelengths of hydrogen in the visible regime could be fit to the relation (the Balmer formula)

$$\lambda_n = 364.5\frac{n^2}{n^2-4} \text{ nm}; \quad n = 3, 4, \ldots \tag{2.14}$$

In fact, other groups of lines in H-spectra were fitted to other expressions. For example, we have the following:
Paschen Series:

$$\lambda_n = 820.1\frac{n^2}{n^2-3^2} \text{ nm}; \quad n = 4, 5, \ldots \tag{2.15}$$

Lyman Series:

$$\lambda_n = 91.35\frac{n^2}{n^2-1} \text{ nm}; \quad n = 2, 3, \ldots \tag{2.16}$$

Other atoms were found to have spectra which satisfied similar relations. In Fig. 2.7 we show series of lines observed in atomic spectra of hydrogen.

The observation of atomic spectra showed that the electrons inside atoms occupy well-defined discrete energy levels which have a special relationship to each other.

An important new development in our understanding of atomic structure came from Bohr who, like Rutherford, assumed that the nucleus was essentially a point particle and the electrons spun around the nucleus, just as planets orbit the sun. *However, unlike the planets, the electrons can only go around in orbits in which*

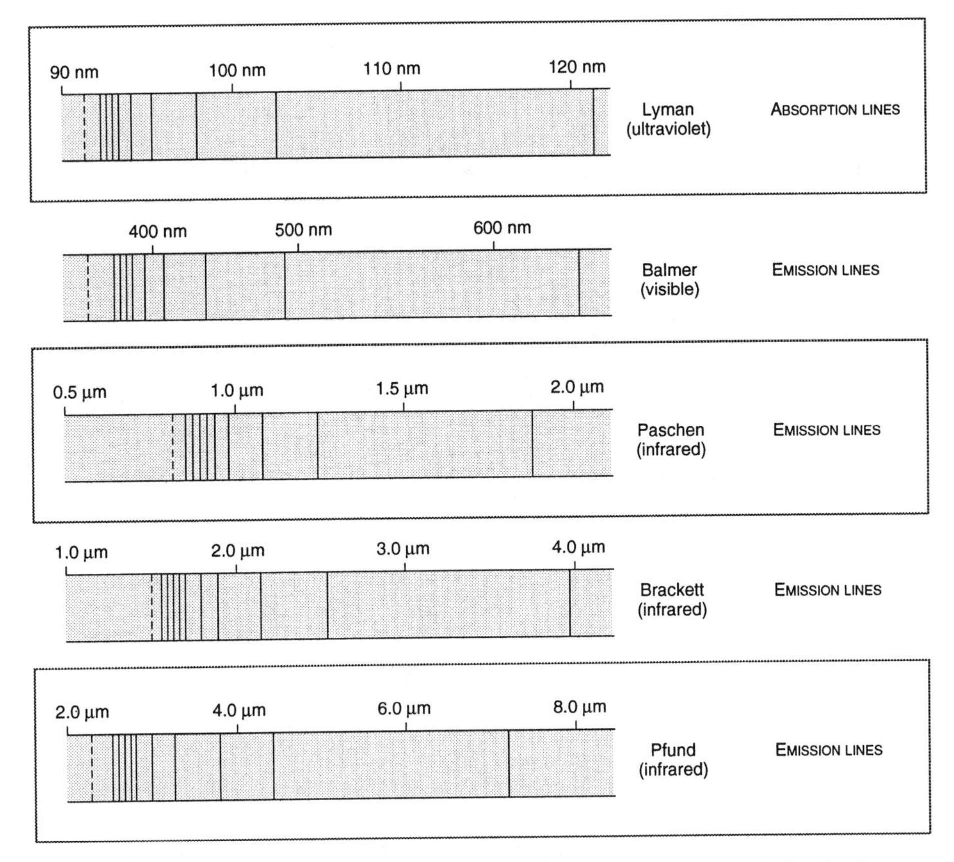

Figure 2.7: Emission and absorption lines in hydrogen. There is a regularity in the spacings of the spectral lines and the lines get closer as they reach the upper limit of each series (denoted by the dashed lines).

the angular momentum was an integral multiple of $\hbar$. If r is the radius of an orbit, we must have

$$m_0 v r = n\hbar; n = 1, 2 \ldots \tag{2.17}$$

By proposing this postulate, Bohr made a profound intellectual leap. He was able to fit the emission and absorption spectra of the hydrogen atom and he was able to explain why the electron does not radiate continuously even though it is orbiting the nucleus. The electron cannot radiate unless it *jumps from one allowed orbit to another.*

Based on the postulate that the electron orbits were *quantized* Bohr was able to calculate the *allowed energies* of electrons in an atom. Equating the centripetal force and the Coulombic force we get

$$F = \frac{1}{4\pi\epsilon_0}\frac{e^2}{r^2} = \frac{m_0 v^2}{r} \tag{2.18}$$

The kinetic energy is now

$$K = \frac{1}{2} m_0 v^2 = \frac{1}{8\pi\epsilon_0} \frac{e^2}{r} \tag{2.19}$$

The potential energy is the Coulombic energy

$$U = -\frac{1}{4\pi\epsilon_0} \frac{e^2}{r} \tag{2.20}$$

and the total energy is

$$E = -\frac{1}{8\pi\epsilon_0} \frac{e^2}{r} \tag{2.21}$$

From Eqns. 2.18 and 2.17 we have

$$(m_0 v) v = \frac{e^2}{4\pi\epsilon_0} \text{ or } v = \frac{e^2}{4\pi\epsilon_0} \cdot \frac{1}{n\hbar} \tag{2.22}$$

From this equation and Eqn. 2.17, we have, for the allowed orbit radii,

$$r_n = \frac{4\pi\epsilon_0}{m_0 e^2} (n\hbar)^2 \tag{2.23}$$

Substituting this equation in Eqn. 2.21 for the electron energy, we have

$$\begin{aligned} E_n &= -\frac{m_0 e^4}{32\pi^2 \epsilon_0^2 \hbar^2} \frac{1}{n^2}; \quad n = 1, 2, \ldots \\ &= -\frac{13.6}{n^2} \text{ eV} \end{aligned} \tag{2.24}$$

The allowed energy levels are shown in Fig. 2.8. The allowed radii of the orbits are

$$r_n = \frac{4\pi\epsilon_0 \hbar^2}{m_0 e^2} n^2 = a_0 n^2 \tag{2.25}$$

where

$$a_0 = \frac{4\pi\epsilon_0 \hbar^2}{m_0 e^2} = 0.529 \text{ Å} \tag{2.26}$$

Based on his model Bohr was able to provide a model which was consistent with observations made on the hydrogen atom. For example, the emission lines resulted from an electron jumping from a higher energy level to a lower energy level, as shown in Fig. 2.8b. Absorption lines resulted from reverse transitions. However, this model was not able to explain the spectra of atoms with more than one electron. Also, the hypothesis of restricted angular momentum leaves an uneasy feeling, even though the model explains the hydrogen atom spectra. So the puzzle remained and was only resolved with the development of quantum mechanics.

The experiments mentioned and many others seemed to suggest that there was a *duality* in nature. Particles sometimes behaved as waves and vice versa. It is a great tribute to the geniuses of the time that they were able to resolve these issues and develop a formalism which explained these remarkable new results and yet *continued to explain phenomena that were classical in nature.*

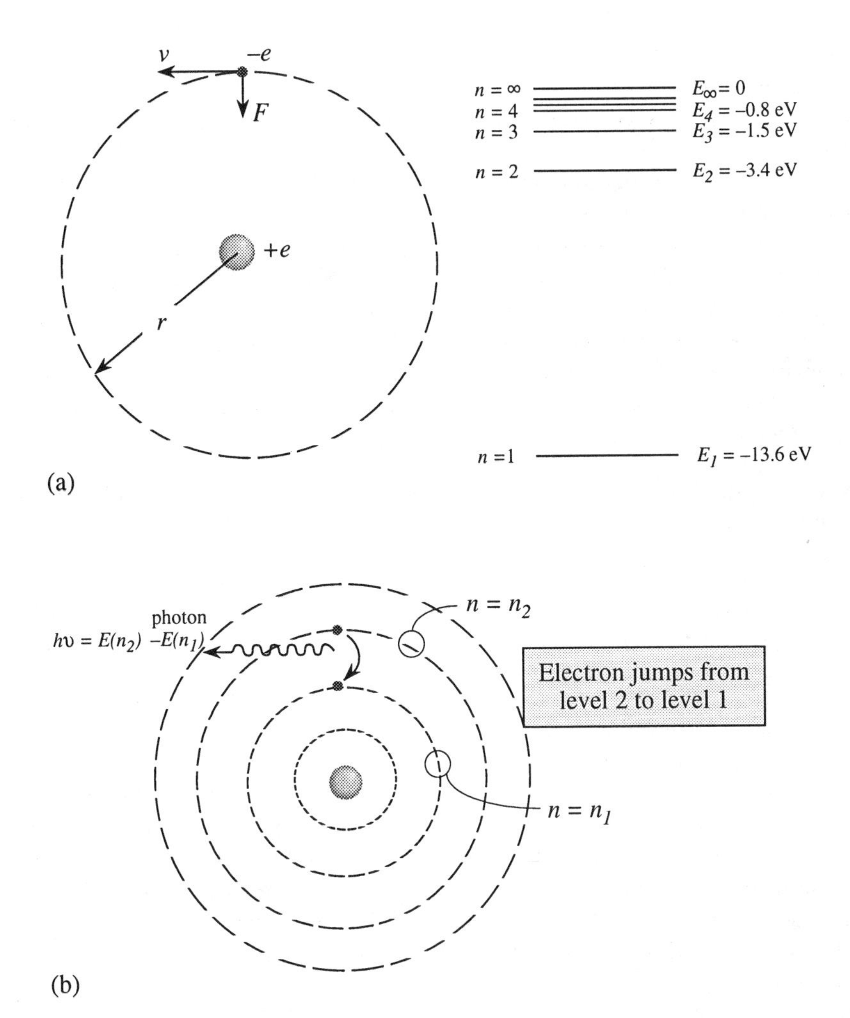

Figure 2.8: (a) A conceptual picture of the Bohr model along with the energy levels for a hydrogen atom; and b) discrete spectral lines are explained by transitions of the electron from one level to another.

2.3 WAVE-PARTICLE DUALITY: A HINT IN OPTICS

Today we know that classical physics tells us that most optical phenomena can be understood on the basis of Maxwell equations, which describe light as waves. This understanding was not always so biased in favor of waves. In the history of optics it is well known that there was considerable disagreement on the nature of light, i.e., whether light was made up of particles (Newton's corpuscular theory) or of waves (Huygens' theory). There were several phenomena, such as reflection and refraction, that could be understood by both approaches. Thus Snell's law, which tells us how light bends when it goes from one medium to another, is explained quite well by treating light as particles which move in straight lines in a given medium. The branch of optics known as geometric optics does not invoke any wave aspect of light and is quite successful in many areas. Of course, these phenomena are also explained equally well by wave optics.

Newton's corpuscular theory was eventually discarded when experiments showing interference effects (first done by Thomas Young) were demonstrated. It thus appears that wave optics encompasses geometric optics, but goes beyond it. An examination of the wave equation for light shows that (see Fig. 2.9) geometric optics (which ignores the wave nature of light) adequately describes optical phenomenon if the characteristic distance L, over which refractive index in the media changes, is large compared to the wavelength of light. *Thus the wave nature of light is only revealed when the characteristic dimensions are smaller than the wavelength of light.*

One can now ask the following questions:

- Is it possible that objects that appear as particles to us are really waves, but under *most* conditions their wave nature is not revealed?

- Is it possible that equations we see in classical mechanics are a *small wavelength* limit to a more general equation, just like geometric optics is a small-wavelength limit to wave optics?

These questions are valid and did occur to physicists of the nineteenth century. However, at that time, there was no experimental evidence that required that one should, indeed, look for a more general equation to describe particles. Once the experimental results, some of which have been described here, started to appear, physicists were forced to generalize classical mechanics and develop wave mechanics (i.e., quantum mechanics) to describe particles. Just as geometric optics is adequate when $L \gg \lambda$, classical mechanics is adequate when the "wavelength" of a particle is small compared with dimensions over which potential energy in a problem changes significantly.

2.3.1 De Broglie's Hypothesis

A brilliant hypothesis was put forward by de Broglie, according to which a particle was given a wave character by assigning a wavelength to it. The wavelength is given by

$$\lambda = \frac{h}{p} \tag{2.27}$$

where p is the particle momentum. For most situations encountered by us in our daily life the wavelength of particles is extremely small. It was later understood that when the wavelength of a particle is small compared with the distance over which potential energy changes, classical concepts are quite valid. This is just like the case encountered in optics, where geometric optics is valid if the distance over which refractive index changes is large compared with the wavelength of light.

There are cases where the potential energy changes over distances comparable or smaller than the de Broglie wavelength of a particle. In such cases the wave

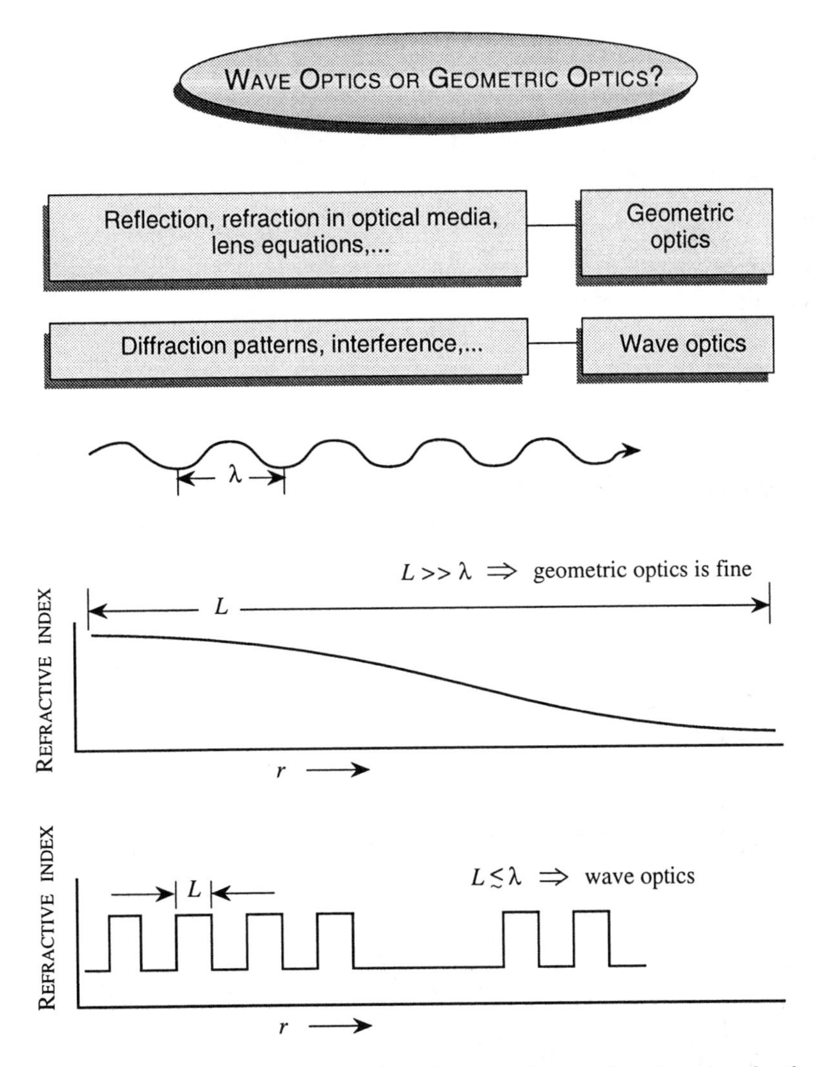

Figure 2.9: The importance of the comparison between the wavelength and scale of spatial variation in refractive index determines if geometric or wave optics is needed.

character of the particle gets manifested and quantum mechanics is needed. These ideas are illustrated schematicaly in Fig. 2.10.

2.4 SCHRÖDINGER EQUATION

We have seen that in order to describe the behavior of waves we need to solve an appropriate wave equation. The solutions of the wave equation (a differential equation) give us the *allowed* forms of the wave. The question that we now ask is the following: If particles have a wave-like behavior, what is the nature of the equation they satisfy? This question was first answered by Schrödinger and led to

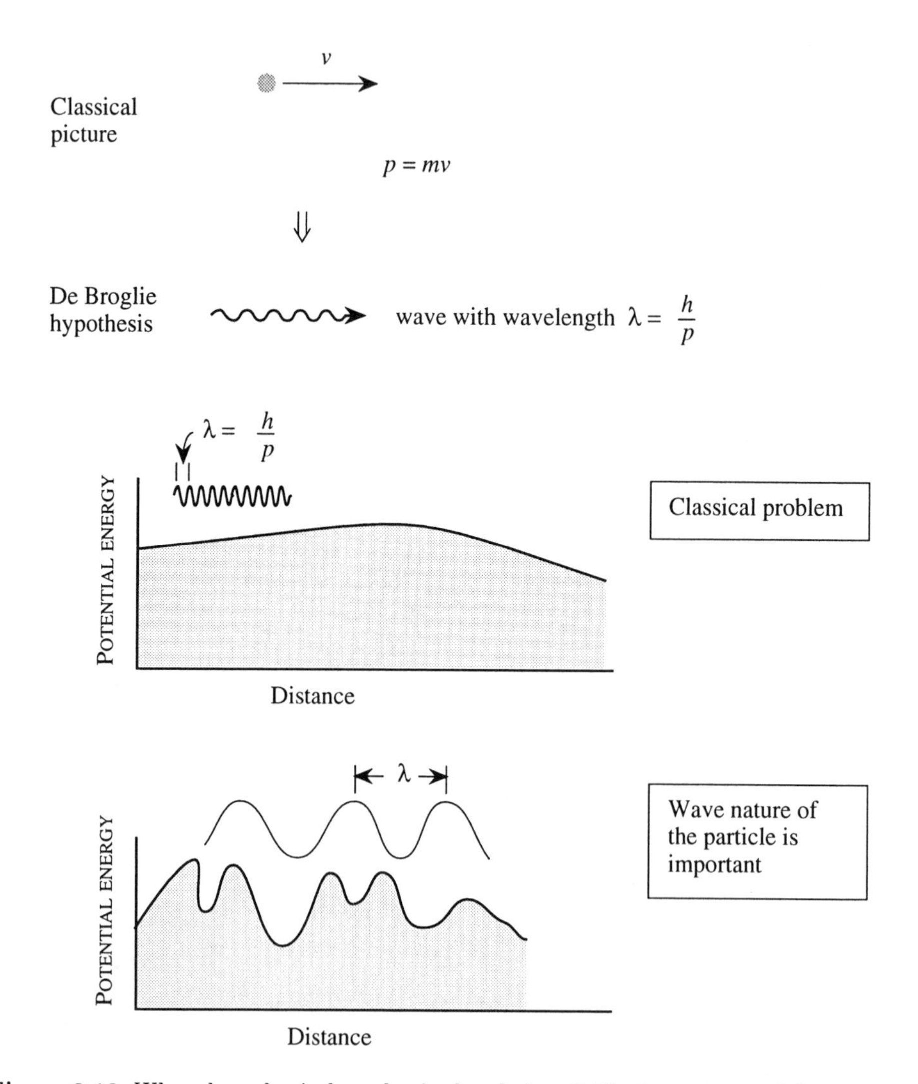

Figure 2.10: When does classical mechanics break down? The importance of the comparison between the de Broglie wavelength and scale of spatial variation in potential energy determines if quantum mechanics is needed.

the equation bearing his name.

It is important to note that the Schrödinger equation describing the non-relativistic behavior of particles cannot be derived from any fundamental principles—just like Newton's equation cannot be derived. It is an equation that gives us solutions that can explain experimentally observed physical phenomena.

A clue to the form of the Schrödinger equation comes from the relation between a particle momentum and its wavelength as given by de Broglie. According to this relation, for a particle in free space the kinetic energy is (replacing the momentum by wavelength or wavevector)

$$K = \frac{p^2}{2m} = \frac{\hbar^2 k^2}{2m}; \quad k = \frac{2\pi}{\lambda} \tag{2.28}$$

If we examine the identity

$$K + U = E$$

where U is the potential energy and E is the total energy, we can write a *wave equation* for a wave with amplitude ψ

$$(K + U)\psi = E\psi$$

In case the particle is free (i.e., the potential energy is zero) we have

$$K\psi = E\psi$$

Using a hint from Eqn. 2.28 for the form of the kinetic energy, we write the kinetic energy as an operator (say, for a one-dimensional case)

$$K = -\frac{\hbar^2}{2m}\frac{d^2}{dx^2} \tag{2.29}$$

We now see that the wave equation takes the form

$$-\frac{\hbar^2}{2m}\frac{d^2}{dx^2}\psi = E\psi$$

and the general solution to the equation is

$$\psi = Ae^{ikx} + Be^{-ikx}$$

The kinetic energy is

$$K = E = \frac{\hbar^2 k^2}{2m}$$

which is consistent with the relation we get from the de Broglie relation.

The Schrödinger equation is written by expressing the kinetic energy as a second-order operator, so that the identity

$$(K + U)\psi = E\psi$$

becomes

$$\boxed{\left[-\frac{\hbar^2}{2m}\nabla^2 + V(r)\right]\psi = E\psi} \tag{2.30}$$

To write the time dependence of the particle-wave we use the analogy for the phase of the particle-wave which goes as

$$\psi(t) \sim e^{iwt}$$

We can write the time-dependence of the wave as

$$\psi(t) \sim \exp\left(-\frac{iEt}{\hbar}\right) \tag{2.31}$$

so that the knowledge of ψ at any time allows one to predict the value of ψ at all times via the equation

$$\frac{\partial\psi}{\partial t} = \frac{-i}{\hbar}E\psi$$

or

$$\boxed{i\hbar\frac{\partial\psi}{\partial t} = E\psi} \tag{2.32}$$

Note that in this development, the quantity h or $\hbar$ has been introduced to define the proportionality between the energy and the particle-wave frequency. In classical physics, this quantity is assumed to be zero. As noted earlier in this chapter, experiments carried out in the early twentieth century showed that $\hbar$ was not zero but had a value of 1.055×10^{-34} J.s.

It is important to emphasize that the derivation given above *does not constitute a proof for the Schrödinger equation.* Only experiments could determine the validity of such an extension.

In our "derivation" of the Schrödinger equation we see that the observable properties of the particle such as momentum and energy appear as *operators* operating on the particle-wavefunction. From Eqns. 2.27 and 2.30 we see the operator form of these observables:

$$\begin{aligned} p_x &\rightarrow -i\hbar\frac{\partial}{\partial x} \\ p_y &\rightarrow -i\hbar\frac{\partial}{\partial y} \\ p_z &\rightarrow -i\hbar\frac{\partial}{\partial z} \end{aligned} \tag{2.33}$$

$$E \rightarrow i\hbar\frac{\partial}{\partial t} \tag{2.34}$$

This observation that physical observables are to be treated as operators is quite generic in the quantum description. The energy operator is called the *Hamiltonian* in quantum mechanics.

EXAMPLE 2.1 Calculate the wavelength associated with a 1 eV (a) photon, (b) electron, and (c) neutron.

(a) The relation between the wavelength and the energy of a photon is

$$\begin{aligned} \lambda_{ph} &= \frac{hc}{E} = \frac{(6.6 \times 10^{-34}\ \text{J.s})(3 \times 10^{8}\ \text{m/s})}{(1.6 \times 10^{-19}\ \text{J})} = 1.24 \times 10^{-6}\ \text{m} \\ &= 1.24\ \mu\text{m} \end{aligned}$$

(b) The relation between the wavelength and energy for an electron is ($k = 2\pi/\lambda$)

$$\lambda_e = \frac{h}{\sqrt{2m_0 E}} = \frac{6.6 \times 10^{-34} \text{ J.s}}{[2(0.91 \times 10^{-30} \text{ kg})(1.6 \times 10^{-19} \text{ J})]^{1/2}}$$
$$= 12.3 \text{ Å}$$

(c) For the neutron using the same relation, we have

$$\lambda_n = \lambda_e \left(\frac{m_0}{m_n}\right)^{1/2} = \lambda_e \left(\frac{1}{1824}\right)^{1/2} = 0.28 \text{ Å}$$

The wavelengths of different "particles" play an important role when these particles are used to "see" atomic phenomena in a variety of microscopic techniques.

2.5 WAVE AMPLITUDE

When we solve a wave equation describing electromagnetic waves, we get a solution that gives the amplitude of the fields and the frequency of vibration. Similarly, when we solve the Schrödinger equation for a given potential energy term, we get a set of solutions $\{E_n, \psi_n\}$, which give us the allowed energy E_n of the particle along with the wavefunction ψ_n. The wavefunction that is introduced must provide a complete quantum mechanical description of the behavior of a particle of mass m in a potential energy V. We now develop a mathematical and physical interpretation of the wave amplitude. This interpretation must be consistent with the experimental observations.

Let us briefly recall the meaning of the wave amplitude $\phi(r,t)$ in the wave equations describing sound waves or light waves. The energy of the wave at a particular point in space and time is related to $|\phi(r,t)|^2$. It is "more likely" that the wave is found in regions where ϕ is large. Thus the quantity $|\phi(r,t)|^2$ represents some sort of probability function for the wave. A similar interpretation has to be developed for a wave ψ describing a particle of mass m. The interpretation has to be statistical in nature. It is thus natural to regard ψ as the measure of finding the particle at a particular point and space. We assume, therefore, that the product of ψ and its complex conjugate ψ^* is the probability density

$$P(r,t) = \psi^*(r,t)\psi(r,t) = |\psi(r,t)|^2 \tag{2.35}$$

Thus the probability of finding the particle in a region of volume $dx\ dy\ dz$ around the position r is simply

$$\boxed{P(r,t)\ dx\ dy\ dz = |\psi(r,t)|^2\ dx\ dy\ dz} \tag{2.36}$$

2.5.1 Normalization of the Wavefunction

If we assume that the particle is confined to a certain volume V, we know that the probability of finding it somewhere in the region must be unity, so that one must have the following condition satisfied:

$$\int_V |\psi(r,t)|^2\ d^3r = 1 \tag{2.37}$$

The volume over which the integral is carried out is sometimes arbitrary and the coefficient of the wavefunction must be chosen to satisfy the normalization integral. The normalization factor does not change the fact that ψ is a solution of the Schrödinger equation due to the homogeneous (linear) form of the equation in ψ. Thus if ψ is a solution, and if A is a constant, then $A\psi$ is also a solution.

Note that once a normalization is done for a wavefunction, the normalization is valid for all times. This can be shown by considering the time derivative of the total integrated probability and by using the Schrödinger equation:

$$\begin{aligned}
\frac{\partial}{\partial t}\int_V P(r,t)\, d^3r &= \int_V \left(\psi^*\frac{\partial\psi}{\partial t} + \frac{\partial\psi^*}{\partial t}\psi\right) d^3r \\
&= \frac{i\hbar}{2m}\int_V \left[\psi^*\nabla^2\psi - (\nabla^2\psi^*)\psi\right] d^3r \\
&= \frac{i\hbar}{2m}\int_V \nabla\cdot\left[\psi^*\nabla\psi - (\nabla\psi^*)\psi\right] d^3r \\
&= \frac{i\hbar}{2m}\int_A \left[\psi^*\nabla\psi - (\nabla\psi^*)\psi\right]_n dA
\end{aligned} \tag{2.38}$$

Here we have used the time-dependent Schrödinger equation to represent $\frac{\partial\psi}{\partial t}$ and $\frac{\partial\psi^*}{\partial t}$, and used Green's theorem (using partial integration) to represent a volume integral over V to a surface integral over the boundary area of A. The term $[\,]_n$ in the surface integral represents the normal to the surface element dA.

Writing

$$\mathrm{S}(r,t) = \frac{\hbar}{2im}\left[\psi^*\nabla\psi - (\nabla\psi^*)\psi\right] \tag{2.39}$$

we have

$$\frac{\partial}{\partial t}\int_V P(r,t)\, d^3r = -\int_V \nabla\cdot\mathrm{S}\, d^3r = -\int_A \mathrm{S}_n\, dA \tag{2.40}$$

To proceed further, we need to identify the boundary conditions that the wavefunction satisfies at the boundary of the volume V. Two classes of problems occur in nature: (i) problems where the potential energy term $V(r,t)$ is such that the wavefunction is localized strongly in some region inside the volume V and goes to zero at the boundaries; (ii) problems where the wavefunction extends over all space and periodic boundary conditions can be imposed on the wavefunction. In either case, the integral of Eqn. 2.40 vanishes so that the normalization integral is constant in time.

2.5.2 Probability Current Density

In classical physics, if we have a flow of particles, we can define the particle current density as the particle flow per second per unit area. In quantum mechanics a similar concept can be developed in a statistical sense. We have shown in Eqn. 2.40 that we have the differential relation

$$\frac{\partial P(r,t)}{\partial t} + \nabla\cdot\mathrm{S}(r,t) = 0 \tag{2.41}$$

This relation is similar to that obtained in continuity equations in classical physics and is associated with the conservation of flow of a fluid of density P and current density S, with no source or sink terms. We therefore interpret $S(r,t)$ as the probability current density. Note that since the momentum is associated with the operator $-i\hbar\nabla$, the velocity is associated with $(\hbar/im)\nabla$. We can see from Eqn. 2.39 that

$$\boxed{S(r,t) = \text{Real part of}\left(\psi^* \frac{\hbar}{im}\nabla\psi\right)} \tag{2.42}$$

It may be noted, however, that it is not possible to measure the current precisely at a given point in space because of the uncertainty in the values of velocity and position. However, the probability current density is a useful concept if there is little r dependence in the problem and an accurate velocity measurement is possible.

2.5.3 Expectation Values

In classical physics, when a measurement is made on a particle, the outcome is information on the particle energy position, momentum, etc. We need to know how to relate a physical observation to what we calculate by solving the Schrödinger equation. In view of the probabilistic nature of quantum mechanics one can only define the "expectation value" of an observable in a physical measurement. Using the probability function $P(r,t) = |\psi(r,t)|^2$, we define the *expectation value* for the measurement of the position of the particle as

$$\langle r\rangle = \int rP(r,t)\; d^3r = \int \psi^*(r,t) r\psi(r,t)\; d^3r \tag{2.43}$$

For the expectation value along different axes we have

$$\begin{aligned}\langle x\rangle &= \int \psi^* x\psi\; d^3r\\ \langle y\rangle &= \int \psi^* y\psi\; d^3r\\ \langle z\rangle &= \int \psi^* z\psi\; d^3r\end{aligned}$$

In all these equations, ψ is normalized to the volume of the integral.

In a similar manner, the expectation value for the potential energy of the particle is

$$\langle V\rangle = \int \psi^*(r,t)V(r,t)\psi(r,t)\; d^3r \tag{2.44}$$

To find the expectation values of momentum and energy, we need to express them in their operator form. We note that the expectation value of energy is

$$\langle E\rangle = \left\langle \frac{p^2}{2m}\right\rangle + \langle V\rangle \tag{2.45}$$

or, in the form of the differential operators,

$$\left\langle i\hbar\frac{\partial}{\partial t}\right\rangle = \left\langle -\frac{\hbar^2}{2m}\nabla^2\right\rangle + \langle V\rangle \tag{2.46}$$

This equation is consistent with Schrödinger equations with the following expectation values:

$$\langle E \rangle = \int \psi^* i\hbar \frac{\partial \psi}{\partial t} \, d^3r \tag{2.47}$$

$$\langle p \rangle = \int \psi^* (-i\hbar) \nabla \psi \, d^3r \tag{2.48}$$

The different momentum components are

$$\langle p_x \rangle = \int \psi^* (-i\hbar) \frac{\partial \psi}{\partial x} \, d^3r$$

$$\langle p_y \rangle = \int \psi^* (-i\hbar) \frac{\partial \psi}{\partial y} \, d^3r$$

$$\langle p_z \rangle = \int \psi^* (-i\hbar) \frac{\partial \psi}{\partial z} \, d^3r$$

Once we solve a Schrödinger equation, we will get one or more allowed wavefunctions. The expressions given above can then be used to determine the expectation values of various physical observables in each of these allowed states. Since all quantities calculated from the wavefunction which can be related to direct physical meaning have a form involving $\psi^*\psi$ product, it is clear that the wavefunction is only determined within a phase factor of the form $e^{i\alpha}$, where α is a real number. This indeterminacy in the wavefunction is unimportant, since it has no effect on physical results.

A very important and useful aspect of quantum mechanics is the principle of superposition of states which is directly related to the quantum mechanics wave equation being linear in the wavefunction ψ. If $\psi_1(q)$ is a state leading to measurement R_1 and $\psi_2(q)$ is a state leading to R_1, then every function of the form $a_1\psi_1 + a_2\psi_2$ (where c_1 and c_2 are constants) gives a state in which the results of measurements are known from the knowledge ψ_1 and ψ_2.

2.6 WAVES, WAVEPACKETS, AND UNCERTAINTY

In Chapter 1, Section 1.6, we have discussed the uncertainty that exists in wave phenomena between wavevector and the position of the wave. This uncertainty is not of concern in classical physics when we deal with particles. For example, we have no problem defining *precisely* any combinations of physical observables of the particle. However, we have seen that in quantum mechanics we must describe the properties of particles via a wave equation. This description will inevitably bring in an uncertainty in the precision with which we can simultaneously define certain physical observables.

In Chapter 1, Section 1.6 we have discussed (see Fig. 1.7) how a wave packet is constructed out of plane waves with different wave vectors. As pointed out there, we have an uncertainty relation, given by

$$\Delta k \, \Delta x \approx 1 \tag{2.49}$$

Similarly we have an uncertainty relation between the frequency and time

$$\Delta\omega \; \Delta t \approx 1 \tag{2.50}$$

A more precise treatment of the uncertainty shows that for the particle momentum-position and energy-time uncertainty we have the following relations, known as the Heisenberg uncertainty relations (using $p = \hbar k$; $E = \hbar\omega$):

$$\boxed{\Delta p_x \cdot \Delta x \geq \hbar/2} \tag{2.51}$$

$$\boxed{\Delta E \Delta t \geq \hbar/2} \tag{2.52}$$

Note that for the momentum-position, the uncertainty only exists between p_i, r_i i.e., between p_x, x, p_y, y and p_z, z. There is no uncertainty in the simultaneous measurement of, say, p_x and y.

2.6.1 Equations of Motion: The Ehrenfest Theorem

How does this *particle-wavepacket* respond to external forces? From our discussion in Section 2.3, we expect that if the potential energy of the particle changes negligibly over the dimensions of the wavepacket, the quantum equation of motion should agree with the classical equations.

Let us consider the rate of change of momentum of the particle, using the operator form of the momentum,

$$\begin{aligned}
\frac{d}{dt}\langle p_x\rangle &= -i\hbar\frac{d}{dt}\int \psi^*\frac{\partial\psi}{\partial x}\; d^3r \\
&= -i\hbar\left(\int \psi^*\frac{\partial}{\partial x}\frac{\partial\psi}{\partial t}\; d^3r + \int \frac{\partial\psi^*}{\partial t}\frac{\partial\psi}{\partial x}\; d^3r\right) \\
&= -\int \psi^*\frac{\partial}{\partial x}\left(-\frac{\hbar^2}{2m}\nabla^2\psi + V\psi\right)\; d^3r \\
&\quad + \int\left(-\frac{\hbar^2}{2m}\nabla^2\psi^* + V\psi^*\right)\frac{\partial\psi}{\partial x}\; d^3r
\end{aligned} \tag{2.53}$$

By using integration by parts, the terms involving the particle mass vanish, and we get

$$\begin{aligned}
\frac{d}{dt}\langle p_x\rangle &= -\int \psi^*\left[\frac{\partial}{\partial x}(V\psi) - V\frac{\partial\psi}{\partial x}\right]\; d^3r \\
&= -\int \psi^*\frac{\partial V}{\partial x}\psi\; d^3r
\end{aligned}$$

or

$$\boxed{\frac{d}{dt}\langle p_x\rangle = \left\langle -\frac{\partial V}{\partial x}\right\rangle} \tag{2.54}$$

The spatial derivative of the potential energy is the force applied, and thus one obtains a classical-like equation for the rate of change of the expectation value of the momentum.

The result above showing that the properties of a wavepacket can be determined from classical equations is called the Ehrenfest theorem. This is an extremely useful theorem in applied quantum mechanics, especially as used in describing various devices. In such cases an appropriate Schrödinger equation is solved to describe the allowed energy and momentum states of a particle. *Once the quantum problem is solved, the response of the particle to slowly varying external forces can be treated as if the particle is obeying classical equations.* This approach is widely used when we use quantum mechanics in describing devices such as semiconductor transistors and lasers.

EXAMPLE 2.2 In quantum mechanics the term $[A, B] = AB - BC$ is called the commutator of A and B. This commutator is not zero in quantum mechanics. Using the operator form of momentum $p = \frac{\hbar}{i}\frac{d}{dx}$, show that the following relations exist:

$$
\begin{aligned}
[p, x] &= -i\hbar \\
[p, x^n] &= -in\hbar x^{n-1}, \quad n > 1 \\
[p, A] &= -i\hbar \frac{dA}{dx}
\end{aligned}
$$

where A is a function of x.

Let $f(x)$ be an arbitrary function of x. Then, applying the commutator to $f(x)$, we have

$$
\begin{aligned}
[p, x]f(x) &= pxf(x) - xpf(x) \\
&= -i\hbar \left[f(x) + x\frac{d}{dx}f(x) - x\frac{d}{dx}f(x)\right] \\
&= -i\hbar f(x)
\end{aligned}
$$

Thus we get

$$[p, x] = -i\hbar$$

To verify the second equality, we note that

$$
\begin{aligned}
[p, x^2] &= px^2 - x^2 p \\
&= px^2 - x^2 p + xpx - xpx \\
&= x[px - xp] + [px - xp]x \\
&= -2i\hbar x
\end{aligned}
$$

By performing this procedure repeatedly, we get the second relation. Finally, for the third relation, we can consider A as a polynomial in x and the third result follows from the second one.

EXAMPLE 2.3 In an experiment known as the Frank and Hertz experiment, electrons are raised to an excited state by colliding with a beam of electrons. It is found that, even if the beam of electrons is monoenergetic, after the *collision with the atomic electrons* the energy of the electrons in the beam has a certain spread. Explain this spread. In a particular experiment the energy spread is found to be 10^{-6} eV. What is the lifetime of the excited state?

The electronic levels in an atom have a certain lifetime which produces an uncertainty in the energy needed to excite the electrons

$$\Delta E \ \Delta t = \frac{\hbar}{2}$$

For the given case

$$\Delta t = \frac{\hbar}{2\Delta E} = \frac{1.05 \times 10^{-34} \text{ J.s}}{2\left(10^{-6} \times 1.6 \times 10^{-19} \text{ J}\right)} = 3.28 \times 10^{-10} \text{ s}$$

EXAMPLE 2.4 Use the uncertainty relation to evaluate the ground state energy of the hydrogen atom.

Let us assume that the electron is confined to a radius r_o in the hydrogen atom. We will find the value of r_o that gives the lowest energy. The uncertainty principle tells us that the momentum associated with the electron is

$$p \sim \frac{\hbar}{r_o}$$

The electron-proton system energy is

$$\begin{aligned} E(r_o) &= \frac{p^2}{2m} + V(r) \\ &= \frac{\hbar^2}{2mr_o^2} - \frac{e^2}{4\pi\epsilon_o r_o} \end{aligned}$$

This is minimized when

$$\begin{aligned} r_o &= \frac{4\pi\epsilon_o\hbar^2}{me^2} \\ &= \frac{4\pi\left(8.84 \times 10^{-12} \text{ F/m}\right)\left(1.05 \times 10^{-34} \text{ J.s}\right)^2}{\left(9.1 \times 10^{-31} \text{ kg}\right)\left(1.6 \times 10^{-19} \text{ C}\right)^2} \\ &= 5.24 \times 10^{-11} \text{ m} \\ &= 0.524 \text{ Å} \end{aligned}$$

With this value of r_o, the energy becomes

$$\begin{aligned} E_1 &= -\frac{1}{2}\frac{me^4}{(4\pi\epsilon_o)^2\hbar^2} \\ &= -\frac{\left(9.1 \times 10^{-31} \text{ kg}\right)\left(1.6 \times 10^{-19} \text{ C}\right)^4}{2\left(4\pi \times 8.84 \times 10^{-12} \text{ F/m}\right)^2\left(1.05 \times 10^{-34} \text{ J.s}\right)^2} \\ &= -2.16 \times 10^{-18} \text{ J} \\ &= -13.5 \text{ eV} \end{aligned}$$

These results are quite accurate, considering the crude approximations used.

2.7 HOW DOES ONE SOLVE THE SCHRÖDINGER EQUATION?

2.7.1 What Are We Trying to Do?

Now that we have written down the Schrödinger equation which describes particle behavior, we will spend most of this textbook solving this equation for various cases and exploiting the results. What exactly do we mean by solving the Schrödinger equation and how do we go about obtaining the solutions? In classical mechanics, as outlined in Fig. 2.11, when we "solve" a problem for one or more particles, under the influence of some forces, we are describing the trajectory of the particle or particles. Questions like, "Is the particle forbidden from having certain energies?" "Is the particle momentum restricted to certain well-defined *quantized* values?" etc., simply do not arise for the classical particles.

If we examine classical physics for waves, as shown in Fig. 2.11, the question of allowed and forbidden solutions is very relevant. So is the question on allowed wavelengths being *quantized*. For example, when we solve Maxwell equations for electromagnetic waves in a dielectric waveguide, only certain modes are allowed to propagate in the guide. These modes have well-defined spatial electric and magnetic fields and well-defined frequencies. In such problems when we say we have "solved" the problem, it means that we know which modes are allowed, what the field profiles are, what the relation between the frequency and wavelength is, etc.

In quantum mechanics, when we solve a problem for a particle, we address not only how the particle responds to external forces acting on it, but also if the *wave nature of the particle places certain restrictions on the particle properties.* Usually we first address the question: What are the allowed states of the particle-wave for the given problem? Then we address the question: How do the states respond to the forces present? There are, however, some cases where the entire problem of the allowed states and the effect of forces is addressed simultaneously.

The Schrödinger equation in space is a second-order differential equation. It takes various forms, depending upon the nature of the potential energy term. Depending upon the form the equation takes, one can choose from a broad arsenal of mathematical tools. Many problems of interest can be solved analytically, since the Schrödinger equation takes a form which has already been solved by mathematicians. Other problems require numerical approaches which may require sophisticated computer software. In this book we will not spend our energy on the mathematical details of how differential equations are solved. We will provide solutions and focus on understanding the physical implications. We will now give an overview of some classes of problems that are of importance.

The Schrödinger equation becomes considerably simplified if the potential energy has no time-dependence. It is then possible to write the wavefunction solution as a product of the time-dependent and space-dependent terms:

$$\Psi(r,t) = f(t)\psi(r) \tag{2.55}$$

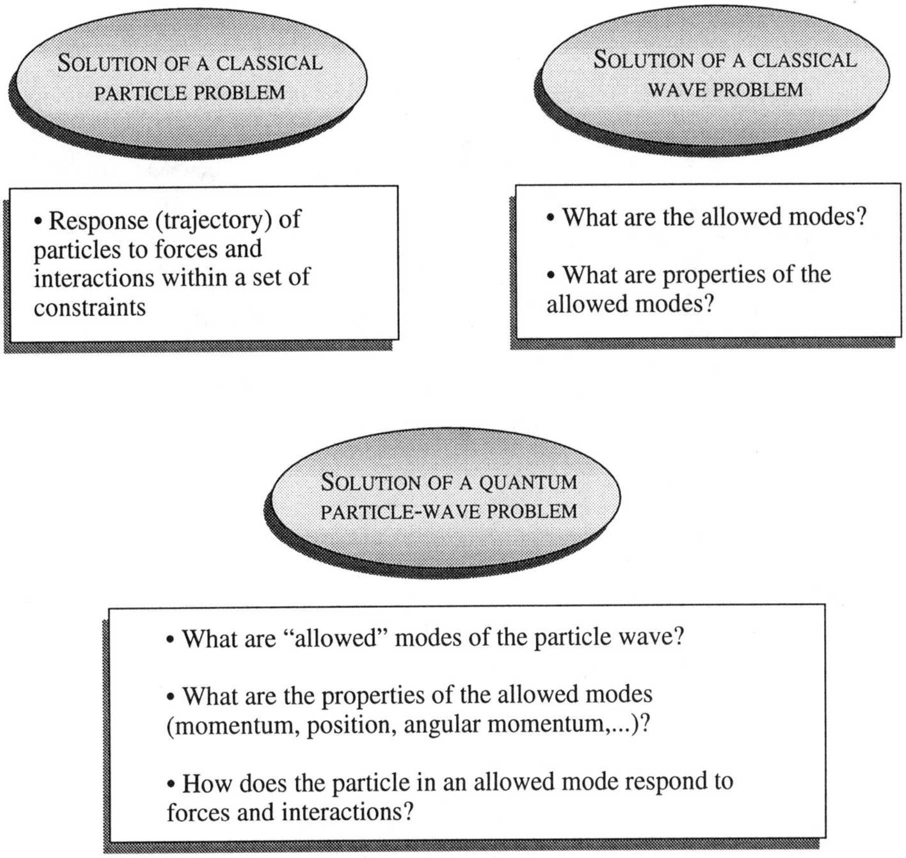

SOLUTION OF A CLASSICAL PARTICLE PROBLEM

- Response (trajectory) of particles to forces and interactions within a set of constraints

SOLUTION OF A CLASSICAL WAVE PROBLEM

- What are the allowed modes?
- What are properties of the allowed modes?

SOLUTION OF A QUANTUM PARTICLE-WAVE PROBLEM

- What are "allowed" modes of the particle wave?
- What are the properties of the allowed modes (momentum, position, angular momentum,...)?
- How does the particle in an allowed mode respond to forces and interactions?

Figure 2.11: A description of what a "solution of a problem" means in classical physics for particles and waves and in quantum mechanics for the "particle-wave."

where we have for the time-dependent term

$$i\hbar \frac{df}{dt} = Ef \tag{2.56}$$

and for the space-dependent part

$$\left[-\frac{\hbar^2}{2m} \nabla^2 + V(r) \right] \psi(r) = E\psi(r) \tag{2.57}$$

The solution for the time-dependent part f is straightforward:

$$\boxed{f(t) = C \exp \frac{iEt}{\hbar}} \tag{2.58}$$

where C is an arbitrary constant. *The energy E has to be obtained from the solution of the space-dependent Schrödinger equation.* Equations of the form given by Eqn. 2.57 are called eigenvalue equations and E, the allowed energies, are called

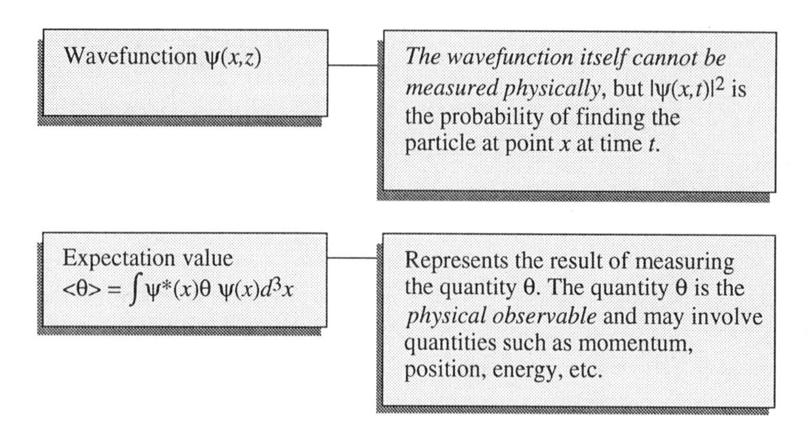

Figure 2.12: Key relations between the solutions of Schrödinger equation and physical observations.

eigenvalues while $\Psi(r,t)$ is the eigenfunction. The problem is solved when the set $\{E_n, \Psi_n(r,t)\}$ of eigenvalues E_n and eigenfunctions $\Psi_n(r,t)$ are known. The subscript n represents the various allowed values and may represent discrete or continuous eigenvalues. Once the eigenfunctions are known, we can evaluate the expectations values of various physical observables (see Fig. 2.12)

For the cases discussed here, i.e., $V(r)$ has no time-dependence, the eigenfunction is called a *stationary state* since $|\Psi(r,t)|^2$ has no time-dependence. In Fig. 2.13 we show some of the important approaches used for solving Schrödinger equation when there is no time dependence in the Hamiltonian.

Problems with Simple Potential Energy Terms

It may turn out that for some eigenvalue equations, the potential energy is simple and the problem can be solved analytically. In many cases, the potential energy term is so simple that it may be possible to "guess" the solution by simple intuition. Problems that fall in this category are:

(i) The free particle problem, where the potential energy term is zero or uniform in space:

$$V(r) = 0$$

(ii) The "quantum well" problem, where

$$\begin{aligned} V(x) &= V_o \quad -a \le x \le a \\ &= 0 \quad \text{otherwise} \end{aligned}$$

(iii) The periodic potential where we have a spatial periodicity defined by a vector **R**

$$V(\boldsymbol{r} + \mathbf{R}) = V(\boldsymbol{r})$$

and $V(\boldsymbol{r})$ itself is simple like the quantum well problem.

In these problems, the solutions are quite simple and can be obtained analytically or with easy numerical techniques. Many of the problems addressed in

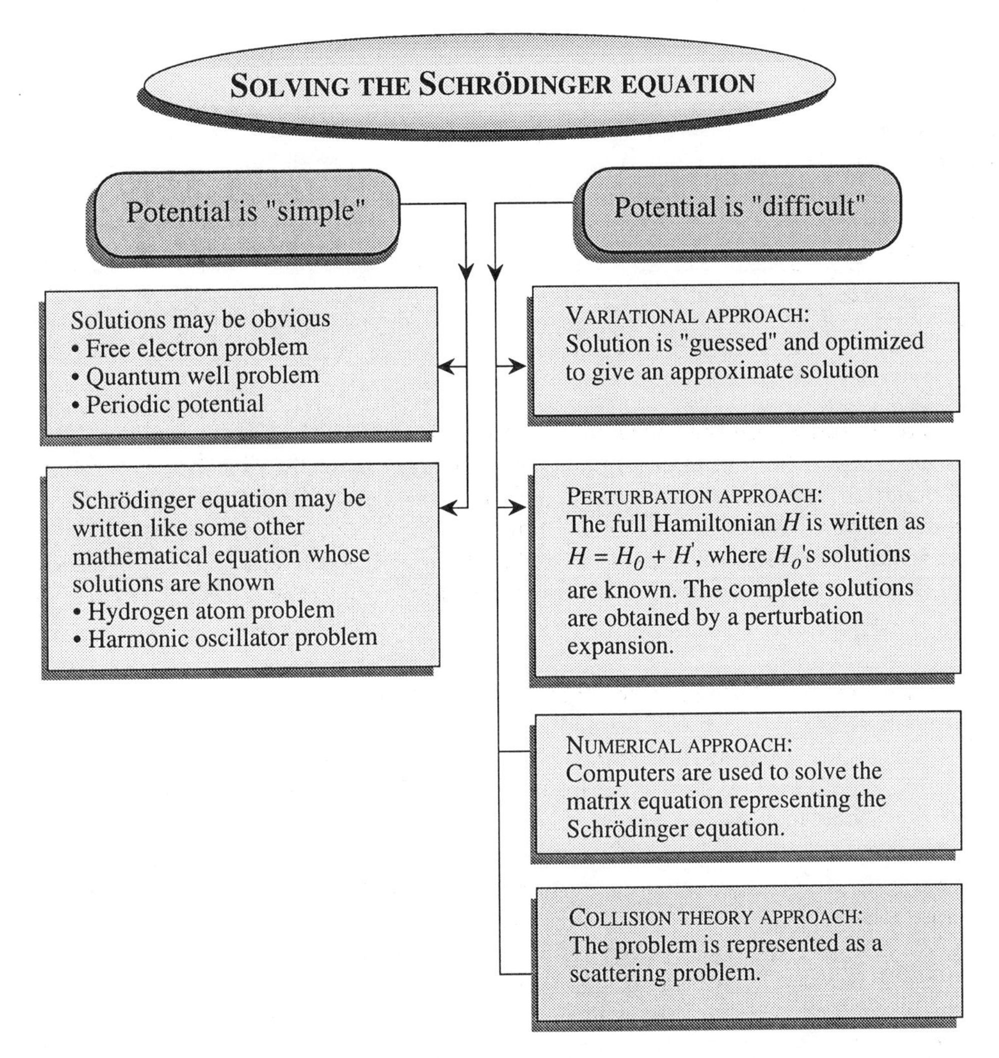

Figure 2.13: Approaches used to address Schrödinger equation when there is no time dependence.

Chapter 4 fall into this category.

In other problems it may be that the potential energy term is simple enough to obtain analytical results, but one has to go through a certain amount of mathematics to solve the problem. Typically, the Schrödinger equation is transformed by a simple manipulation into a form in which it looks like a well-studied differential equation. This equation may then have solutions that are well-known mathematical functions, such as Hermite functions, Legendre functions, Bessel functions, etc. Examples of problems in quantum mechanics where such an approach works are: (i) The harmonic oscillator problem where the potential energy has the form

$$V(x) = \frac{Cx^2}{2}$$

This is a very important problem in quantum mechanics with implications in understanding the properties of atomic vibrations in crystals, electromagnetic field quantization, and a host of other important problems. This problem is discussed in Chapter 5.
(ii) The hydrogen atom problem in which the potential energy is simply the Coulombic energy between an electron and a proton,

$$V(r_e - r_p) = \frac{e^2}{4\pi\epsilon_o |r_e - r_p|}$$

where r_e and r_p are the coordinates of the electron and the proton. This problem is also a central one in applied quantum mechanics. Its study allows us to build the foundation upon which one can discuss properties of isolated atoms as well as of a collection of atoms as in a solid. Hydrogen atom problem physics is also used to understand problems of dopants in materials as well as excitonic effects—problems of great technical importance. The hydrogen atom problem is discussed in Chapter 5.

Problems with Difficult Potential Energy Terms
In many cases the problems are too complicated to be solved with such ease. This is especially true for applied problems where one has to deal with realistic potential energy terms that are quite complex. In such cases, a range of tools have been developed to address the problem. In most cases, the solutions obtained are not exact, but can be obtained with sufficient accuracy for many applications. We will briefly list a few of the approaches used for the solution of "difficult" Schrödinger equations.
• **Variational approach:** A useful approach to solving difficult Schrödinger equations is the variational approach. In this approach one "guesses" the general form of the wavefunction (Gaussian, exponential, etc.) with a few parameters. The parameters are then varied to minimize the expectation value of the energy of the system. In some cases the minimization can be carried out analytically. In other cases, it may require a computer. The variational approach is discussed in Chapter 7.
• **Perturbation approach:** The perturbation approach is an extremely important technique used to solve quantum mechanics problems. In this approach, the Hamiltonian H describing the problem is written in the form

$$H = H_0 + H'$$

where H_0 is a simple Hamiltonian whose solutions are easy to calculate. If H' is small, the perturbation approach is useful and involves finding the solutions of H in terms of the solutions of H_0. Various orders of perturbation theory are developed, depending upon the level of accuracy needed. This method is discussed in Chapter 7.
• **Numerical approach:** The numerical approach is very useful in real-life problems where the problem is usually too complicated to be solved by the techniques

mentioned so far. This approach relies upon powerful computers. The Schrödinger equation is written in terms of a large matrix eigenvalue equation and matrix techniques are used to solve the problem. Powerful math libraries exist to solve such problems, and it is possible to solve a $10,000 \times 10,000$ matrix equation without too much difficulty! So important are the numerical techniques in modern applied quantum mechanics that every student should take some time to visit the local computer center and scan through the documentation in math libraries, such as IMSL and EISPACK. Once the use of the appropriate subroutines is mastered, the student will find his or her fear of quantum mechanics evaporating!

• **Collision theory approach:** Certain kinds of time-independent Hamiltonian problems are best described by the collision theory. These problems may be physically represented as a particle traveling in a region where its Hamiltonian is H_0 and scattering off a localized potential $V(r)$. Thus the particle is initially in a state $\Psi_i(r,t)$ and after scattering has a finite probability of being in a different state. The solution of such a problem involves the calculation of the probability of scattering from one state to another. The wavefunctions $\Psi(r,t)$ are thus not eigenfunctions of the full Hamiltonian and are not stationary states. The $\Psi(r,t)$ are solutions of the H_0 part of the Hamiltonian and the potential $V(r)$ causes the scattering of the particle from one state to another. The scattering problem is extremely important in applied quantum mechanics and determines important macroscopic properties of materials such as resistance, mobility, speed of transistors, etc. The collision problem is treated by perturbation approaches described in Chapter 8.

2.8 SOME MATHEMATICAL TOOLS FOR QUANTUM MECHANICS

In this section, we will review some of the important mathematical tools used to solve the Schrödinger equation. Some of these tools simply involve the boundary conditions that are satisfied or can be imposed on the wavefunctions, and others involve important mathematical concepts.

2.8.1 Boundary Conditions on the Wavefunction

Continuity Conditions

The Schrödinger equation is a second-order differential equation in spatial coordinates. Such an equation can always be integrated if one knows the value of the wavefunction and its derivative at some point in space. An important requirement in this regard is that the wavefunction be finite and single valued at every point in space. This ensures that the solution is unique and the probability density is finite and continuous. Another restriction is imposed on the gradient of the wavefunction and comes from requiring the particle probability current density to be continuous. If the particle mass does not change across the boundary, this requires the wavefunction gradient to be continuous. The statement about particle mass changing may seem redundant since the particle mass is a constant in non-relativistic physics. However, we will see later that in crystalline materials an electron can behave "as if" it has an effective mass m^* which is different from its free space mass. In such a case, when the electron moves from one region to another, its "effective mass" changes, and the current density, not the wavefunction gradient, is continuous. The two conditions on the wavefunction and its derivative can be summarized as

$$\psi \mid_{x+} = \psi \mid_{x-} \tag{2.59}$$

$$\left. \frac{1}{m} \frac{\partial \psi}{\partial x} \right|_{x+} = \left. \frac{1}{m} \frac{\partial \psi}{\partial x} \right|_{x-} \tag{2.60}$$

Bound State Problem

In addition to the aforementioned restrictions, depending upon the nature of the problem, very often certain conditions can be imposed on the nature of the wavefunction at the boundaries of a volume. In some problems *the nature of the potential energy is such that the potential energy is low over some region of space and high at the boundaries of a volume.* In such cases, the wavefunction goes toward zero at the boundaries, provided the region of spaces considered is large enough. One can then impose the boundary conditions on the "bound" state problem

$$\psi \mid_{\text{boundaries}} = 0 \tag{2.61}$$

This requirement is equivalent to saying that the potential energy goes to infinity at the boundaries since the infinite potential requires the wavefunction to be zero. The approach used is shown schematically in Fig. 2.14.

Periodic Boundary Conditions

In some classes of problems, one imposes "periodic boundary conditions" on the wavefunction. The wavefunction is normalized to a volume V. Even though we focus our attention on a finite volume V, the wave can be considered to spread in all space as we conceive the entire space to be made up of identical cubes of sides L, as shown in Fig. 2.15. Then

$$\begin{aligned} \psi(x, y, z + L) &= \psi(x, y, z) \\ \psi(x, y + L, z) &= \psi(x, y, z) \\ \psi(x + L, y, z) &= \psi(x, y, z) \end{aligned} \tag{2.62}$$

2.8.2 Basis Functions

There are some functions in mathematics which have the property that any wavefunction can be expressed in terms of a properly weighted combination of these functions. Such functions are called *basis functions.* For example, if $\phi_n(x)$ represents a set of basis functions, a general function $G(x)$ can be written as

$$G(x) = \sum_n a_n \phi_n(x) \tag{2.63}$$

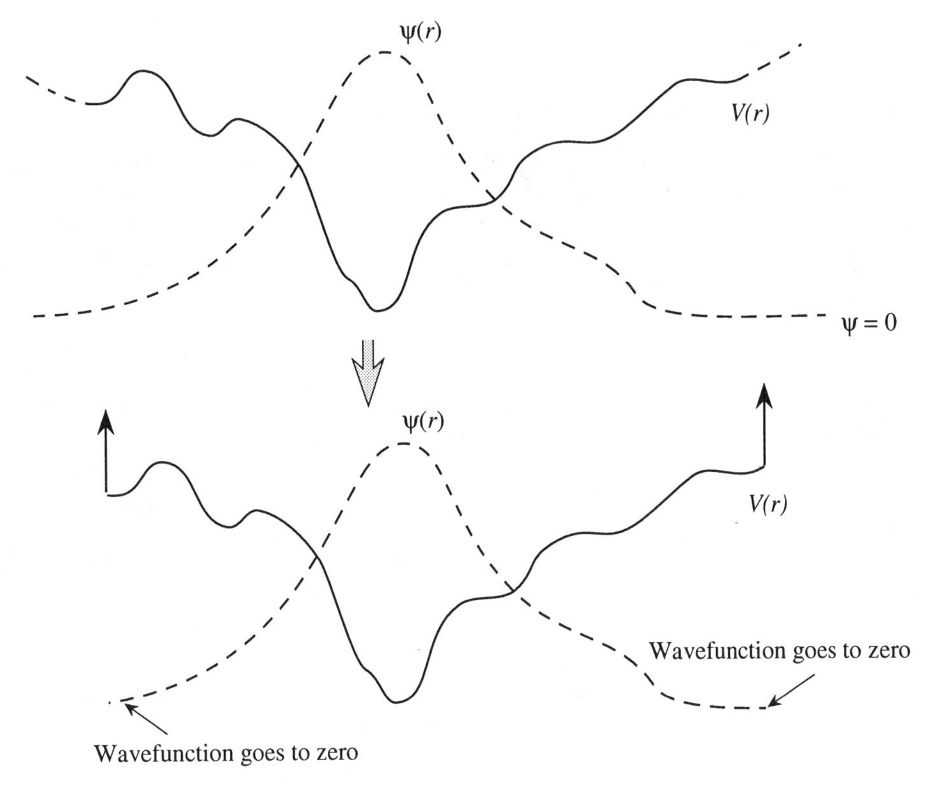

Figure 2.14: In many problems the form of the potential is such that one can require the wavefunction to go to zero at the boundaries. This is equivalent to requiring the potential energy to go to infinity at the boundary.

where a_n are the expansion coefficients of $G(x)$. The index n could represent discrete or continuous values.

The set of basis funtions must obey certain key requirements. The first requirement is that the functions should be complete. By completeness we mean that the basis functions should be such that any general function should be expressible in terms of the basis functions. In other words, the relation given by Eqn. 2.63 should be possible no matter what $G(x)$ is.

Another important requirement on the basis function set is that it should be orthonormal. This means that the basis functions should be orthogonal to each other and should be normalized. This is expressed as

$$\int \phi_m^*(r)\phi_n(r)\ d^3r = \delta_{mn} \tag{2.64}$$

The orthogonality of the basis functions simply ensures that each function is independent of others (and thus not redundant).

There are a large number of basis function sets that can be used in mathematics. Which is an appropriate one to use for a particular problem? A great deal of insight is necessary to choose the "best" basis set since the complexity of the quantum mechanics solutions can be greatly influenced by this choice.

2.8.3 Dirac δ-Function and Dirac Notation

An extremely useful mathematical concept for handling the properties of eigenfunctions and basis sets used in quantum mechanics is the δ-function introduced by Dirac. The δ-function is defined by the relation

$$\delta(x) = 0 \quad \text{if } x \neq 0$$
$$\int \delta(x)\, dx = 1 \tag{2.65}$$

The range of the integral includes the $x = 0$ point. An equivalent definition of the δ-function is that if we consider an arbitrary function $f(x)$ which is continuous at $x = 0$, then

$$\boxed{\int f(x)\delta(x)\, dx = f(0)} \tag{2.66}$$

where the integration range includes the $x = 0$ point. As shown schematically in Fig. 2.16a, the δ-function is an extremely singular function which is zero everywhere except at $x = 0$. At $x = 0$, its value is so large that the area under the curve defined by it on the x-axis is unity.

Several mathematical functions can be used in a limiting form to represent the δ-function. A particularly useful form is the function

$$\frac{\sin gx}{\pi x}$$

where g is a positive real function. As shown in Fig. 2.16b, the function has a value g/π at $x = 0$ and oscillates with decreasing amplitude with a period $2\pi/g$ as $|x|$ increases. Moreover, regardless of the value of g, we have the equality

$$\int_{-\infty}^{\infty} \frac{\sin gx}{\pi x} dx = 1$$

Thus the limit of this function as g goes to infinity has the proper properties of δ-function. We thus have

$$\delta(x) = \lim_{g \to \infty} \frac{\sin gx}{\pi x} \tag{2.67}$$

This mathematical relation is used quite often in quantum mechanics for applications in scattering problems.

Dirac Notation

A very elegant notation to represent frequently encountered quantities in quantum mechanics was introduced by Dirac. In this notation, we represent a wavefunction and its complex state by

$$\psi_n \rightarrow |n\rangle$$
$$\psi_n^* \rightarrow \langle n| \tag{2.68}$$

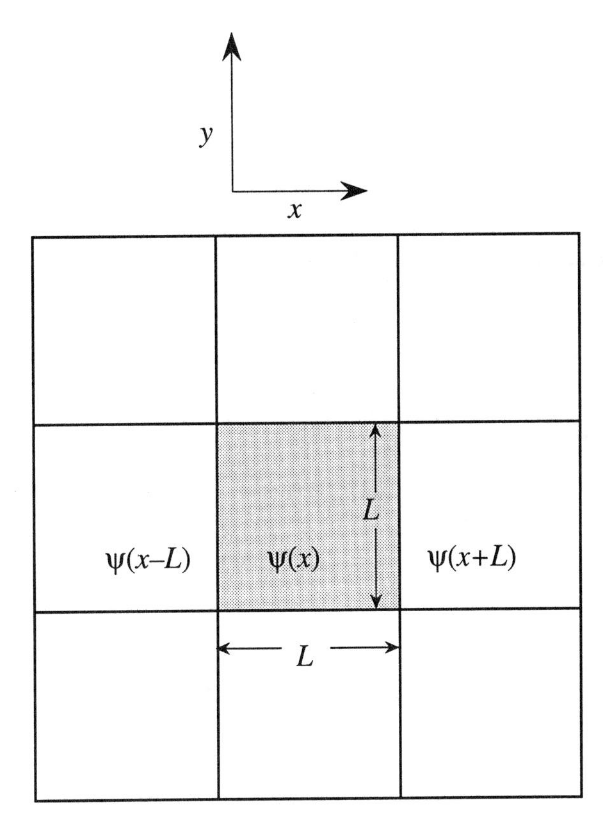

Figure 2.15: A schematic showing space filled with identical cubes of volume L^3. Periodic boundary conditions allow one to focus on a single cube, but have wavefunctions filling the entire space.

The representations are called bra and ket, respectively. The notation for products commonly appearing in quantum mechanics are

$$\begin{aligned}\int \psi_n^* \psi_m d^3r &\rightarrow \langle n \mid m \rangle \\ \int \psi_n^* Q \psi_m d^3r &\rightarrow \langle n|Q|m\rangle \end{aligned} \tag{2.69}$$

where Q is a quantum mechanical operator.

EXAMPLE 2.5 Prove that $x\delta'(x) = -\delta(x)$.

To prove this relation, we use an approach in which we multiply both sides by a continuous function $f(x)$ and integrate over x. Thus, integrating by parts, we have

$$\begin{aligned}\int f(x)x\delta'(x)\, dx &= -\int \delta(x)\frac{d}{dx}[xf(x)]\, dx \\ &= -\int \delta(x)[f(x) + xf'(x)]\, dx\end{aligned}$$

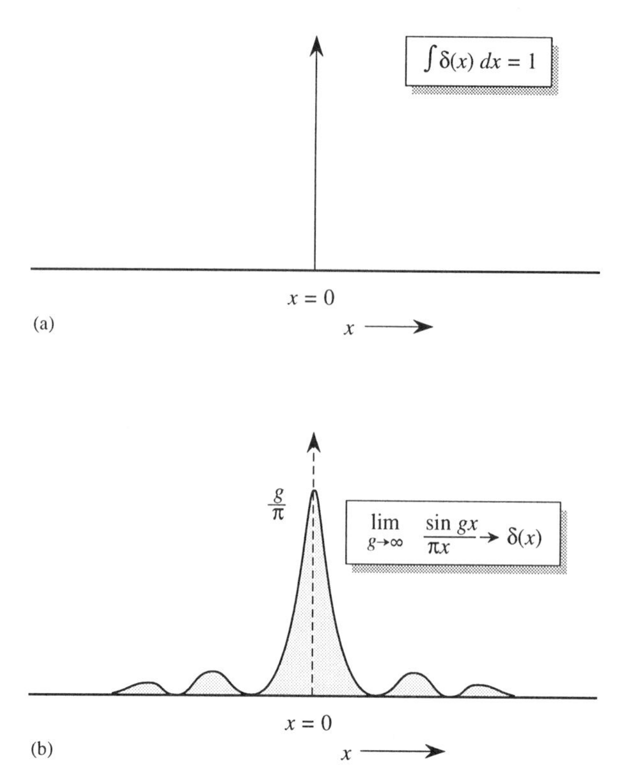

Figure 2.16: (a) The δ-function is an extremely singular function which can be mathematically represented by a limit function as shown in (b).

$$= -\int f(x)\delta(x)\, dx$$

where we have used the property $x\delta(x) = 0$. Comparing the left and right sides, we get

$$x\delta'(x) = -\delta(x)$$

2.9 CHAPTER SUMMARY

In this chapter we have established the essential mathematical framework on which quantum mechanics is based. The remaining part of this text will apply this framework to a number of physical problems of interest. The summary tables (Tables 2.1–2.3) cover key findings and topics studied in this chapter.

2.10 PROBLEMS

Problem 2.1 The surface of the sun has a temperature of 6000 K. Calculate the wavelength at which the sun emits its peak radiancy.

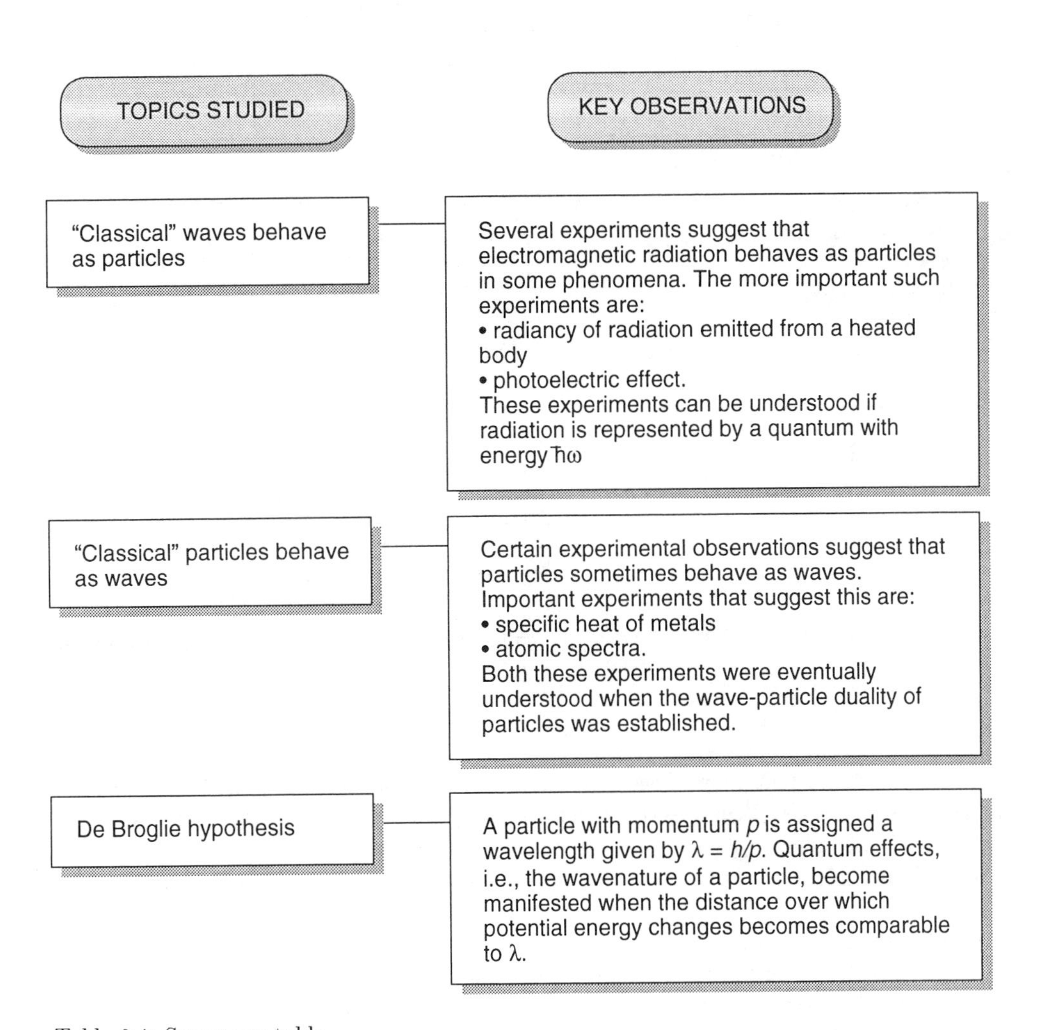

Table 2.1: Summary table

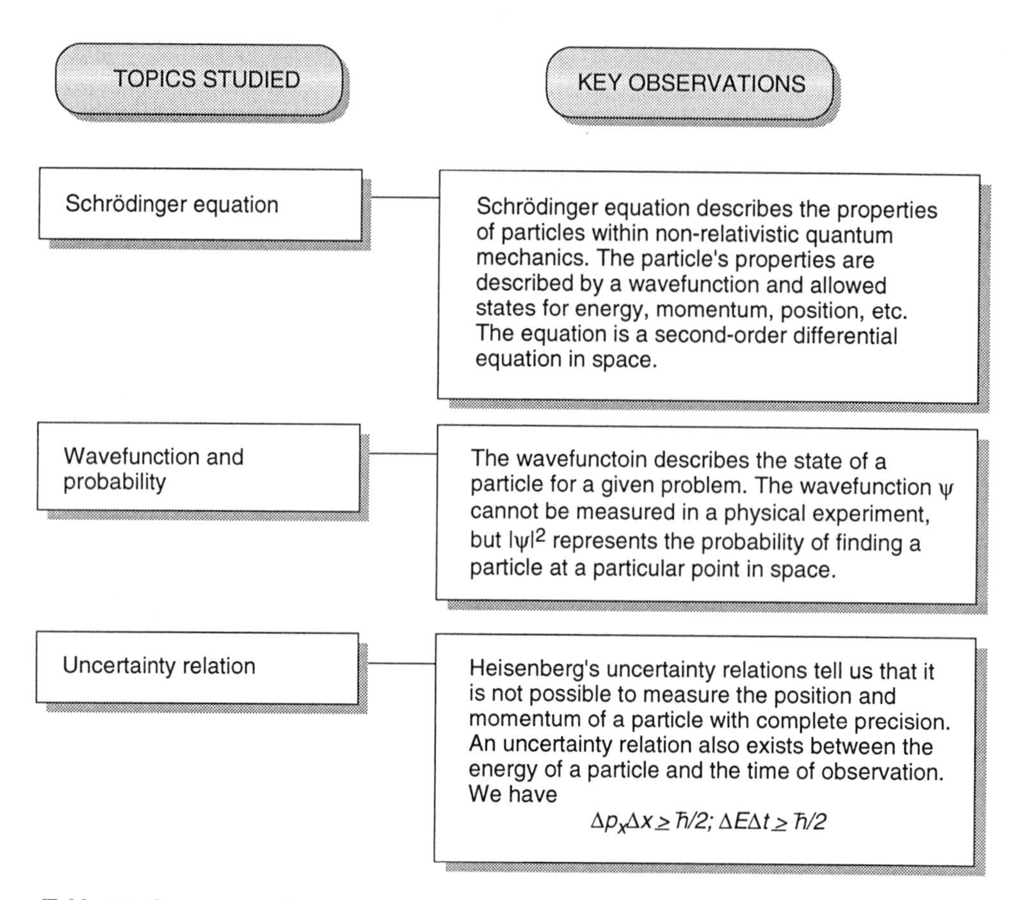

Table 2.2: Summary table

Problem 2.2 The deep space radiation has a spectrum corresponding to a temperature of 2.7 K. What is the peak wavelength of this radiation? Calculate the energy of the photon corresponding to this peak wavelength.
Problem 2.3 Calculate the peak wavelength emitted by a healthy human. Can we see this radiation?
Problem 2.4 Estimate the power radiated by a typical human body.
Problem 2.5 The human eye response is limited between wavelengths of 1000 Å and 4400 Å. Calculate the blackbody temperature for which the peak wavelength coincides with these limits.
Problem 2.6 Calculate the cutoff wavelength for the photoelectric effect for an aluminum sample.
Problem 2.7 A 1 W light source produces radiation in a uniform spatial distribution. The wavelength of light is 5500 Å. Calculate the number of photons per second striking an area 20 cm $\times$ 10 cm located 1 m from the source.
Problem 2.8 A metal surface is found to have a photoelectric cutoff wavelength of 3250 Å. Calculate the stopping potential if the surface is illuminated with light of wavelength 2000 Å.

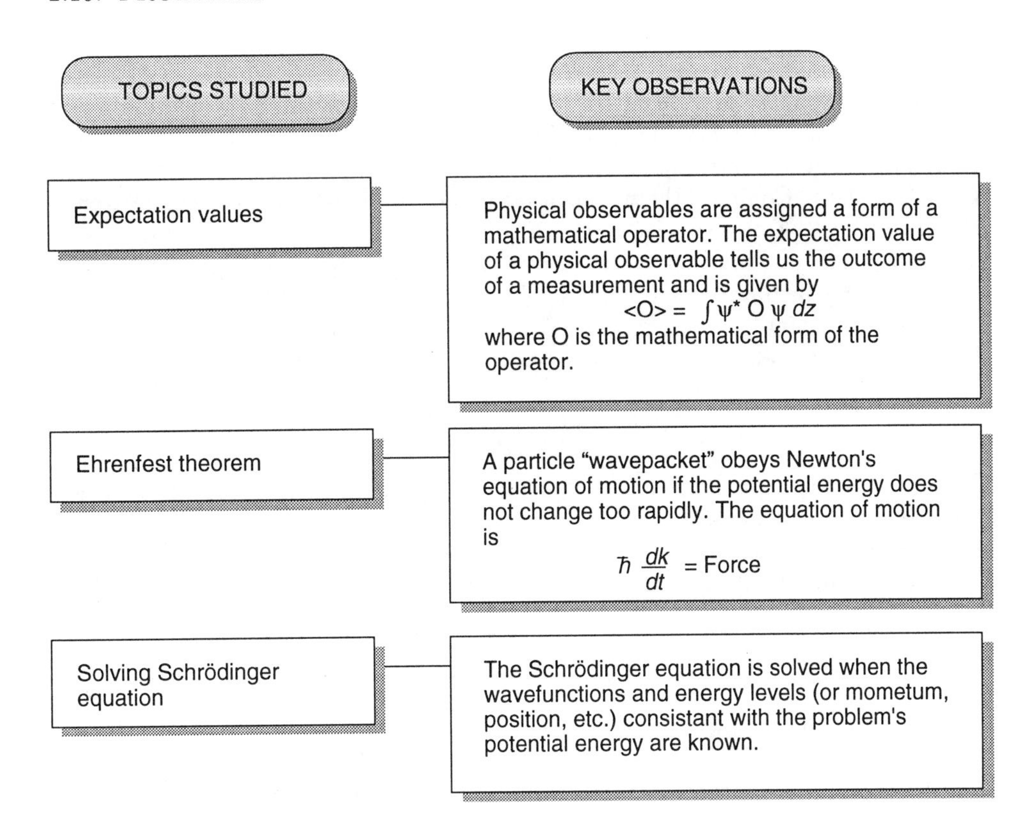

Table 2.3: Summary table

Problem 2.9 The surface of copper is illuminated with light of wavelength 2200 Å. Calculate the stopping potential for a photoelectric experiment.

Problem 2.10 It is found that when a sample of sodium is illuminated with light of a wavelength 3200 Å the stopping potential is 1.6 V. When it is illuminated with light of a wavelength 2800 Å, the stopping potential is 2.15 V. Calculate the work function of sodium.

Problem 2.11 Use the energy levels calculated in the Bohr model for hydrogen to discuss what levels are involved in producing the Balmer and Paschen series.

Problem 2.12 Discuss how the Lyman series would be produced using the Bohr model for a hydrogen atom.

Problem 2.13 Calculate the speed of an electron in the $n = 3$ level of the hydrogen atom.

Problem 2.14 Calculate the radii and speed of an electron in the $n = 1$ and $n = 3$ levels in a hydrogen like atom with a nuclear charge Ze. Assume that $Z = 2$ and $Z = 3$.

Problem 2.15 A gas containing hydrogen atoms is illuminated with light of a wavelength 500 Å. Calculate the kinetic energy of the emitted electrons if the atoms are in the ground state.

Problem 2.16 Calculate the de Broglie wavelength of the following:

i) A 1000 kg car moving at 100 km/h.
ii) An organic molecule of mass 10^{-24} kg moving at 10 cm/s.
iii) An electron with kinetic energy of 10 eV.
iv) A proton with kinetic energy of 1 MeV.

Problem 2.17 Compare the wavelengths of a 1 eV photon and a 1 eV electron.

Problem 2.18 Consider electrons with kinetic energy of 10 eV moving in *i)* a medium where the smallest distance over which the potential energy changes is 1.0 μ m; *ii)* a medium where this distance is 1.0 Å. Which problem will display "quantum behavior?"

Problem 2.19 Consider an electron wavefunction given by the general form

$$\psi = A \sin \frac{n\pi x}{L}; n = 1, 2, 3$$

Calculate the factor A (normalization factor) if the wavefunction is to be normalized between $x = 0$ and $x = L$.

Problem 2.20 Consider an electron wavefunction describing an electron state extending from $-W/2$ to $W/2$

$$\psi(x) = \int \frac{2}{W} \cos \frac{n\pi x}{W}; n = 1, 3, 5$$

Calculate the energy, momentum and position expectation value of an electron in state $n = 1$ and $n = 3$.

Problem 2.21 Estimate the minimum velocity of a ball confined to a region of length 10 cm.

Problem 2.22 In beta decay electrons emerge from a nucleus. The diameter of a typical nucleus is 10^{-14} m. Use the uncertainty principle to estimate the kinetic energy of an electron emerging in a beta decay.

Problem 2.23 The size of a typical atom is 1 Å. Use the uncertainty relation to estimate the energy of electrons in an atom.

Problem 2.24 Alpha particles (He nuclei) are emitted from atoms in radioactive decay at energies of $\sim$ 5 MeV. Use this information and the uncertainty principle to estimate the size of a nucleus.

CHAPTER 3

PARTICLES THAT MAKE UP OUR WORLD

Chapter at a Glance

3.1 INTRODUCTION

We have seen in the previous chapter that particles must be described by the Schrödinger equation. Once the Schrödinger equation is solved we get information on the *allowed* eigenvalues (i.e., energies, momenta, angular momenta, etc.) and the allowed wavefunctions corresponding to these eigenvalues. When we use the term allowed we mean that the particle can only have these values and all other values are forbidden.

While solving the Schrödinger equation is an important first step in understanding a quantum problem, it is only one component in describing how an actual experiment will proceed. In Fig. 3.1 we show the three steps involved in predicting the outcome of a physical experiment. Once the Schrödinger equation is solved one has to choose how particles are actually distributed among the allowed states, i.e., determine the *occupation of the states.* Finally, once the occupation is determined at some point in time, one has to determine how the system responds to forces or perturbations.

In this chapter we will discuss how particles are distributed among the allowed states. We will assume that we are dealing with a large number of particles that are in equilibrium at a temperature T. In Chapter 1, Section 1.7, we have discussed this problem within classical physics. Here we will see how things change as a result of applying quantum concepts.

3.2 NATURE OF PARTICLES

In classical and quantum physics we often deal with problems with many identical particles. The particles may be electrons, atoms, molecules, etc. The fact that particles are identical has a profound influence in their quantum treatment. In classical physics, identical particles do not lose their particular identity, in spite of having properties indistinguishable from others. For example, if two identical particles have a collision, we can, in classical physics, follow each particle throughout the collision process with complete precision and, therefore, identify which particle is scattering where. In quantum mechanics this seemingly simple task cannot be accomplished. The reason for this is not difficult to identify since the quantum phenomena are defined by probabilities and we have an uncertainty relation that does not allow us to precisely follow a particle's path. In this chapter we will study the consequences of this indistinguishability on physical phenomena. In particular, we will discuss the identical particles known as *fermions* and those known as *bosons.*

In classical mechanics an important quantity used to describe properties of particles is the angular momentum given by

$$\mathbf{L} = \mathbf{r} \times \mathbf{p} \tag{3.1}$$

In quantum mechanics one has an analogous concept except that, while in classical mechanics the allowed values of angular momentum are continuous, in quantum mechanics the values are quantized. We have discussed in Chapter 2 how Bohr postulated the quantization of angular momentum to develop his atomic model. In quantum mechanics the values of $\mathbf{L}$ are quantized to $0, \hbar, 2\hbar, 3\hbar$, etc. Unlike the case in classical physics, particles in quantum mechanics exhibit another property—spin. The spin is an intrinsic property of particles, much like charge, and has no classical analog in terms of an actual motion or spinning of the particle about an axis. It is a purely quantum property associated with the particle.

Spin can take integral and half-integral (in units of $\hbar$) values in contrast to orbital angular momentum, which is quantized in integral units only. *Particles with integral spin values are called bosons, while those with half-integral spin are fermions.* This seemingly simple distinction has a profound effect on the physical properties of systems containing these particles.

Particles such as electrons, protons, and neutrons are fermions. Particles like mesons (π-meson, K-meson) are bosons. We will see in this chapter that in quantum mechanics, just as particles behave like waves, waves also behave as particles. For example, electromagnetic radiation behaves as particles known as photons. Photons are bosons.

It is important to note that while electrons are fermions, since their spin is $\frac{1}{2}\hbar$, it is possible for electrons to form a "complete pair" in some materials. The composite can have integral spin and the pair then behaves as a boson. This change over from fermion to boson-like behavior is manifested in electrical properties of materials. For example, when electrons behave as fermions, materials have finite resistance. Under certain conditions, when electrons couple together to form pairs (called Cooper pairs), the material becomes a superconductor and loses its resistance completely.

3.3 SIMPLE EXPERIMENT WITH IDENTICAL PARTICLES

To illustrate the differences between classical particles, bosons, and fermions, we examine the scattering of particles. Two particles, A and B, scatter from each other and we measure their scattering rates by placing detectors as shown in Fig. 3.2. We will look for the angular distribution of the probability that *some* particle arrives at detector D_1. We will work in the center-of-mass system and denote by $f(\theta)$ the amplitude that the particle A is scattered by an angle θ. Since the particles are identical, a particle can arrive at the detector in two different ways, as shown in Fig. 3.2. As shown, the particles are simply exchanged in these two possibilities. The amplitude of scattering for the second case is $f(\pi - \theta)$. *The following is observed experimentally*:

- If the particles are *distinguishable*, the probability of *some* particle appearing at D_1 is

$$|f(\theta)|^2 + |f(\pi - \theta)|^2 \tag{3.2}$$

- If the particles are *indistinguishable and are bosons* (e.g., α-particles, photons, mesons), the probability of one of the particles appearing at D_1 is

$$|f(\theta) + f(\pi - \theta)|^2 \tag{3.3}$$

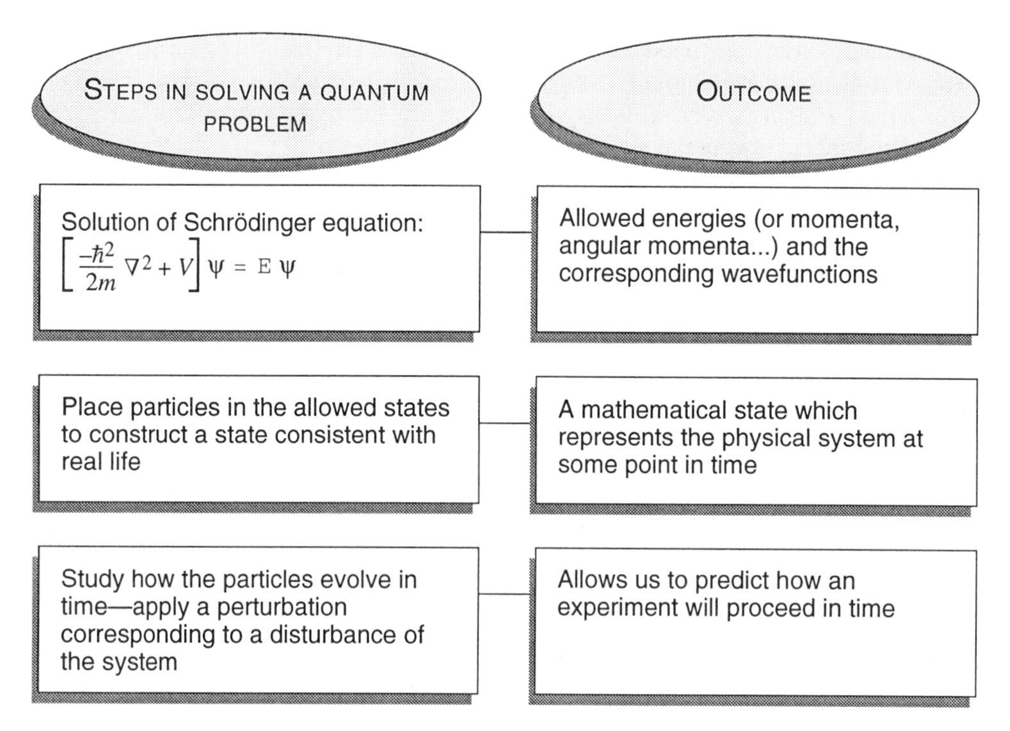

Figure 3.1: Steps taken to solve a general quantum problem.

- If the particles are *identical but are fermions* (e.g., electrons, neutrinos, protons, neutrons), the probability is

$$|f(\theta) - f(\pi - \theta)|^2 \tag{3.4}$$

Note that if the two particles are electrons but have different spins, and the scattering is not supposed to alter the spin, then the probability is given by the *distinguishable* case.

In general, the direct and exchange states would have an amplitude of scattering $f(\theta)$ and $\exp(i\delta)f(\pi - \theta)$, respectively. If one carries out the exchange again from the exchanged state the new amplitude is $\exp(2i\delta)f(\theta)$. But this is equal to $f(\theta)$ so that $\exp(i\delta)$ is $+1$ or -1. It was shown by Pauli that for fermions the choice of the phase factor is -1 and for bosons it is $+1$. In fact, in general, the wavefunction describing a set of fermions must be such that it is *antisymmetric* (changes sign) to any permutation or interchange of two particles. Similarly, for bosons the wavefunction is *symmetric* to such a permutation.

3.3.1 Pauli Exclusion Principle

In the previous subsection we have seen that the amplitude that two identical fermions scatter and go into a detector at angle θ is given by Eqn. 3.4. It can be seen that if the two particles are in the same state the amplitude given by the equation

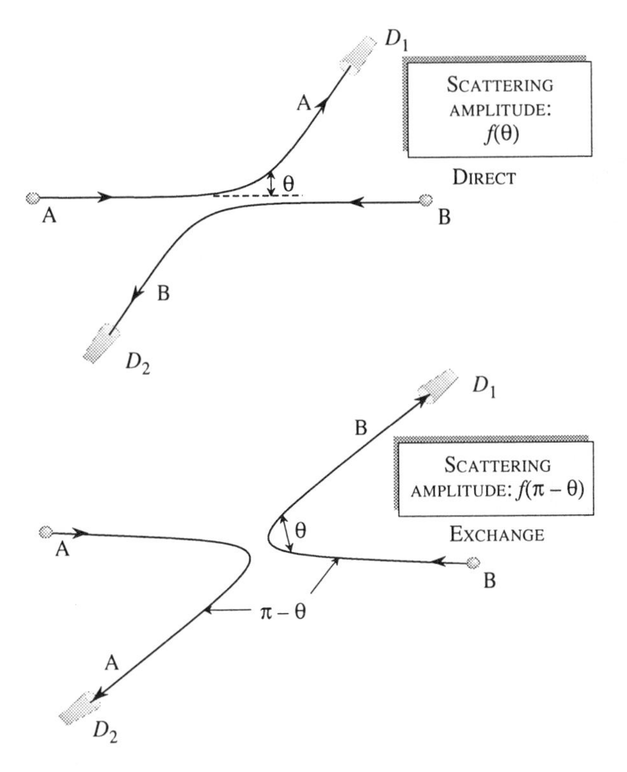

Figure 3.2: Scattering of two particles in the center-of-mass system. Detector D_1 is able to detect any particle being scattered at the angle. In the first process, particle A is scattered into detector D_1, and in the second process, the particles are exchanged.

above is zero! This is just a complicated way to state a most important finding in quantum mechanics—the Pauli exclusion principle. According to this principle two (or more) identical fermions cannot occupy the same allowed state. Thus for example, if an electron with a certain spin direction occupies an energy level, we cannot place another electron with the same spin in that state. This is illustrated in Fig. 3.3.

The Pauli exclusion principle is extremely central to our understanding of materials and devices. It is the basis for understanding the periodic chart, chemical properties of elements, and it answers why some materials are insulators and some are conductors. It also forms the basis for how ordinary metals become superconductors at low temperatures.

Particles that are bosons do not have any restriction on how many of them can occupy a given allowed state. Any number of bosons can occupy a given state. As we discussed earlier, photons, mesons, etc., are bosons. It is also possible for electrons (or other fermions) to form a coupled state where they can have integral spin. In that case these coupled states act as bosons and are not restricted by the Pauli exclusion principle. In some materials electrons do form pairs (called Cooper pairs) at low temperatures. A consequence of this is superconductivity.

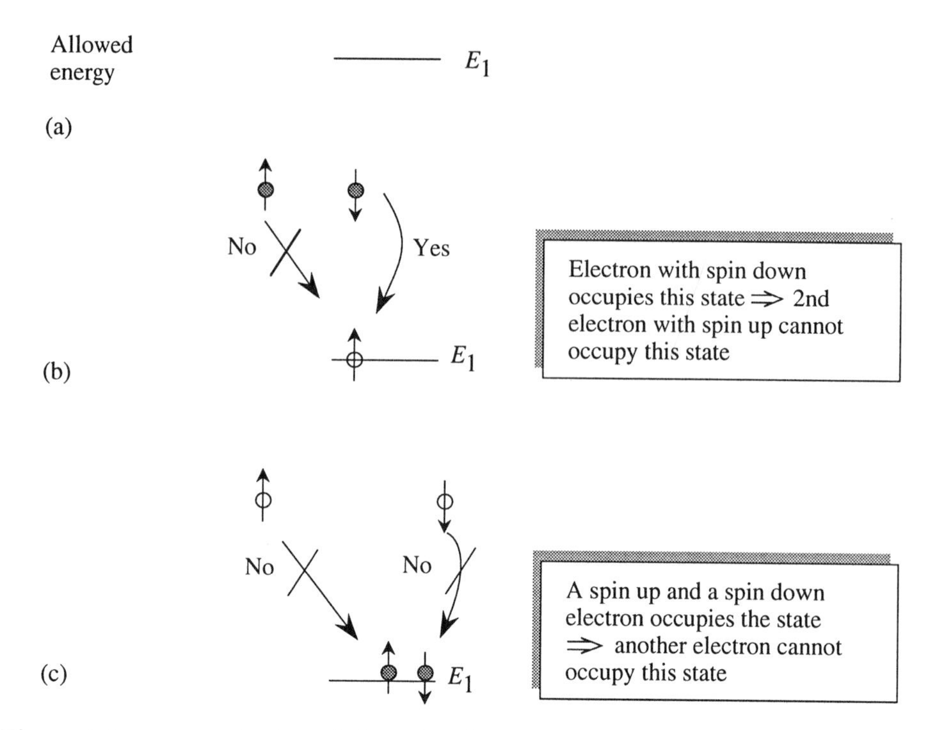

Figure 3.3: An example of restrictions placed on electrons occupation of an allowed state by the Pauli exclusion principle. (a) An allowed energy level, (b) two electrons with same spin cannot occupy the energy level; only one can, and (c) two electrons with opposite spin can occupy the level. Another electron is forbidden from further occupying the state.

3.4 WHEN WAVES BEHAVE AS PARTICLES

In Chapter 2, Section 2.2, we discussed some experiments which suggest that under some conditions classical *waves* such as electromagnetic waves behave as particles. For example, the classical wave picture of electromagnetic radiation does not agree with the photoelectric experiment or with the blackbody radiation spectral density. We also saw that classical particles sometimes behave as waves. The initial development of quantum mechanics, both historically and in this text, focused on the physics of entities which in the classical limit are particles. For example, the Schrödinger equation is the wave equation for electrons and protons—objects that have a certain mass and are "particles." These classical particles are given a wave nature so that, depending upon the problem at hand, their behavior is more particle-like or more wave-like. This treatment for classical particles may be called the "first" quantization.

It is natural to expect that entities that are waves in classical physics (electromagnetic waves, vibrations of a stretched string, etc.) should also be given a treatment in quantum mechanics which naturally allows them to have a particle-wave duality. For example, we know that electromagnetic radiation behaves as "particles" (with a well-defined energy $\hbar\omega$) in some instances. In such cases, it is appropriate to

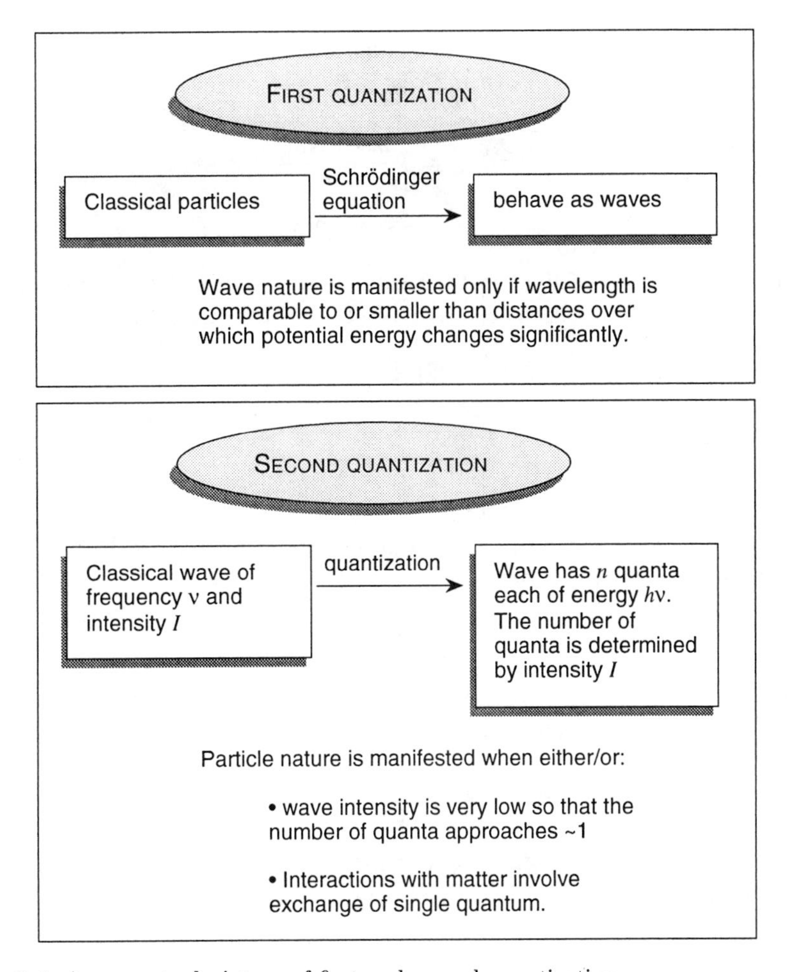

Figure 3.4: A conceptual picture of first and second quantization.

define electromagnetic radiation not by its field amplitudes E and H, but by how many "photons" or radiation quanta are present in a given mode. Such a "particle" description is particularly important when we discuss the physics of the interaction of the electromagnetic radiation with matter, and where the number of the photons can change as a result.

The procedure to represent the classical waves by a particle description is known as the *second quantization*. In Fig. 3.4 we show a schematic of the quantum treatment in the first and second quantizations.

3.4.1 Second Quantization of the Radiation Field

As an example of how clasical waves are treated as particles we will (very superficially) consider the quantization of electromagnetic radiation. The classical wave is described by its frequency ν and the amplitude of vibration. The amplitude is directly related to the intensity of the radiation. For example, the time-averaged (see Chapter 1, Section 1.3) Poynting vector **S** representing the optical power is

$$\begin{aligned}\langle \mathbf{S}\rangle_{\text{time}} &= \hat{k}\frac{2vk^2\,|\mathbf{A}_0|^2}{\mu_0} \\ &= 2v\epsilon\omega^2\,|\mathbf{A}_0|^2\,\hat{k}\end{aligned} \tag{3.5}$$

where v is the velocity of light in the medium ($= c/\sqrt{\tilde{\epsilon}}$) and $\hat{k}$ is a unit vector in the direction of $\mathbf{k}$. Here $\tilde{\epsilon}$ is the relative dielectric constant. We have

$$|\mathbf{k}| = \omega/v \tag{3.6}$$

The energy density is then

$$\left|\frac{\mathbf{S}}{v}\right| = 2\epsilon\omega^2\,|\mathbf{A}_0|^2 \tag{3.7}$$

In second quantization the wave properties are described by quanta of energy. Each quantum has an energy is

$$E = \hbar\omega = h\nu \tag{3.8}$$

The number of quanta, n_{ph}, are now related to the energy density of the wave. The energy density of the photon field is

$$\frac{n_{ph}\hbar\omega}{V} \tag{3.9}$$

We have already calculated the energy density in Eqn. 3.7 above. Equating these two, we get

$$\begin{aligned}|A_0|^2 &= \frac{\hbar n_{ph}}{2\epsilon\omega V} \\ \frac{n_{ph}}{V} &= \frac{\langle S\rangle}{v}\cdot\frac{1}{\hbar\omega}\end{aligned} \tag{3.10}$$

The approach mentioned for electromagnetic field can be used for other wave fields as well. This leads to a number of *particles* which are given special names. Fig. 3.5 illustrates some of the particles and their classical wave forms.

3.5 CLASSICAL AND QUANTUM STATISTICS

In our discussion of the solutions of the Schrödinger equation so far in this text, we have not indicated whether particles actually occupy the allowed states. In an actual physical system, not only are the allowed states important, but so is the occupation of the states. In this section we will discuss this occupation probability or distribution function for systems with a very large number of particles. In the spirit of second quantization, these particles need not be electrons or protons alone, but could also be photons, phonons, magnons, etc.

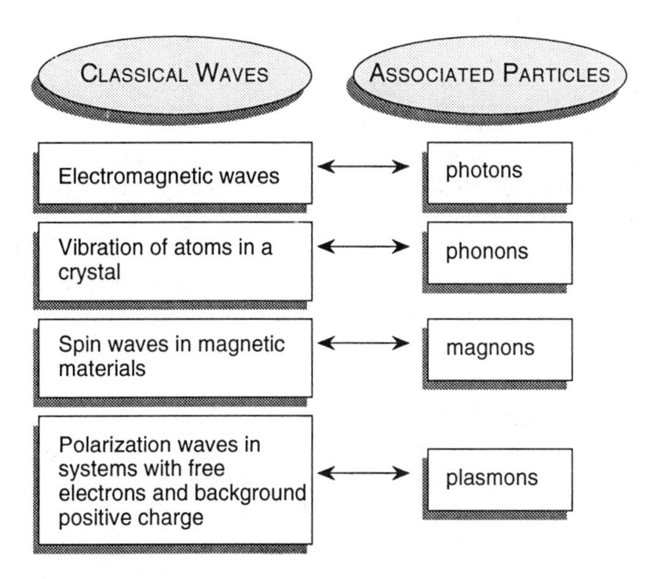

Figure 3.5: Some important classical waves and the particles that result after second quantization.

The interaction and properties of a large number of particles is described through statistical mechanics. We will discuss the problem of the distribution function at thermodynamic equilibrium, i.e., where a system is maintained at a temperature T and no external force is applied to the particles.

To find the distribution function of a system of particles we need to exploit two important properties (once the allowed states are known): (i) the distribution should be such that the free energy of the system is a minimum; (ii) the distribution should respect any special property of the particles. The first requirement says that the free energy F, where

$$F = U - TS \tag{3.11}$$

should be minimized. Here U is the total energy of the system of particles and S is the entropy of the system. Entropy is a measure of the system disorder. As can be seen from Eqn. 3.11, minimization of the internal energy would like to produce an ordered system, where all particles occupy the lowest possible energy, while the entropy term would like the particles to be distributed as randomly as possible with no regard to their energies. The entropy is given by

$$S = k_B \ln W \tag{3.12}$$

where W is the total degeneracy (i.e., how many different ways the particles can be arranged) of the particular distribution.

In addition to the minimization of the free energy, the distribution must satisfy the following constraints, which lead to different distribution functions:

(i) *Classical or Boltzmann distribution*: The particles are distinguishable from each other. No other constraint is placed on the particles.

(ii) *Fermi-Dirac distribution*: The particles are indistinguishable and only one particle can occupy a particular state. Particles such as electrons obey this constraint (the Pauli exclusion principle).
(iii) *Bose-Einstein distribution*: The particles are indistinguishable and can occupy the same available state. Particles such as photons and phonons (lattice vibrations) obey the resulting distribution function.

We start the problem with the knowledge that the available states are defined by the sequence $\{\epsilon_i, g_i\}$ where ϵ_i is an allowed energy with a degeneracy g_i. We wish to distribute N particles among these allowed states. The key issue of interest is the total degeneracy W, which gives us the entropy term.

Classical or Boltzmann Distribution

In a classical distribution, we can distinguish among the various particles. To find the total degeneracy of a distribution where N_1 particles are in state ϵ_1, N_2 in state $\epsilon_2, \ldots$, as shown in Fig. 3.6, we need to find the following values:
(a) How many ways can N particles be arranged so that N_1 particles are in state ϵ_1, N_2 in ϵ_2, N_3 in ϵ_3, etc.?
(b) Within each energy ϵ_i with degeneracy g_i, in how many distinct ways can N_i particles be arranged?

The total degeneracy is the product of the numbers calculated in (a) and (b) above. Let us first consider the number involved in (a). We will answer the following question: In how many ways can N_1 particles be chosen from N particles $(N = N_1 + N_2 + \cdots)$?

- The first particle can be chosen in N ways.
- The second particle can be chosen in $(N - 1)$ ways.
- The third particle can be chosen in $(N - 2)$ ways.

$\vdots$

- The N_1^{th} particle can be chosen in $(N - N_1 + 1)$ ways.

Thus the total number is

$$P_1^{'} = N(N-1)(N-2)\cdots(N-N_1+1) = \frac{N!}{(N-N_1)!} \tag{3.13}$$

However, in this counting we have counted as distinct ways the sequence in which the particles have been chosen. Thus, for example, the second particle can be really chosen in $\frac{N-1}{2}$ ways, the third one in $\frac{N-1}{3}$, etc., to avoid double or triple counting. This gives us the number

$$P_1 = \frac{N!}{N_1!(N-N_1)!} \tag{3.14}$$

The same holds for the ways N_2 particles (for the state ϵ_2) can be chosen (remember that we have $(N - N_1)$ particles left now). Thus

$$P_2 = \frac{(N-N_1)!}{N_2!(N-N_1-N_2)!} \tag{3.15}$$

etc.

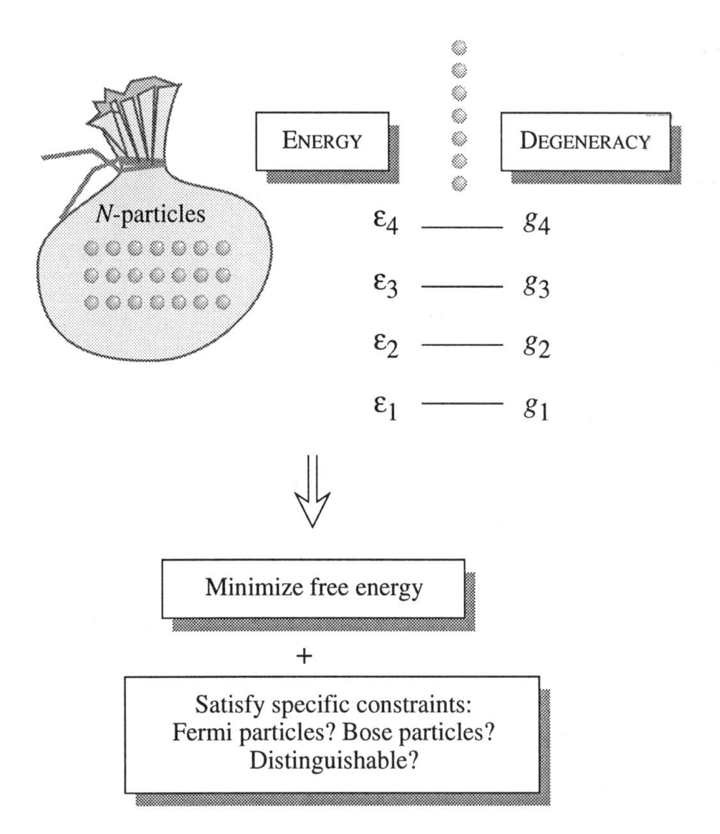

Figure 3.6: The N particles are to be distributed into the available states in a manner such that the free energy is minimized and any special constraints on the particles are to be respected.

Now within each energy level, the particles can be arranged in different ways to produce new configurations. The number of ways N_1 particles can be arranged in g_1 states is simply, $g_1 \times g_1 \times g_1 \cdots$ (N_1 times) $= (g_1)^{N_1}$. The total degeneracy of the distribution is thus

$$\begin{aligned}
W &= \left\{ \frac{N!}{N_1!(N-N_1)!}(g_1)^{N_1} \right\} \left\{ \frac{(N-N_1)!}{N_2!(N-N_1-N_2)!}(g_2)^{N_2} \right\} \cdots \\
&= \frac{N!(N-N_1)!(N-N_1-N_2)!(N-N_1-N_2-N_3)!\ldots(g_1)^{N_1}(g_2)^{N_2}\ldots}{N_1!(N-N_1)!N_2!(N-N_1-N_2)!\ldots} \\
&= \frac{N!(g_1)^{N_1}(g_2)^{N_2}(g_3)^{N_3}\ldots}{N_1!N_2!N_3!\ldots} \\
&= N!\prod_{i=1}^{N} \frac{(g_i)^{N_i}}{N_i!}
\end{aligned} \tag{3.16}$$

The free energy of the classical system is

$$F = U - TS = \sum_i \epsilon_i N_i - k_B T \ \ln \ W \tag{3.17}$$

We must minimize this, along with the constraint that the total number of particles

$$N = \sum_i N_i \tag{3.18}$$

is a constant. Let us examine the term $\ln W$. To estimate this, we use Stirling's formula

$$\ln \ (x!) = x \ \ln \ x - x \tag{3.19}$$

From Eqns. 3.16 and 3.19, we have

$$\begin{aligned} \ln W &= \ln \ N! + \sum_i (N_i \ \ln \ g_i - \ln \ N_i!) \\ &= N \ \ln \ N - N + \sum_i (N_i \ \ln \ g_i - N_i \ \ln \ N_i + N_i) \end{aligned} \tag{3.20}$$

We note that at equilibrium we must have $\delta F = 0$ and $\delta N = 0$. Using the result of Eqn. 3.20 in Eqn. 3.17, and using Eqn. 3.18, we have

$$\begin{aligned} \delta F &= \sum_i \epsilon_i \delta N_i - k_B T \left\{ \sum_i (\ln \ g_i) \delta N_i - (\ln \ N_i) \delta N_i \right\} = 0 \\ \delta N &= \sum_i \delta N_i = 0 \end{aligned} \tag{3.21}$$

To simultaneously satisfy both these requirements, we use the principle of Lagrange multipliers where we multiply the second equation by $-A$ and add it to the first equation to obtain

$$\sum_i \left[-k_B T \ \ln \left(\frac{g_i}{N_i} \right) + \epsilon_i - A \right] = 0 \tag{3.22}$$

This must hold for each level i and we have

$$\ln \frac{g_i}{N_i} = \frac{\epsilon_i - A}{k_B T} \tag{3.23}$$

or

$$N_i = g_i \ \exp - \left(\frac{\epsilon_i - A}{k_B T} \right)$$

The occupation probability f is now given by

$$\boxed{f(\epsilon_i) = \frac{N_i}{g_i} = \exp - \left(\frac{\epsilon_i - A}{k_B T} \right)} \tag{3.24}$$

The quantity A (called the *chemical potential*) is obtained from the condition

$$N = \sum_i N_i = \sum_i g_i \exp - \left(\frac{\epsilon_i - A}{k_B T} \right) \tag{3.25}$$

The distribution function given by Eqn. 3.24 is known as *the Boltzmann distribution* and it is applicable to classical particles or to particles under conditions where $f(\epsilon_i)$ is much smaller than unity.

Fermi-Dirac Distribution

As with the Boltzmann distribution, the key to the Fermi-Dirac distribution is to calculate W, the total degeneracy of a particular distribution. Now the particles are indistinguishable and each state can be occupied only by one particle. Let us find the ways N_1 particles can be distributed in g_i states.

Particle 1 can be placed in g_1 ways.
Particle 2 can be placed in $\frac{g_1-1}{2}$ ways.
Particle 3 can be placed in $\frac{g_1-2}{3}$ ways.

⋮

The total number is thus

$$P_1 = \frac{g_1(g_1-1)(g_1-2)\cdots(g_1-N_1+1)}{N_1!} = \frac{g_1!}{N_1!(g_1-N_1)!} \tag{3.26}$$

The total degeneracy of the distribution is now

$$W = \prod_i \frac{g_i!}{N_i!(g_i-N_i)!} \tag{3.27}$$

Once again, we must have $\delta F = 0$ and $\delta N = 0$, which gives the two equations

$$\sum_i \epsilon_i \delta N_i - \sum_i k_B T \{(-\ln N_i + \ln(g_i - N_i))\,\delta N_i\} = 0$$

$$\sum_i \delta N_i = 0 \tag{3.28}$$

As before, we multiply the second equation by $-A$, add to the first equation, and obtain for any energy level ϵ_i,

$$\epsilon_i - k_B T \ell n \left(\frac{g_i - N_i}{N_i}\right) - A = 0 \tag{3.29}$$

or

$$\ell n \left(\frac{g_i - N_i}{N_i}\right) = \frac{\epsilon_i - A}{k_B T} \tag{3.30}$$

which gives

$$N_i = \frac{g_i}{\exp\left[\frac{\epsilon_i - A}{k_B T}\right] + 1} \tag{3.31}$$

The occupation probability f for the general energy ϵ is now

$$\boxed{f(\epsilon) = \frac{N(\epsilon)}{g(\epsilon)} = \frac{1}{\exp\left[\frac{\epsilon - A}{k_B T}\right] + 1}} \tag{3.32}$$

which is the Fermi-Dirac distribution function. The term A is usually denoted by μ, which is known as the chemical potential. The chemical potential is also called the Fermi level in many applied problems. The occupation function has a value of 0.5 at the Fermi level. It is important to note that if $E - A \gg k_B T$, the unity in the denominator of the Fermi distribution function can be neglected and we obtain the Boltzmann distribution function.

Bose-Einstein Distribution
In this distribution we assume that the particles are indistinguishable, and one can place any number of them in any given state. By arguments similar to those used earlier, it can be shown that

$$W = \prod_i \frac{(N_i + g_i - 1)!}{N_i!(g_i - 1)!} \tag{3.33}$$

which leads to a probability distribution of the form

$$\boxed{f(\epsilon) = \frac{1}{\exp\left[\frac{\epsilon - A}{k_B T}\right] - 1}} \tag{3.34}$$

which is applicable to integral spin particles (photons, phonons, mesons, etc.).

3.6 STATISTICS OF DISCRETE ENERGY LEVELS

In the previous section we have seen how electrons are distributed in thermodynamic equilibrium. We will now consider a special case where the allowed states in a problem are made up of a continuum of allowed states and a discrete state as shown in Fig. 3. 7. This problem is of great importance in our later discussion of doping in materials.

Let us assume that there are N_d impurity levels with energy E_d each with a degeneracy of g_d. Let us assume that an electron can either occupy the impurity level or be in the continuum. Also *only one electron can occupy the level although the electron can be in anyone of g_d states.* We denote by N_d^0 the number of impurities which have an electron. The question we will address is: What is the occupation probability for the energy levels E_d? Thus we want to know the ratio of N_d^0 and N_d.

The number of ways N_d^0 electrons can be distributed among N_d states is

$$W(N_d^0, N_d) = \frac{N_d!}{N_d^0!(N_d - N_d^0)!} \tag{3.35}$$

Within each impurity level the electron can take any of g_d possibilities. Thus the total degeneracy for distributing N_d^0 electrons among N_d impurity levels is

$$W_{imp} = g_d^{N_d^0} \frac{N_d!}{N_d^0!(N_d - N_d^0)!} \tag{3.36}$$

Let us say that the total number of electrons in the continuum band and the impurity levels are n_{tot}, i.e.,

$$n_{tot} = N_d^0 + \sum_i N_i \tag{3.37}$$

where the subscript i represents the states in the continuum. The total energy of the electronic system is now

$$U = E_{tot} = N_d^0 E_d + \sum_i N_i \epsilon_i \tag{3.38}$$

The total degeneracy of the system is the product of the degeneracy arising from the electrons in the impurity levels and the degeneracy arising from the distribution in the continuum band. Using the results obtained previously for the continuum band we get

$$\begin{aligned} W_{tot} &= W_{imp} \times W_{band} \\ &= g_d^{N_d^0} \frac{N_d!}{N_d^0!(N_d - N_d^0)!} \prod_i \frac{g_i!}{N_i!(g_i - N_i)!} \end{aligned} \tag{3.39}$$

To find the equilibrium state we must minimize the free energy

$$F = U - k_B T ln W_{tot}$$

subject to the condition

$$N = N_d^0 + \sum_i N_i$$

Using the conditions

$$\begin{aligned} \delta F &= 0 \\ \delta n &= 0 \end{aligned}$$

we use the method of Lagrangian multipliers, i.e., we multiply the second equation by $-A$ and add it to the first. This gives

$$\begin{aligned} d\Bigg(&\sum_i \epsilon_i N_i + N_d^0 E_d - k_B T(N_d^0 ln g_d + ln N_d! - ln N_d^0! - ln[N_d - N_d^0]!) \\ &+ \sum_i [ln g_i! - ln N_i! - ln(g_i - N_i)! - A N_d^0 - A \sum_i N_i \Bigg) = 0 \end{aligned} \tag{3.40}$$

Maximizing the expression with respect to N_d^0 gives

$$k_B T[ln g_d - ln N_d^0 + ln(n_d - N_d^0)] + (E_d - A) = 0$$

which gives

$$\frac{N_d^0}{N_d} = \frac{1}{1 + \frac{1}{g_d}\exp(\frac{E_d - A}{k_B T})} \tag{3.41}$$

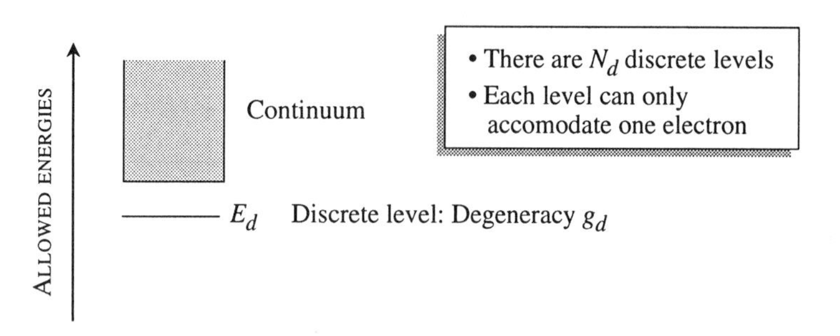

Figure 3.7: A schematic of a discrete allowed energy level separated by an energy E_d from a continuum of allowed energies.

Writing A as the Fermi level E_F we get the distribution function

$$f(E_d) = \frac{N_d^0}{N_d} = \frac{1}{1 + \frac{1}{g_d}\exp(\frac{E_d - E_F}{k_B T})} \tag{3.42}$$

If we maximize Eqn. 3.40 with respect to N_i we get the usual Fermi-Dirac function

$$f(\epsilon) = \frac{1}{1 + \exp(\frac{\epsilon - E_F}{k_B T})}$$

3.7 BLACKBODY SPECTRAL DENSITY

In Chapter 2, Section 2.2, we have mentioned that experimental observation of the radiation emanating from a blackbody at temperature T was a critical factor in the development of modern physics. We have mentioned that classical physics gives a spectral distribution which is reasonable at long wavelengths, but is quite inaccurate at small wavelengths. The resolution of the problems lies in treating radiation as particles of energy $\hbar\omega$ (or hv) and using Bose-Einstein statistics for the distribution among allowed states.

To calculate the spectral density we need to find the density of allowed modes of radiation in a volume. Let us consider the general form of radiation to be described by plane waves of the form

$$u = u_0 \exp\left[i\left(k_x x + k_y y + k_z z\right)\right] \tag{3.43}$$

If we now assume that the radiation is confined in a volume $L \times L \times L (= V)$, the wave goes to zero at the boundaries of this volume. The waves can then be represented by standing waves of the form $\sin k_x x \sin k_y y \sin k_z z$ with the restrictions for the allowed k_x, k_y, k_z values given by

$$\begin{aligned} k_x &= \frac{n\pi}{L} \quad n = 1, 2, 3 \ldots \\ k_y &= \frac{m\pi}{L} \quad m = 1, 2, 3 \ldots \\ k_z &= \frac{\ell\pi}{L} \quad \ell = 1, 2, 3 \ldots \end{aligned} \tag{3.44}$$

Note that since L is much larger than the wavelength of the radiation, the spacing between adjacent allowed k-values is very small. The k-space volume per allowed state is, as shown in Fig. 3.8a,

$$\Delta\Omega = \left(\frac{\pi}{L}\right)^3 \tag{3.45}$$

We now consider a k-space volume Ω. The number of allowed states in this volume are

$$\Delta N = \frac{\Omega}{\pi^3} \cdot L^3 \tag{3.46}$$

To calculate the density of allowed modes per unit volume between photon energies E and $E + dE$ we find the k-space volume between k and $k + dk$ as shown in Fig. 3.8b. This volume (in the positive quadrant only) is

$$\Omega = \frac{1}{8} \cdot 4\pi k^2 \Delta k \tag{3.47}$$

The density of allowed modes (per unit volume) between k and $k+dk$ is, by definition, $g(k)dk$, where $g(u)$ is the density of modes. Thus for a given polarization of light we have

$$g_p(k)\Delta dk = \frac{\pi}{2} \cdot \frac{k^2 \Delta k}{\pi^3} = \frac{k^2 \Delta k}{2\pi^2} \tag{3.48}$$

The relation between the photon energy and photon wavevector is

$$E = \hbar\omega = \hbar ck \tag{3.49}$$

where c is the velocity of light. We now have, for the density of modes (expressing k in terms of photon energy),

$$g_p(E)dE = g_p(k)dk = \frac{E^2 dE}{2\pi^2 c^3 \hbar^3} \tag{3.50}$$

We account for the two independent polarizations of the electromagnetic radiation to get the total density of states as

$$g(E)dE = \frac{E^2 dE}{\pi^2 c^3 \hbar^3} \tag{3.51}$$

The occupation number of each mode is given by the Bose-Einstein expression

$$n(E) = n(\hbar\omega) = \frac{1}{\exp\left(\hbar\omega/k_B T\right) - 1} \tag{3.52}$$

The spectral density is now

$$u(E)dE = n(\hbar\omega)Eg(E)dE = \frac{E^3 dE}{\pi^2 c^3 \hbar^3 \left(\exp(\hbar\omega/k_B T) - 1\right)} \tag{3.53}$$

This is the Planck distribution function for blackbody radiation. We can also express this in terms of energy density between wavelength λ and $\lambda + d\lambda$. This gives

$$u_\lambda d_\lambda = 8\pi hc \frac{\lambda^{-5} d\lambda}{e^{hc/\lambda k_B T} - 1} \tag{3.54}$$

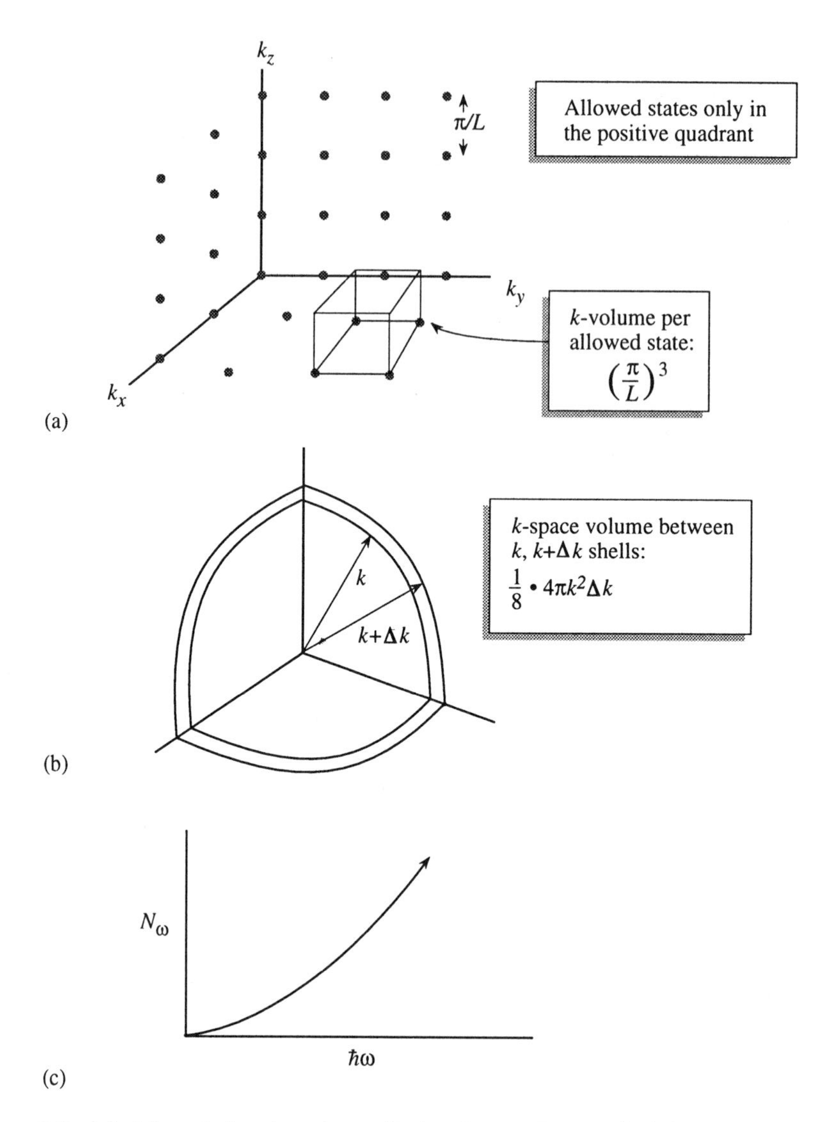

Figure 3.8: (a) Allowed k-points for radiation in a volume L^3, (b) geometry used to calculate k-states between $k, k + dk$, and (c) dependence of photon density of states on photon energy.

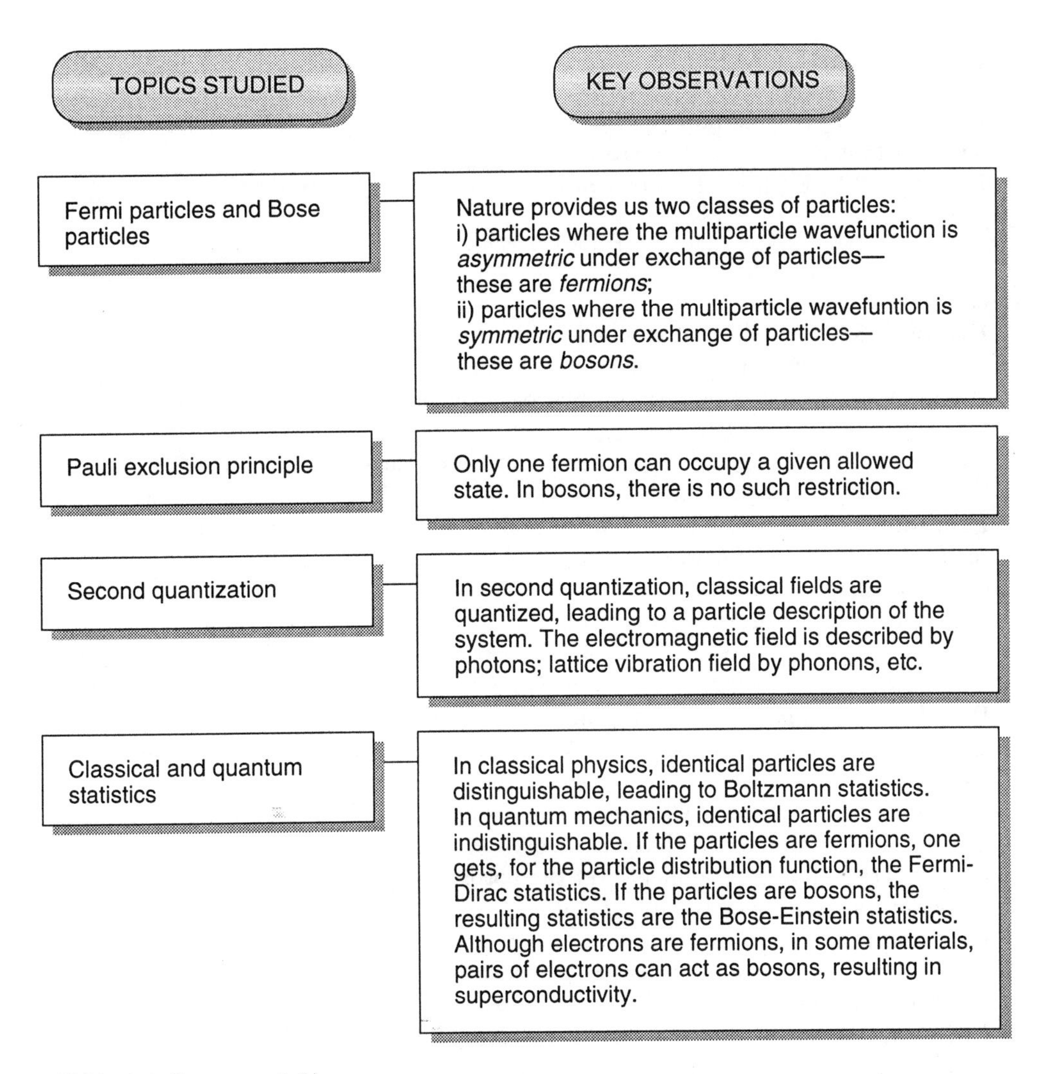

Table 3.1: Summary table

The radiancy $R(\lambda)$ discussed in Chapter 2 and u_λ are related by

$$R(\lambda) = u_\lambda \frac{c}{4} \tag{3.55}$$

3.8 CHAPTER SUMMARY

The topics discussed in this chapter are summarized in Table 3.1.

3.9 PROBLEMS

Problem 3.1 Discuss how the Bose-Einstein distribution function differs for Bose particles that have finite and zero mass.
Problem 3.2 Evaluate the error produced by Stirling's formula

$$\ell n(x!) \approx x\ell nx - x$$

as a function of x.
Problem 3.3 Use the Boltzmann distribution function to determine the ratio of the density of oxygen molecules in the atmosphere at the earth's surface and at 10 km. Also calculate this ratio for hydrogen molecules. Assume that the temperature is independent of height.
Problem 3.4 Discuss the conditions under which classical and Fermi-Dirac distribution functions become similar.
Problem 3.5 Use the Planck distribution for blackbody radiation and show that the peak wavelength (for which the radiancy is a maximum) is given by Wien's law (given in Chapter 2).
Problem 3.6 Integrate Planck's blackbody radiation law to obtain the Stefan-Boltzmann equation given in Chapter 2 for total power radiated by a blackbody. Use the integral $\int_0^\infty x^3dx/(e^x-1) = \pi^4/15$. (Power density is energy density $\times c/4$.)
Problem 3.7 Find the relative number of hydrogen atoms in the $n = 2, 3$, and 4 states with reference to the $n = 1$ state in the solar chromosphere where the temperature is 5000 K. The degeneracy of the states are given by $2n^2$. You may use the energy values given in Chapter 2 and assume that the occupation of the $n = 1$ state is close to unity and that of the excited states is very small. Why does the Balmer series appear prominently in the absorption spectrum of the sun, in spite of the low occupation of the $n = 2$ state?
Problem 3.8 A laser emits at $\lambda = 1.3$ μm with a power of 10 mW. The emitting area is 1 μm$\times$10 μm and the laser linewidth is $\Delta\lambda = 5$ Å. Estimate the photon number in this laser.
Problem 3.9 The temperature of the sun is 6000 K. Calculate the photon number for photons of wavelength 1.3 μm and 1.55 μm. These wavelengths are used in long-haul optical communications. Typical lasers used for communication applications have photon occupation numbers of $\sim$ 100. How hot will a blackbody have to be to have photon numbers suitable for communication?
Problem 3.10 Estimate the power emitted by a typical human body in the visible regime (between 4400 Å and 7000 Å).
Problem 3.11 A typical laser has photon occupation numbers of about 100 to 1000. Yet such lasers operate at room temperature. Can Planck's distribution equation for blackbody radiation explain this?

CHAPTER 4

PARTICLES IN PERIODIC POTENTIALS

CHAPTER AT A GLANCE

MAGIC OF TECHNOLOGY

• Scientists in a gene-splicing factory have developed a new *bio-material.* This one is for a self-dissolving suture for certain surgeries. They are not quite sure if the material has the right structure their computer model predicts. An electron diffraction experiment is conducted and after some data analysis the structure of the molecule is displayed on a computer screen. The image looks exactly like the computer model. Two years of research has paid off!
• A semiconductor manufacturing house is developing a new material to build lasers that can emit blue light. The crystal growers want to *see how the crystal grows at the atomic level.* An electron gun is positioned. The image made by the electrons on a fluorescent screen modulates as the crystal grows. Using the human eye the scientists are able to see how each monolayer of material grows. Each monolayer is only ~ 3 Å thick!
• In a modern integrated chip factory aluminum lines are used to connect switches much like highways in a city. Thin layers of silicon dioxide are used to insulate the devices. The device itself is made from silicon.

One of the greatest successes of modern science has been to explain why aluminum is a conductor, silicon dioxide is an insulator and silicon can be used as a switch. The hardware for the entire computer and communication industry is based on this basic understanding. In this chapter we will examine the concepts upon which the technologies mentioned above are based.

4.1 INTRODUCTION

In Chapters 2 and 3 we established the formal mathematical framework for quantum mechanics. We will now start our journey to apply this framework to problems of interest in applied physics and engineering. In this chapter we will address several important "simple" problems that have far-reaching consequences for applied physics and engineering. The problems are outlined in Fig. 4.1. Also shown are the implications of the understanding developed for different applied problems. As we address each problem we will highlight how modern engineering and applied physics problems are benefiting from the solutions.

Some of the problems chosen for this chapter are simple in the sense that the potential energy of the problem is such that it is possible to obtain analytical results. These problems provide us physical insight, useful in understanding more complicated problems which may require numerical techniques.

4.2 FREE PARTICLE PROBLEM AND DENSITY OF STATES

The free particle problem where the potential energy term is zero (or spatially constant) seems like a rather trivial problem, but the concepts developed are quite useful in a number of important problems. It will be shown later that when a particle moves in a perfectly periodic potential (in space) the solutions to the Schrödinger equation have a form very similar to the solutions in free space. Thus many concepts

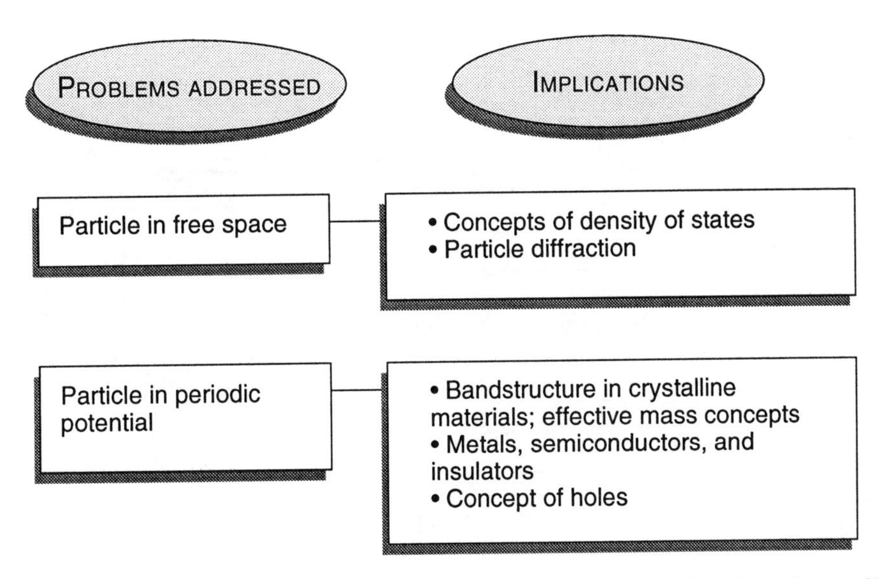

Figure 4.1: The problems addressed in this chapter and the implications for applied problems.

developed for the free-space particle can be applied to the description of electrons in crystalline media. In particular, the concept of density of states developed here is widely used.

Let us consider the Schrödinger equation for a free particle of mass m. The time-independent equation for the background potential equal to V_0 is

$$\frac{-\hbar^2}{2m}\left(\frac{\partial^2}{\partial x^2}+\frac{\partial^2}{\partial y^2}+\frac{\partial^2}{\partial z^2}\right)\psi(r)=(E-V_0)\psi(r) \tag{4.1}$$

A general solution of this equation is

$$\psi(r)=\frac{1}{\sqrt{V}}e^{\pm ik\cdot r} \tag{4.2}$$

and the corresponding energy is

$$E=\frac{\hbar^2k^2}{2m}+V_0 \tag{4.3}$$

where the factor $\frac{1}{\sqrt{V}}$ in the wavefunction occurs because we wish to have one particle per volume V or

$$\int_V d^3r \mid \psi(r) \mid^2 = 1 \tag{4.4}$$

We assume that the volume V is a cube of side L.

In classical mechanics the energy momentum relation for the free particle is $E = p^2/2m + V_0$, and p can be a *continuous variable*. The quantity $\hbar k$ seems to be replacing p in quantum mechanics. Due to the wave nature of the electron,

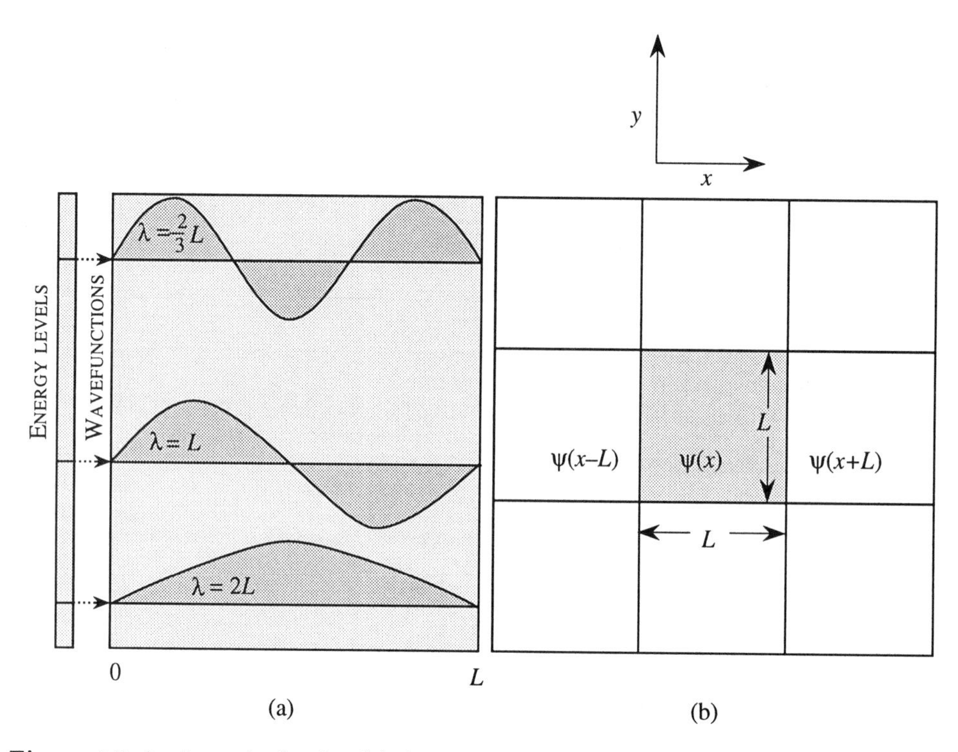

Figure 4.2: A schematic showing (a) the stationary boundary conditions leading to standing waves and (b) the periodic boundary conditions leading to exponential solutions with the electron probability equal in all regions of space.

in a finite volume, k is not continuous but discrete. To correlate with physical conditions we may want to describe, two kinds of boundary conditions are imposed on the wavefunction. In the first one the wavefunction is considered to go to zero at the boundaries of the volume, as shown in Fig. 4.2a. In this case, the wave solutions are standing waves of the form $\sin(k_x x)$ or $\cos(k_x x)$, etc., and k-values are restricted to positive values:

$$k_x = \frac{\pi}{L}, \frac{2\pi}{L}, \frac{3\pi}{L} \cdots \tag{4.5}$$

In Chapter 3, Section 3.7 we have used the stationary boundary conditions to obtain the density of modes of electromagnetic radiation. Here we will use a different set of conditions. Periodic boundary conditions are shown in Fig. 4.2b. Even though we focus our attention on a finite volume V, the wave can be considered to spread in all space as we conceive the entire space was made up of identical cubes of sides L. Then

$$\begin{aligned} \psi(x, y, z+L) &= \psi(x, y, z) \\ \psi(x, y+L, z) &= \psi(x, y, z) \\ \psi(x+L, y, z) &= \psi(x, y, z) \end{aligned} \tag{4.6}$$

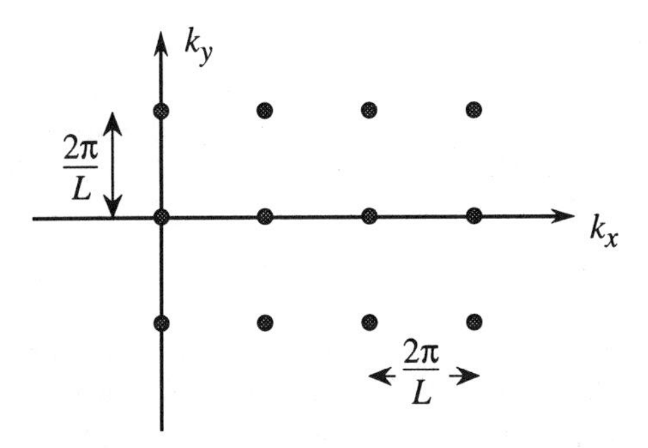

Figure 4.3: k-Space volume of each electronic state. The separation between the various allowed components of the k-vector is $\frac{2\pi}{L}$.

Because of the boundary conditions the allowed values of k are (n are integers—positive and negative)

$$k_x = \frac{2\pi n_x}{L}; \quad k_y = \frac{2\pi n_y}{L}; \quad k_z = \frac{2\pi n_z}{L} \tag{4.7}$$

If L is large, the spacing between the allowed k-values is very small. Also it is important to note that the results one obtains for properties of the particles in a *large volume are independent of whether we use the stationary or periodic boundary conditions.* It is useful to discuss the *volume in k-space that each electronic state occupies.* As can be seen from Fig. 4.3, this volume is (in three dimensions)

$$\left(\frac{2\pi}{L}\right)^3 = \frac{8\pi^3}{V} \tag{4.8}$$

If Ω is a volume of k-space, the number of electronic states in this volume is

$$\boxed{\frac{\Omega V}{8\pi^3}} \tag{4.9}$$

4.2.1 Density of States for a Three-Dimensional System

We will now use the discussion of the previous subsection to derive the extremely important concept of density of states. Although we will use the periodic boundary conditions to obtain the density of states, the stationary conditions lead to the same result, as long as the space under consideration is large, compared to the wavelength of the particle.

The concept of density of states is extremely powerful, and important physical properties in materials such as optical absorption, transport, etc., are intimately dependent upon this concept. Density of states is the number of available electronic states *per unit volume per unit energy* around an energy E. If we denote the density of states by $N(E)$, the number of states in a unit volume in an energy interval dE around an energy E is $N(E)dE$. To calculate the density of states, we need to know the dimensionality of the system and the energy versus k relation that the particles obey. We will choose the particle of interest to be the electron, since in most applied problems we are dealing with electrons. Of course, the results derived can be applied to other particles as well. For the free electron case we have the parabolic relation

$$E = \frac{\hbar^2 k^2}{2m_0} + V_0$$

The energies E and $E + dE$ are represented by surfaces of spheres with radii k and $k + dk$, as shown in Fig. 4.4. In a three-dimensional system, the k-space volume between vector k and $k + dk$ is (see Fig. 4.4a) $4\pi k^2 dk$. We have shown in Eqn. 4.9 that the k-space volume per electron state is $(\frac{2\pi}{L})^3$. Therefore, the number of electron states in the region between k and $k + dk$ is

$$\frac{4\pi k^2 dk}{8\pi^3} V = \frac{k^2 dk}{2\pi^2} V$$

Denoting the energy and energy interval corresponding to k and dk as E and dE, we see that the number of electron states between E and $E + dE$ per unit volume is

$$N(E)\, dE = \frac{k^2 dk}{2\pi^2}$$

Using the E versus k relation for the free electron, we have

$$k^2 dk = \frac{\sqrt{2} m_0^{3/2} (E - V_0)^{1/2} dE}{\hbar^3}$$

and

$$N(E)\, dE = \frac{m_0^{3/2} (E - V_0)^{1/2} dE}{\sqrt{2}\pi^2 \hbar^3} \tag{4.10}$$

We have seen in Chapter 3 that a particle is characterized by an *internal angular momentum value called spin*, which can take values $0, \hbar/2, \hbar, 3\hbar/2 \ldots$. An electron is a fermion and can have two possible states with a given energy. The electron can have a spin state $\hbar/2$ or $-\hbar/2$. Accounting for spin, the density of states obtained is simply multiplied by 2

$$\boxed{N(E) = \frac{\sqrt{2} m_0^{3/2} (E - V_0)^{1/2}}{\pi^2 \hbar^3}} \tag{4.11}$$

4.2.2 Density of States in Sub-Three-Dimensional Systems

Let us now consider a 2D system, a concept that has become a reality with use of quantum wells. The two-dimensional density of states is defined as the number of available electronic states *per unit area per unit energy* around an energy E. Similar arguments as used in the derivation show that the density of states for a parabolic band (for energies greater than V_0) is (see Fig 4.4b)

$$\boxed{N(E) = \frac{m_0}{\pi \hbar^2}} \tag{4.12}$$

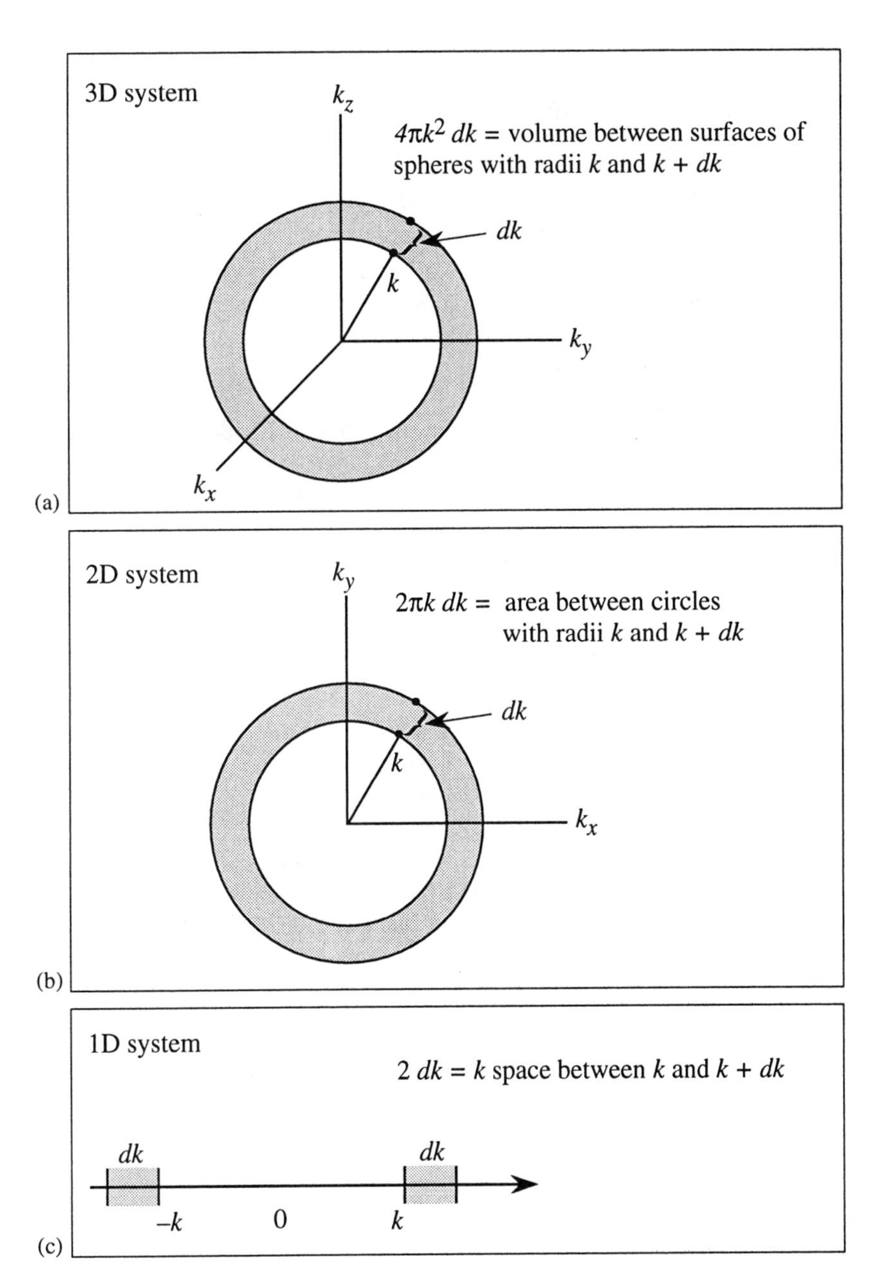

Figure 4.4: Geometry used to calculate density of states in three, two, and one dimensions. By finding the k-space volume in an energy interval between E and $E + dE$, one can find the number of allowed states.

The factor of 2 resulting from spin has been included in this expression.

Finally, we can consider a one-dimensional system often called a "quantum wire." The one-dimensional density of states is defined as the number of available electronic states *per unit length per unit energy* around an energy E. In a 1D system or a "quantum wire" the density of states is (including spin) (see Fig. 4.4c)

$$\boxed{N(E) = \frac{\sqrt{2}m_0^{1/2}}{\pi\hbar}(E - V_0)^{-1/2}} \tag{4.13}$$

Notice that as the dimensionality of the system changes, the energy dependence of the density of states also changes. As shown in Fig. 4.5, for a three-dimensional system we have $(E - V_0)^{1/2}$ dependence, for a two-dimensional system we have no energy dependence, and for a one-dimensional system we have $(E - V_0)^{-1/2}$ dependence.

We will see later in the next section that when a particle is in a periodic potential, its wavefunction is quite similar to the free particle wavefunction. Also, the particle responds to external forces as if it is a free particle *except that its energy-momentum relation is modified by the presence of the periodic potential.* In some cases it is possible to describe the particle energy by the relation

$$E = \frac{\hbar^2 k^2}{2m^*} + E_0 \tag{4.14}$$

where m^* is called the effective mass in the material. The expressions derived for the free electron density of states *can then be carried over to describe the density of states for a particle in a crystalline material (which has a periodic potential) by simply replacing m_0 by m^*.*

EXAMPLE 4.1 Calculate the density of states of electrons in a 3D system and a 2D system at an energy of 0.1 eV. Assume that the background potential is zero.

The density of states in a 3D system (including the spin of the electron) is given by (E is the energy in Joules)

$$\begin{aligned} N(E) &= \frac{\sqrt{2}(m_0)^{3/2}E^{1/2}}{\pi^2\hbar^3} \\ &= \frac{\sqrt{2}(0.91 \times 10^{-30}\ \text{kg})(E^{1/2})}{\pi^2(1.05 \times 10^{-34}\ \text{J.s})^3} \\ &= 1.07 \times 10^{56} E^{1/2}\ \text{J}^{-1}\ \text{m}^{-3} \end{aligned}$$

Expressing E in eV and the density of states in the commonly used units of $\text{eV}^{-1}\ \text{cm}^{-3}$, we get

$$\begin{aligned} N(E) &= 1.07 \times 10^{56} \times (1.6 \times 10^{-19})^{3/2}(1.0 \times 10^{-6})E^{1/2} \\ &= 6.8 \times 10^{21} E^{1/2}\ \text{eV}^{-1}\ \text{cm}^{-3} \end{aligned}$$

At $E = 0.1$ eV we get

$$N(E) = 2.15 \times 10^{21}\ \text{eV}^{-1}\ \text{cm}^{-3}$$

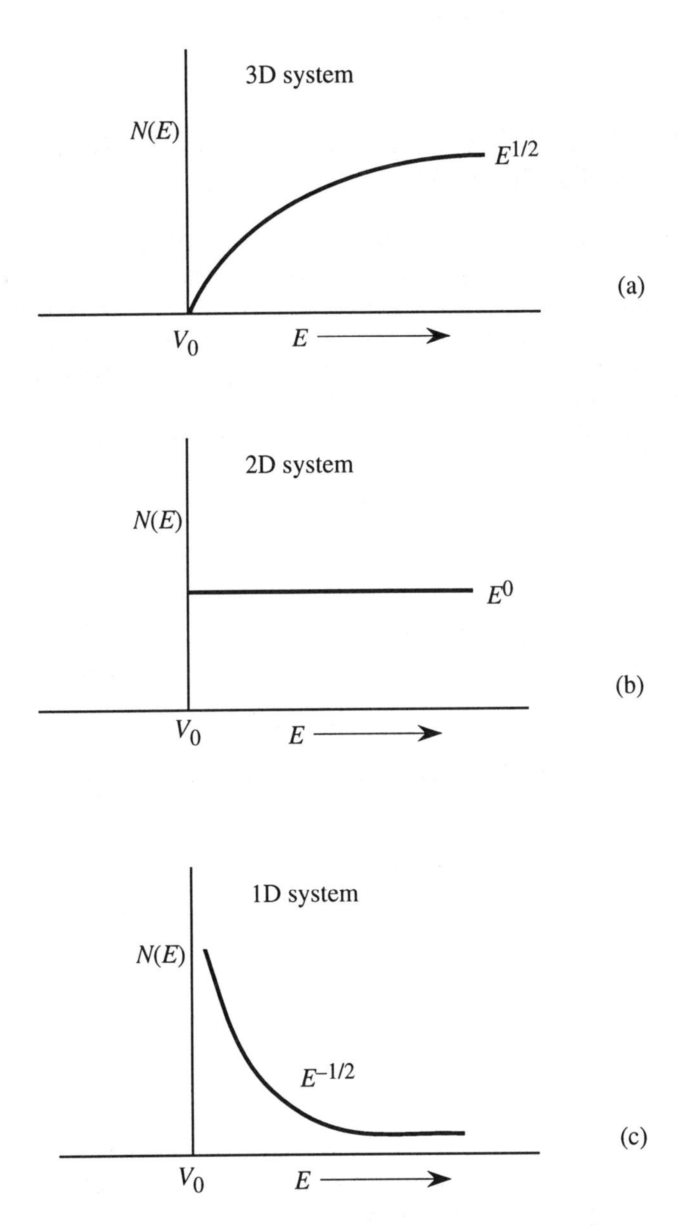

Figure 4.5: Variation in the energy dependence of the density of states in (a) three-dimensional, (b) two-dimensional, and (c) one-dimensional systems. The energy dependence of the density of states is determined by the dimensionality of the system.

For a 2D system the density of states is independent of energy and is

$$N(E) = \frac{m_0}{\pi\hbar^2} = 4.21 \times 10^{14} \text{ eV}^{-1} \text{ cm}^{-2}$$

4.3 PARTICLE IN A PERIODIC POTENTIAL: BLOCH THEOREM

One of the first great successes of quantum theory in applied physics is band theory, which eventually resolved the puzzle of electrical transport in metals, semiconductors, and insulators. The core of modern understanding of electronic and optical properties of solid-state materials is based on the band theory, which describes the properties of electrons in a periodic potential. The periodic potential is due to the periodic arrangement of atoms in a crystal.

The description of the electron in the periodic material has to be via the Schrödinger equation

$$\left[\frac{-\hbar^2}{2m_0}\nabla^2 + U(\mathbf{r})\right]\psi(\mathbf{r}) = \mathbf{E}\psi(\mathbf{r}) \tag{4.15}$$

where $U(\mathbf{r})$ is the background potential seen by the electrons. Due to the crystalline nature of the material, the potential $U(\mathbf{r})$ has the same periodicity, R, as the lattice

$$U(\mathbf{r}) = U(\mathbf{r} + \mathbf{R}) \tag{4.16}$$

If the background potential is zero, the electronic function in a volume V is

$$\psi(\mathbf{r}) = \frac{\mathrm{e}^{\mathbf{ik}\cdot\mathbf{r}}}{\sqrt{\mathbf{V}}}$$

and the electron momentum and energy are

$$\mathbf{p} = \hbar\mathbf{k}$$
$$E = \frac{\hbar^2 k^2}{2m_0}$$

The wavefunction is spread in the entire sample and has equal probability $(\psi^*\psi)$ at every point in space. Let us examine the periodic crystal. We expect the *electron probability to be same in all unit cells of the crystal because each cell is identical.* If the potential was random, this would not be the case, as shown schematically in Fig. 4.6a. This expectation is, indeed, correct and is put in a mathematical form by Bloch's theorem. Bloch's theorem states that the eigenfunctions of the Schrödinger equation for a periodic potential are the product of a plane wave $e^{i\mathbf{k}\cdot\mathbf{r}}$ and a function $u_{\mathbf{k}}(\mathbf{r})$, which has the *same periodicity as the periodic potential.* Thus

$$\boxed{\psi_{\mathbf{k}}(\mathbf{r}) = e^{i\mathbf{k}\cdot\mathbf{r}}u_{\mathbf{k}}(\mathbf{r})} \tag{4.17}$$

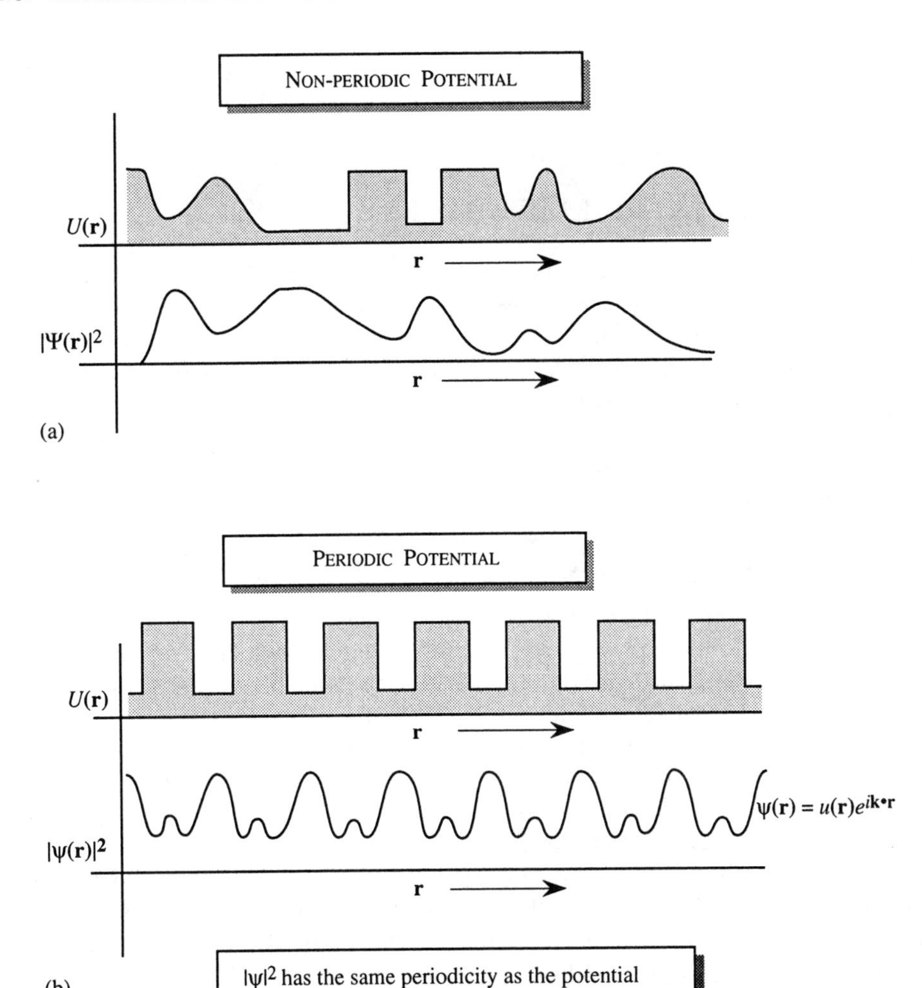

Figure 4.6: (a) Potential and electron probability value of a typical electronic wavefunction in a random material. (b) The effect of a periodic background potential on an electronic wavefunction. In the case of the periodic potential, $|\psi|^2$ has the same spatial periodicity as the potential. This puts a special constraint on $\psi(\mathbf{r})$ according to Bloch's theorem.

is the form of the electronic function. The periodic part $u_{\mathbf{k}}(\mathbf{r})$ has the same periodicity as the crystal; i.e.,

$$\boxed{u_{\mathbf{k}}(\mathbf{r}) = u_{\mathbf{k}}(\mathbf{r} + \mathbf{R})} \tag{4.18}$$

The wavefunction has the property

$$\begin{aligned}\psi_{\mathbf{k}}(\mathbf{r} + \mathbf{R}) &= e^{i\mathbf{k}\cdot(\mathbf{r}+\mathbf{R})} u_{\mathbf{k}}(\mathbf{r} + \mathbf{R}) = e^{i\mathbf{k}\cdot\mathbf{r}} u_{\mathbf{k}}(\mathbf{r}) e^{i\mathbf{k}\cdot\mathbf{R}} \\ &= e^{i\mathbf{k}\cdot\mathbf{R}} \psi_{\mathbf{k}}(\mathbf{r})\end{aligned} \tag{4.19}$$

The wavefunction is illustrated in Fig. 4.6b. Before discussing the solutions of the periodic potential problem, let us take a look at some of the important properties of crystalline materials.

4.4 CRYSTALLINE MATERIALS

Most of the modern information-processing devices are made from crystalline materials, where long range order is present among the atoms. Semiconductors form the most important class of materials used in information-processing technologies. Certain kinds of materials called ferroelectrics and some dielectrics are also used in modern optoelectronics. Recently polymers have started to become important as information-processing materials. We will now discuss some general properties of crystalline materials and will then focus on the specific structural properties of important crystals.

4.4.1 Periodicity of a Crystal

Crystals are made up of identical building blocks, the block being an atom or a group of atoms. While in most crystals the crystalline symmetry is fixed by nature, new advances in crystal growth techniques are allowing scientists to produce artificial crystals with modified crystalline structure. These advances depend upon being able to place atomic layers with exact precision and control during growth. The underlying periodicity of crystals is the key which controls the properties of the electrons inside the material. Thus by altering crystalline structure artificially, one is able to alter electronic properties.

To understand and define the crystal structure, two important concepts are introduced. The *lattice* represents a set of points in space which form a periodic structure. Each point sees an exact similar environment. The lattice is by itself a mathematical abstraction. A building block of atoms called the *basis* is then attached to each lattice point yielding the crystal structure.

An important property of a lattice is the ability to define three vectors $\mathbf{a}_1$, $\mathbf{a}_2$, $\mathbf{a}_3$, such that any lattice point $\mathbf{R}'$ can be obtained from any other lattice point $\mathbf{R}$ by a translation

$$\mathbf{R}' = \mathbf{R} + m_1\mathbf{a}_1 + m_2\mathbf{a}_2 + m_3\mathbf{a}_3 \tag{4.20}$$

where m_1, m_2, m_3 are integers. Such a lattice is called Bravais lattice. The entire lattice can be generated by choosing all possible combinations of the integers m_1, m_2, m_3 . The crystalline structure is now produced by attaching the basis to each of these lattice points.

$$\boxed{\text{lattice} + \text{basis} = \text{crystal structure}} \tag{4.21}$$

The translation vectors $\mathbf{a}_1$, $\mathbf{a}_2$, and $\mathbf{a}_3$ are called primitive if the volume of the cell formed by them is the smallest possible. There is no unique way to choose the primitive vectors. One choice is to pick

System	Number of Lattices	Restrictions on Conventional Cell Axes and Singles
Triclinic	1	$a_1 \neq a_2 \neq a_3$ $\alpha \neq \beta \neq \gamma$
Monoclinic	2	$a_1 \neq a_2 \neq a_3$ $\alpha = \gamma = 90^o \neq \beta$
Orthorhombic	4	$a_1 \neq a_2 \neq a_3$ $\alpha = \beta = \gamma = 90^o$
Tetragonal	2	$a_1 = a_2 \neq a_3$ $\alpha = \beta = \gamma = 90^o$
Cubic	3	$a_1 = a_2 = a_3$ $\alpha = \beta = \gamma = 90^o$
Trigonal	1	$a_1 = a_2 = a_3$ $\alpha = \beta = \gamma < 120^o, \neq 90^o$
Hexagonal	1	$a_1 = a_2 \neq a_3$ $\alpha = \beta = 90^o$ $\gamma = 120^o$

Table 4.1: The 14 Bravais lattices in 3-dimensional systems and their properties

$\mathbf{a}_1$ to be the shortest period of the lattice
$\mathbf{a}_2$ to be the shortest period not parallel to $\mathbf{a}_1$
$\mathbf{a}_3$ to be the shortest period not coplanar with $\mathbf{a}_1$ and $\mathbf{a}_2$.

It is possible to define more than one set of primitive vectors for a given lattice, and often the choice depends upon convenience. The volume cell enclosed by the primitive vectors is called the *primitive unit cell.*

4.4.2 Basic Lattice Types

The various kinds of lattice structures possible in nature are described by the symmetry group that describes their properties. Rotation is one of the important symmetry groups. Lattices can be found which have a rotation symmetry of $2\pi, \frac{2\pi}{2}, \frac{2\pi}{3}, \frac{2\pi}{4}, \frac{2\pi}{6}$. The rotation symmetries are denoted by 1, 2, 3, 4, and 6. No other rotation axes exist, e.g., $\frac{2\pi}{5}$ or $\frac{2\pi}{7}$ are not allowed because such a structure could not fill up an infinite space.

There are 14 types of lattices in three dimensions. These lattice classes are defined by the relationships between the primitive vectors a_1, a_2, and a_3, and the angles α, β, and γ between them. The general lattice is triclinic ($\alpha \neq \beta \neq \gamma, a_1 \neq a_2 \neq a_3$) and there are 13 special lattices. Table 4.1 provides the basic properties of these three-dimensional lattices. We will focus on the cubic lattice, which is the structure taken by all semiconductors.

There are three kinds of cubic lattices: simple cubic, body-centered cubic,

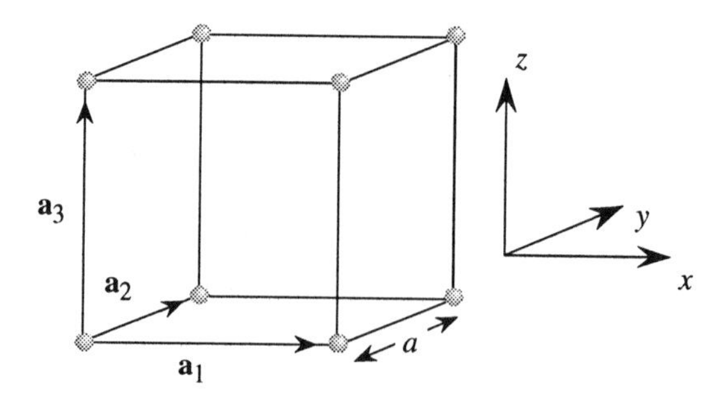

Figure 4.7: A simple cubic lattice showing the primitive vectors. The crystal is produced by repeating the cubic cell through space.

and face-centered cubic.

Simple Cubic: The simple cubic lattice shown in Fig. 4.7 is generated by the primitive vectors

$$a\mathbf{x}, a\mathbf{y}, a\mathbf{z} \tag{4.22}$$

where the $\mathbf{x}$, $\mathbf{y}$, $\mathbf{z}$ are unit vectors.

Body-Centered Cubic: The bcc lattice shown in Fig. 4.8 can be generated from the simple cubic structure by placing a lattice point at the center of the cube. If $\hat{\mathbf{x}}, \hat{\mathbf{y}}$, and $\hat{\mathbf{z}}$ are three orthogonal unit vectors, then a set of primitive vectors for the body-centered cubic lattice could be

$$\mathbf{a}_1 = a\hat{\mathbf{x}},\ \mathbf{a}_2 = a\hat{\mathbf{y}},\ \mathbf{a}_3 = \frac{a}{2}(\hat{\mathbf{x}} + \hat{\mathbf{y}} + \hat{\mathbf{z}}) \tag{4.23}$$

A more symmetric set for the bcc lattice is

$$\mathbf{a}_1 = \frac{a}{2}(\hat{\mathbf{y}} + \hat{\mathbf{z}} - \hat{\mathbf{x}}),\ \mathbf{a}_2 = \frac{a}{2}(\hat{\mathbf{z}} + \hat{\mathbf{x}} - \hat{\mathbf{y}}),\ \mathbf{a}_3 = \frac{a}{2}(\hat{\mathbf{x}} + \hat{\mathbf{y}} - \hat{\mathbf{z}}) \tag{4.24}$$

Face-Centered Cubic: A most important lattice for semiconductors is the *face-centered cubic* (fcc) Bravais lattice. To construct the face-centered cubic Bravais

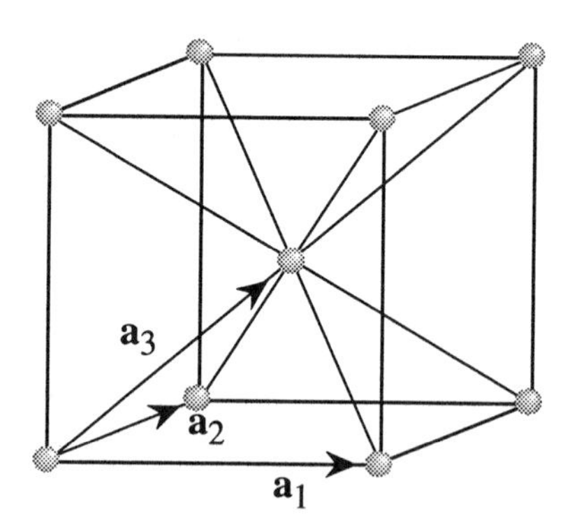

Figure 4.8: The body-centered cubic lattice along with a choice of primitive vectors.

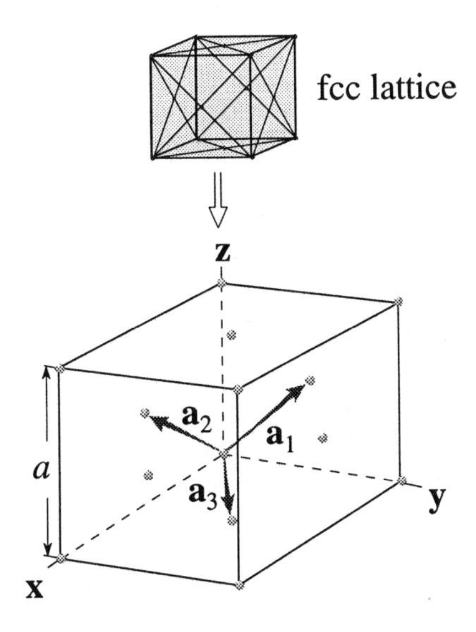

Figure 4.9: Primitive basis vectors for the face-centered cubic lattice.

lattice add to the simple cubic lattice an additional point in the center of each square face (Fig. 4.9).

A symmetric set of primitive vectors for the face-centered cubic lattice (see Fig. 4.9) is

$$\mathbf{a}_1 = \frac{a}{2}(\hat{\mathbf{y}} + \hat{\mathbf{z}}),\ \mathbf{a}_2 = \frac{a}{2}(\hat{z} + \hat{\mathbf{x}}),\ \mathbf{a}_3 = \frac{a}{2}(\hat{\mathbf{x}} + \hat{\mathbf{y}}) \tag{4.25}$$

The face-centered cubic and body-centered cubic Bravais lattices are of great importance, since an enormous variety of solids crystallize in these forms with an atom (or ion) at each lattice site. Essentially all semiconductors of interest for electronics and optoelectronics have fcc structure.

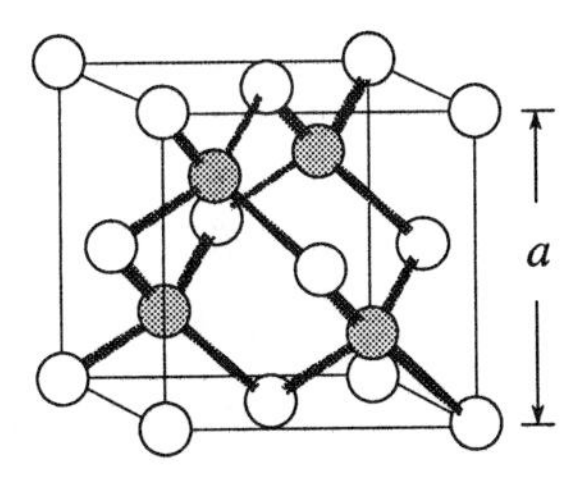

Figure 4.10: The zinc blende crystal structure. The structure consists of the interpenetrating fcc lattices, one displaced from the other by a distance $(\frac{a}{4}\frac{a}{4}\frac{a}{4})$ along the body diagonal. The underlying Bravais lattice is fcc with a two-atom basis. The positions of the two atoms is (000) and $(\frac{a}{4}\frac{a}{4}\frac{a}{4})$.

4.4.3 Diamond and Zinc Blende Structures

Most semiconductors of interest for electronics and optoelectronics have an underlying fcc lattice. However, they have two atoms per basis. The coordinates of the two basis atoms are

$$(000) \; and \; \left(\frac{a}{4}, \frac{a}{4}, \frac{a}{4}\right) \tag{4.26}$$

Since each atom lies on its own fcc lattice, such a two-atom basis structure may be thought of as two inter-penetrating fcc lattices, one displaced from the other by a translation along a body diagonal direction $(\frac{a}{4}\frac{a}{4}\frac{a}{4})$.

Fig. 4.10 gives details of this important structure. If the two atoms of the basis are identical, the structure is called diamond. Semiconductors such as Si, Ge, C, etc., fall in this category. If the two atoms are different, the structure is called the zinc blende structure. Semiconductors such as GaAs, AlAs, CdS, etc., fall in this category. Semiconductors with diamond structure are often called elemental semiconductors, while the zinc blende semiconductors are called compound semiconductors. The compound semiconductors are also denoted by the position of the atoms in the periodic chart, e.g., GaAs, AlAs, InP are called III-V (three-five) semiconductors while CdS, HgTe, CdTe, etc., are called II-VI (two-six) semiconductors.

4.4.4 Hexagonal Close Pack Structure

The hexagonal close pack (hcp) structure is an important lattice structure and many metals have this underlying lattice. Some semiconductors such as BN, AlN, GaN, SiC, etc., also have this underlying lattice (with a two-atom basis). The hcp structure is formed as shown in Fig. 4.11a. Imagine that a close-packed layer of spheres is formed. Each sphere touches six other spheres, leaving cavities, as shown. A second close-packed layer of spheres is placed on top of the first one so that the second layer sphere centers are in the cavities formed by the first layer. The third layer of close-packed spheres can now be placed so that center of the spheres do not fall on the center of the starting spheres (left side of Fig. 4.11a) or coincide with the centers of the starting spheres (right side of Fig. 4.11b). These two sequences, when repeated, produce the fcc and hcp lattices.

In Fig. 4.11b we show the detailed positions of the lattice points in the hcp lattice. The three lattice vectors are a_1, a_2 a_3, as shown. The vector a_3 is denoted by c and the term c-axis refers to the orientation of a_3. In an ideal structure, if $\mid a \mid=\mid a_1 \mid=\mid a_2 \mid$,

$$\frac{c}{a} = \sqrt{\frac{8}{3}} \tag{4.27}$$

In Table 4.2 we show the structural properties of some important materials.

4.4.5 Ferroelectric Crystals

An important class of materials used in solid-state optoelectronics is called ferroelectric crystals. *A ferroelectric material is characterized by the presence of an electric*

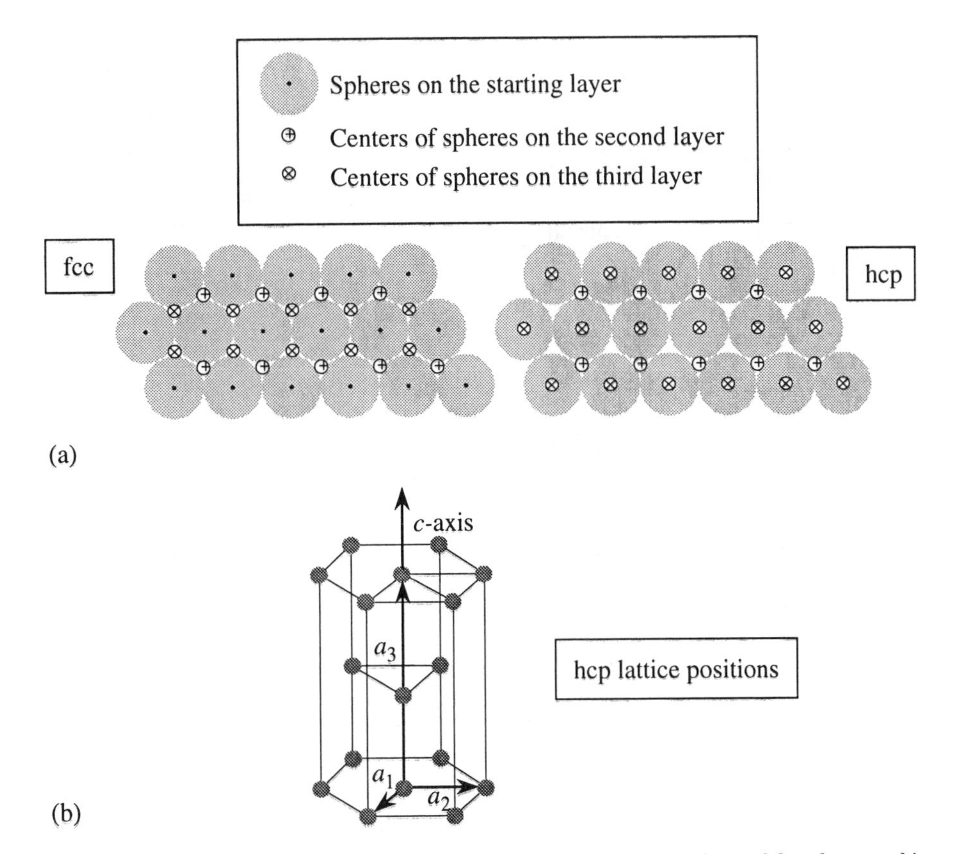

Figure 4.11: (a) A schematic of how the fcc and hcp lattices are formed by close packing of spheres. (b) Arrangement of lattice points on an hcp lattice.

dipole moment even in the absence of an external electric field. Thus in ferroelectric crystals, the center of the positive charge of the crystal does not coincide with the center of the negative charge. If we examine a zinc blende material like GaAs, an important optoelectronic material, we note that due to charge transfer, the Ga atom (from the group III of the periodic chart) carries a slightly negative charge (and is called the cation) while the As atom (from the group V of the periodic chart) has a slightly positive charge (and is called the anion). However, there is no net electric dipole in the crystal in spite of the local charge transfer because in the zinc blende structure the cations and anions are placed so that the center of the positive and negative charges coincides when all the atoms are considered. A net dipole could be present if, for example, the entire Ga sublattice moved with respect to the As sublattice. Such a displacement does not occur in GaAs but does occur in certain materials.

Some important ferroelectric materials include $LiNbO_3$, $LiTaO_3$, KH_2PO_4, $BaTiO_3$, etc. Ferroelectric crystals are produced by two main effects: i) ordering of ions in such a way as to produce a net electric dipole; ii) displacement of one sublattice with respect to another to produce a net dipole. Some ferroelectrics involve

Material	Structure	Lattice Constant (Å)	Density (gm/cm^3)
C	Diamond	3.5668	3.5153
Si	Diamond	5.431	2.329
Ge	Diamond	5.658	5.323
GaAs	Zinc Blende	5.653	5.318
AlAs	Zinc Blende	5.660	3.760
InAs	Zinc Blende	6.058	5.667
GaN	Wurtzite	$a = 3.175; c = 5.158$	6.095
AlN	Wurtzite	$a = 3.111; c = 4.981$	3.255
SiC	Zinc Blende	4.360	3.166
Cd	hcp	$a = 2.98; c = 5.620$	8.65
Cr	bcc	2.88	7.19
Co	hcp	$a = 2.51; c = 4.07$	8.9
Au	fcc	4.08	19.3
Fe	bcc	2.87	7.86
Ag	fcc	4.09	10.5
Al	fcc	4.05	2.7
Cu	fcc	3.61	8.96

Table 4.2: Structure, lattice constant, and density of some materials at room temperature.

the presence of hydrogen or deuterium (the KDP type shown in Table 4.2). The motion of the proton in the hydrogen or deuterium is the cause of the ferroelectric effect.

An important class of ferroelectric crystals falls in the category of perovskite structure shown in Figs. 4.12a and 4.12b. We can understand the basic perovskite crystal by focusing on a material like barium titanate ($BaTiO_3$). The structure is cubic with Ba^{2+} ions at the cube corners. The O^{2-} ions are at the six face centers of the cube while a Ti^{4+} ion is at the body center. In the absence of any deformation, the material does not have any net electric dipole at zero field, but if there is a displacement of ions, a dipole can develop. Fig. 4.12b shows how such a

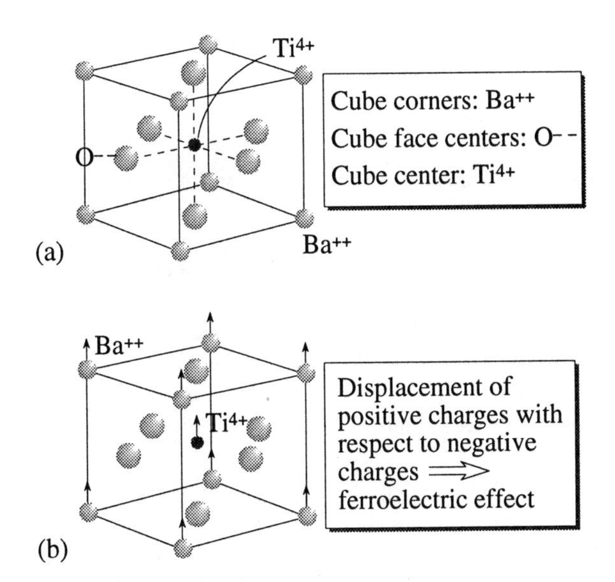

Figure 4.12: (a) The structure of a typical perovskite crystal illustrated by examining barium titanate. (b) The ferroelectric effect is produced by a net displacement of the positive ions with respect to the negative ions .

displacement can arise. In most materials the net displacement is of the order of ~ 0.1 Å to 1 Å.

EXAMPLE 4.2 The crystal barium titanate has a unit cell volume of $(4 \times 10^{-8}\ \text{cm})^3$ and a spontaneous polarization of 26.67 μC cm^{-2}. Calculate the dipole moment of a unit cell and estimate the shift of the positive ions with respect to the negative ions.

The dipole moment of the unit cell is

$$\begin{aligned} p &= (26.67 \times 10^{-6}\ C.\text{cm}^{-2})(64 \times 10^{-24}\ \text{cm}^3) \\ &= 1.71 \times 10^{-27}\ C.\text{cm} \end{aligned}$$

If the displacement of the positive ions, with respect to the negative ions, is δ, the dipole moment of the unit cell will be $6e\delta$. Thus we have

$$\begin{aligned} \delta &= \frac{1.71 \times 10^{-27}\ C.\text{cm}}{6 \times (1.6 \times 10^{-19}\ C)} = 1.78 \times 10^{-9}\ \text{cm} \\ &= 0.178\ \text{Å} \end{aligned}$$

4.4.6 Notation to Denote Planes and Points in a Lattice: Miller Indices

A simple scheme is used to describe lattice planes, directions, and points. For a plane, we use the following procedure:

(1) Define the x, y, z axes (primitive vectors)

(2) Take the intercepts of the plane along the axes in units of lattice constants.
(3) Take the reciprocal of the intercepts and reduce them to the smallest integers.
The notation (hkl) denotes a family of parallel planes.
The notation {hkl} denotes a family of equivalent planes.

To denote directions, we use the smallest set of integers having the same ratio as the direction cosines of the direction.

In a cubic system the Miller indices of a plane are the same as the direction perpendicular to the plane. The notation [ ] is for a set of parallel directions. The notation < > is for a set of equivalent directions. Fig. 4.13 shows some examples of the use of the Miller indices to define planes in a cubic structure.

EXAMPLE 4.3 The lattice constant of silicon is 5.43 Å. Calculate the number of silicon atoms in a cubic centimeter. Also calculate the number density of Ga atoms in GaAs which has a lattice constant of 5.65 Å.

Silicon has a diamond structure which is made up of the fcc lattice with two atoms on each lattice point. The fcc unit cube has a volume a^3. The cube has eight lattice sites at the cube edges. However, each of these points is shared with eight other cubes. In addition, there are six lattice points on the cube face centers. Each of these points is shared by two adjacent cubes. Thus the number of lattice points per cube of volume a^3 is

$$N(a^3) = \frac{8}{8} + \frac{6}{2} = 4$$

In silicon there are two silicon atoms per lattice point. The number density is, therefore,

$$N_{Si} = \frac{4 \times 2}{a^3} = \frac{4 \times 2}{(5.43 \times 10^{-8})^3} = 4.997 \times 10^{22} \text{atoms/cm}^3$$

In GaAs, there is one Ga atom and one As atom per lattice point. The Ga atom density is, therefore,

$$N_{Ga} = \frac{4}{a^3} = \frac{4}{(5.65 \times 10^{-8})^3} = 2.22 \times 10^{22} \text{atoms/cm}^3$$

There are an equal number of As atoms.

EXAMPLE 4.4 In semiconductor technology, a Si device on a VLSI chip represents one of the smallest devices while a GaAs laser represents one of the larger devices. Consider a Si device with dimensions $(5 \times 2 \times 1)$ μm^3 and a GaAs semiconductor laser with dimensions $(200 \times 10 \times 5)$ μm^3. Calculate the number of atoms in each device.

From Example 4.2 the number of Si atoms in the Si transistor is

$$N_{Si} = (5 \times 10^{22} \text{atoms/cm}^3)(10 \times 10^{-12} \text{cm}^3) = 5 \times 10^{11} \text{atoms}$$

The number of Ga atoms in the GaAs laser is

$$N_{Ga} = (2.22 \times 10^{22})(10^4 \times 10^{-12}) = 2.22 \times 10^{14} \text{atoms}$$

An equal number of As atoms are also present in the laser.

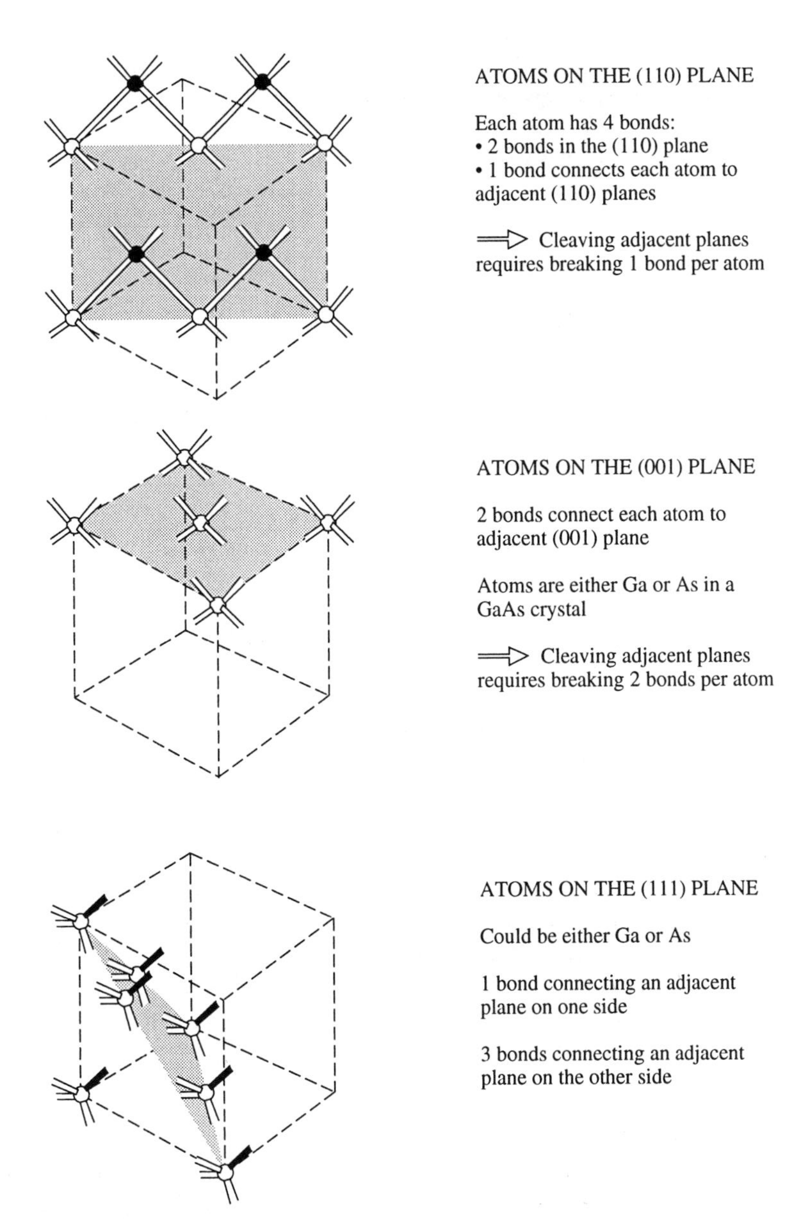

Figure 4.13: Some important planes in the cubic system along with their Miller indices. This figure also shows how many bonds connect adjacent planes. This number determines how easy or difficult it is to cleave the crystal along these planes by cutting the bonds joining the adjacent planes.

EXAMPLE 4.5 Calculate the surface density of Ga atoms on a Ga terminated (001) GaAs surface.

In the (001) surfaces, the top atoms are either Ga or As leading to the terminology Ga terminated (or Ga stabilized) and As terminated (or As stabilized), respectively. A square of area a^2 has four atoms on the edges of the square and one atom at the center of the square. The atoms on the square edges are shared by a total of four squares. The total number of atoms per square is

$$N(a^2) = \frac{4}{4} + 1 = 2$$

The surface density is then

$$N_{Ga} = \frac{2}{a^2} = \frac{2}{(5.65 \times 10^{-8})^2} = 6.26 \times 10^{14} \text{cm}^{-2}$$

EXAMPLE 4.6 Calculate the height of a GaAs monolayer in the (001) direction. In the case of GaAs, a monolayer is defined as the combination of a Ga and As atomic layer. The monolayer distance in the (001) direction is simply

$$A_{m\ell} = \frac{a}{2} = \frac{5.65}{2} = 2.825 \text{ Å}$$

4.4.7 Artificial Structures: Superlattices and Quantum Wells

So far in this chapter we have discussed crystal structures that are present in natural semiconductors. These structures are the lowest free energy configuration of the solid state of the atoms. Since the electronic and optical properties of the semiconductors are completely determined by the crystal structure, scientists have been intrigued with the idea of fabricating artificial structures or superlattices. The key to growing artificial structures with tailorable crystal structure and hence tailorable optical and electronic properties has been the progress in heteroepitaxy, i.e., the growth of one material on another. Heteroepitaxial crystal growth techniques such as molecular beam epitaxy (MBE) and metal-organic chemical vapor deposition (MOCVD) have made a tremendous impact on semiconductor physics and technology. From very high speeds, low-noise electronic devices used for satellite communications to low-threshold lasers for communication, semiconductor devices are being made by these techniques.

MBE or MOCVD are techniques which allow monolayer (~3 Å) control in the chemical composition of the growing crystal. Nearly every semiconductor and many metals have been grown by epitaxial techniques such as MBE and MOCVD.

Since the heteroepitaxial techniques allow one to grow heterostructures with atomic control, one can change the periodicity of the crystal in the growth direction. This leads to the concept of superlattices where two (or more) semiconductors A and B are grown alternately with thicknesses d_A and d_B, respectively. The periodicity of the lattice in the growth direction is then $d_A + d_B$. A $(GaAs)_2$ $(AlAs)_2$ superlattice is illustrated in Fig. 4.14. It is a great testimony to the precision of the new growth techniques that values of d_A and d_B as low as one monolayer have been grown.

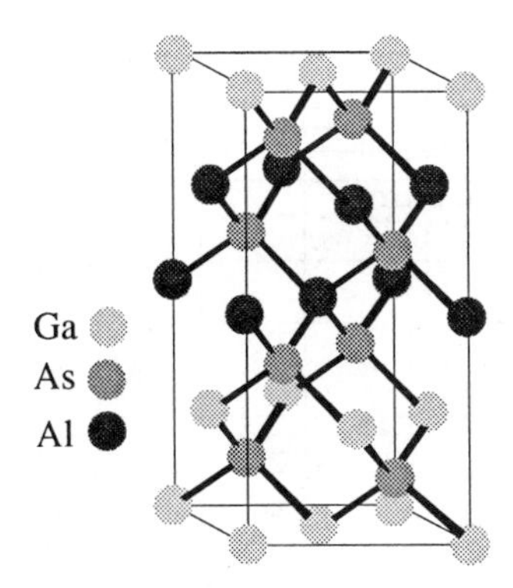

Figure 4.14: Arrangement of atoms in a $(GaAs)_2(AlAs)_2$ superlattice grown along (001) direction.

4.4.8 Wave Diffraction and Bragg's Law

We have described in this section how atoms are arranged in crystalline materials. How do we know all this? We know from optics that to "see" or resolve something the wavelength of light used should be smaller than the size of the object. This applies to seeing atoms in a crystal, too. The wavelength should be ~ 1 Å. For electromagnetic radiation this means the photon energy should be ~ 10 KeV. Such radiation is called x-rays. We can also use particles to see atoms. As we know from the deBroglie wavelength and momentum relation, particles can have very short wavelengths at quite small energies. In Fig. 4.14a we show the relation between wavelength and energy for photons and electrons. Both these particles (and other particles) can be used to "see" atomic positions in materials.

Theory similar to the ones used in optics is used to understand how one obtains the spacing between planes of atoms in crystals. Let us use the geometry shown in Fig. 4.15b. A plane wave is incident on a series of lattice planes separated by a distance d. The plane wave could be due to x-rays or particle-waves. The optical path difference for rays reflected from adjacent planes is $2d \sin \theta$, where θ is the angle of incidence as shown in Fig. 4.15b. Whenever this path difference is an integral multiple of the wavelength, the rays will interfere constructively. This condition, known as Bragg's Law, is

$$2d \sin \theta = n\lambda, \; n = 1, 2 \ldots \tag{4.28}$$

By placing a detector and cataloging the angles at which constructive interference occurs, it is possible to map out all the interplanar separations.

4.5 KRONIG-PENNEY MODEL FOR BANDSTRUCTURE

A useful model for understanding how electrons behave inside crystalline materials is the Kronig-Penney model. Although not a realistic potential for crystals, it allows us to calculate the energy of the electrons as a function of the parameter k that

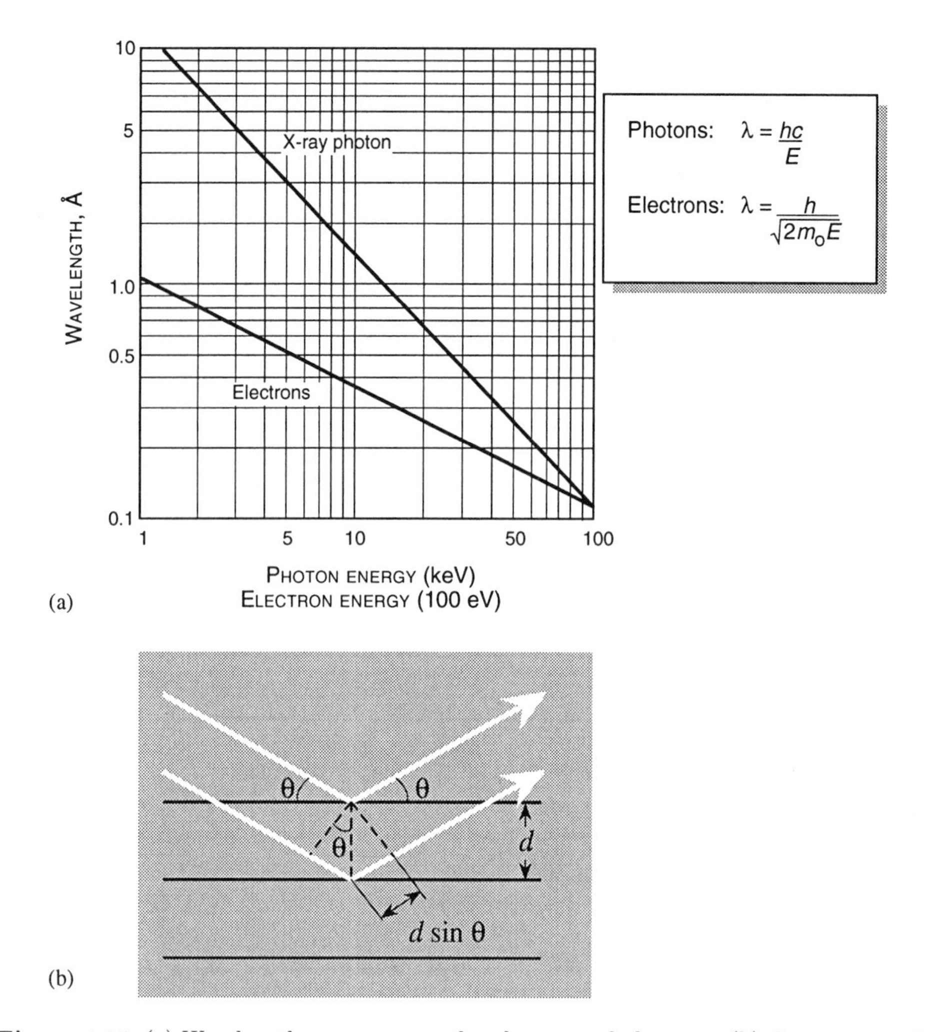

Figure 4.15: (a) Wavelength versus energy for photons and electrons. (b) Geometry used to derive Bragg's law.

appears in Bloch's theorem.

The Kronig-Penney model represents the background periodic potential seen by the electrons in the crystal as a simple potential shown in Fig. 4.16. The one-dimensional potential has the form

$$\begin{aligned} U(x) &= 0 \qquad 0 \leq x \leq a \\ &= U_0 \qquad -b \leq x \leq 0 \end{aligned} \tag{4.29}$$

The potential is repeated periodically as shown in Fig. 4.16 with a periodicity distance d $(= a + b)$. Since the potential is periodic, the electron wavefunction

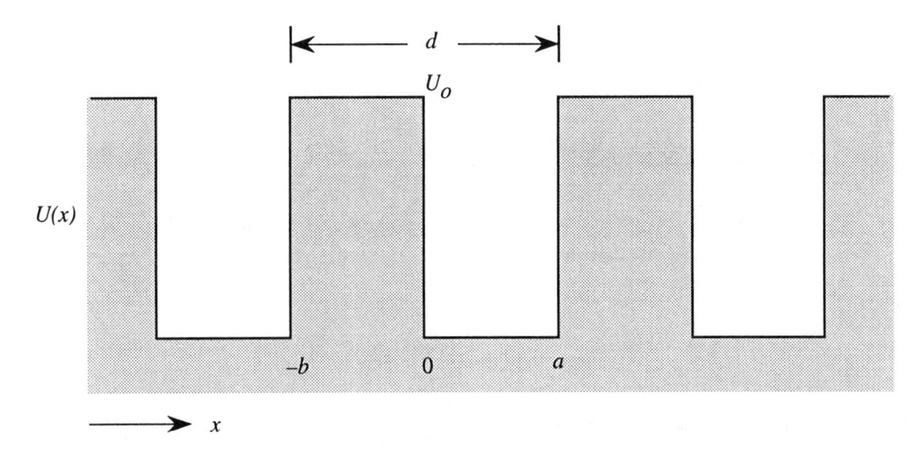

Figure 4.16: The periodic potential used to study the bandstructure in the Kronig-Penney model. The potential varies between 0 and U_0 as shown and has a periodicity of d.

satisfies Bloch's theorem and we may write

$$\psi(x+d) = e^{i\phi}\psi(x) \tag{4.30}$$

where the phase ϕ is written as

$$\phi = k_x d$$

In the region $-b < x < a$, the electron function has the form

$$\psi(x) = \begin{cases} Ae^{i\beta x} + Be^{-i\beta x}, & \text{if } -b < x < 0 \\ De^{i\alpha x} + Fe^{-i\alpha x}, & \text{if } 0 < x < a \end{cases} \tag{4.31}$$

where

$$\beta = \sqrt{\frac{2m_0(E - U_0)}{\hbar^2}}$$
$$\alpha = \sqrt{\frac{2m_0 E}{\hbar^2}} \tag{4.32}$$

Then, in the following period, $a < x < a + d$, from Eqn. 4.30,

$$\psi(x) = e^{i\phi}\begin{cases} Ae^{i\beta(x-d)} + Be^{-i\beta(x-d)}, & \text{if } a < x < d \\ De^{i\alpha(x-d)} + Fe^{-i\alpha(x-d)}, & \text{if } d < x < a + d \end{cases} \tag{4.33}$$

From the continuity conditions for the wavefunction and its derivative at $x = 0$ and at $x = a$, the following system of equations is obtained:

$$\begin{aligned} A + B &= D + F \\ \beta(A - B) &= \alpha(D - F) \\ e^{i\phi}(Ae^{i\beta b} + Be^{-i\beta b}) &= De^{i\alpha a} + Fe^{-i\alpha a} \\ \beta e^{i\phi}(Ae^{i\beta b} - Be^{-i\beta b}) &= \alpha(De^{i\alpha a} - Fe^{-i\alpha a}) \end{aligned} \tag{4.34}$$

Non-trivial solutions for the variables A, B, D, F are obtained only if the determinant of their coefficients vanishes, which gives the condition

$$\begin{aligned}\cos\phi &= \cos\ a\alpha\ \cosh\ b\delta - \frac{\alpha^2-\delta^2}{2\alpha\delta}\sin\ a\alpha\ \sinh\ b\delta, \quad \text{if } 0 < E < U_0 \\ &= \cos\ a\alpha\ \cos\ b\beta - \frac{\alpha^2+\beta^2}{2\alpha\beta}\sin\ a\alpha\ \sin\ b\beta, \quad \text{if } E > U_0 \end{aligned} \tag{4.35}$$

where

$$\delta = \sqrt{\frac{2m_0(U_0-E)}{\hbar^2}} \tag{4.36}$$

The energy E, which appears in Eqn. 4.35 through α, β, and δ, is physically allowed only if

$$-1 \leq \cos\ \phi\ \leq +1$$

Consider the case where $E < U_0$. We denote the right-hand side of Eqn. 4.35 by $f(E)$:

$$\begin{aligned}f(E) &= \cos\left(a\sqrt{\frac{2m_0E}{\hbar^2}}\right)\cosh\left(b\sqrt{\frac{2m_0(U_0-E)}{\hbar^2}}\right) \\ &+ \frac{U_0-2E}{2\sqrt{E(U_0-E)}}\sin\left(a\sqrt{\frac{2m_0E}{\hbar^2}}\right)\sinh\left(b\sqrt{\frac{2m_0(U_0-E)}{\hbar^2}}\right)\end{aligned} \tag{4.37}$$

This function must lie between -1 and $+1$ since it is equal to $\cos\phi$ ($= \cos\ k_xd$). We wish to find the relationship between E and ϕ or E and k_x. In general, one has to write a computer program in which one evaluates $f(E)$ starting from $E = 0$, and verifies if $f(E)$ lies between -1 and 1. If it does, one gets the value of ϕ for each allowed value of E. The approach is shown graphically in Fig. 4.17a. As we can see, $f(E)$ remains between the ± 1 bounds only for certain regions of energies. These "allowed energies" form the allowed bands and are separated by "bandgaps." One can obtain the E versus k relation or the bandstructure of the electron in the periodic structure, as shown in Fig. 4.17b. In the figure, the energies between E_2 and E_1 form the first allowed band, the energies between E_4 and E_3 form the second bandgap, etc.

We note that the $\phi = k_xd$ term on the left-hand side of Eqn. 4.35 appears as a cosine. As a result, if k_xd corresponds to a certain allowed electron energy, then $k_xd + 2n\pi$ is also allowed. This simply reflects a periodicity that is present in the problem. It is customary to show the E–k relation for the smallest k-values. The smallest k-values lie in a region $\pm\pi/d$ for the simple problem discussed here. In more complex periodic structures the smallest k-values lie in a more complicated k-space. The term Brillouin zone is used to denote the smallest unity cell of k-space. If a k-value is chosen beyond the Brillouin zone values, the energy values are simply repeated. The concept of allowed bands of energy separated by bandgaps

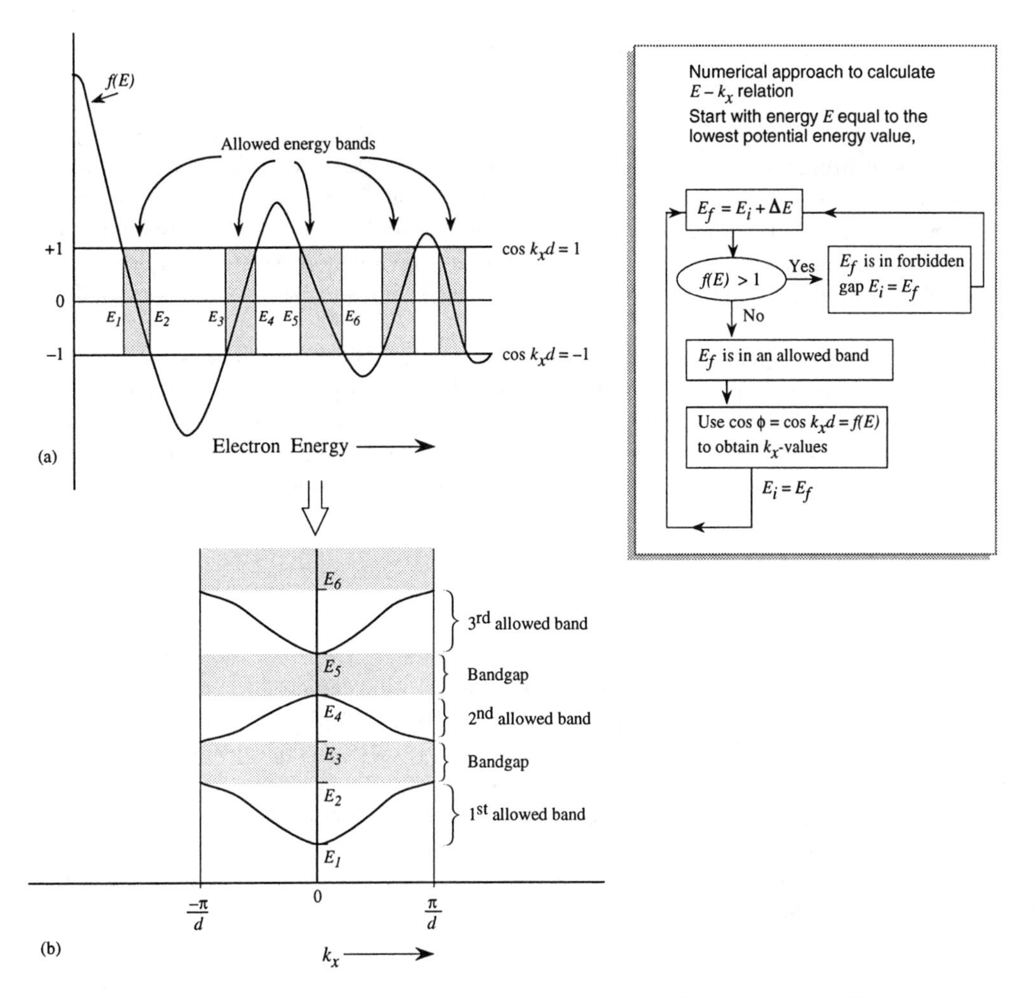

Figure 4.17: (a) The graphical solution to obtain the allowed energy levels. The function $f(E)$ is plotted as a function of E. Only energies for which $f(E)$ lies between $+1$ and -1 are allowed. (b) The allowed and forbidden bands are plotted in the E versus k relation using the results from (a). The inset shows a flow chart of how one can obtain the E-k_x relation.

is central to the understanding of crystalline materials. Near the bandedges it is usually possible to define the electron E–k relation as

$$E = \frac{\hbar^2(k - k_o)^2}{2m^*} \tag{4.38}$$

where k_o is the k-value at the bandedge and m^* is the effective mass. The concept of an effective mass is extremely useful, since it represents the response of the electron-crystal system to the outside world.

4.5.1 Significance of the k-vector

For free electrons moving in space two important laws are used to describe their properties: (i) Newton's second law of motion tells us how the electron's trajectory evolves in presence of an external force; (ii) the law of conservation of momentum allows us to determine the electron's trajectory when there is a collision. As noted in Section 2.4.2, the Ehrenfest theorem tells us that these laws are applicable to particles in quantum mechanics as well. We are obviously interested in finding out what the analogous laws are when an electron is inside a crystal and not in free space.

An extremely important implication of the Bloch theorem is that in the perfectly periodic background potential that the crystal presents, *the electron propagates without scattering.* The electronic state ($\sim \exp(i\mathbf{k}\cdot\mathbf{r})$) is an extended wave which *occupies the entire crystal.* To complete our understanding, we need to derive an equation of motion for the electrons which tells us how electrons will respond to external forces. The equation of motion

$$\frac{d\mathbf{p}}{dt} = \mathbf{F}_{ext} + \mathbf{F}_{int} \tag{4.39}$$

is quite useless for a meaningful description of the electron because it includes the internal forces on the electron. We need a description which does *not* include the evaluation of the internal forces. We will now give a simple derivation for such an equation of motion. The equation of motion also provides a conceptual understanding of the vector $\mathbf{k}$ that has been introduced by the Bloch theorem.

Since the Bloch function is a plane wave which extends over all crystalline space, we examine a localized wavepacket made up of wavefunctions near a particular $\mathbf{k}$-value, as discussed in Sections 1.7 and 2.4. Using our understanding of waves, we can define the group velocity of this wavepacket as

$$\mathbf{v}_g = \frac{d\omega}{d\mathbf{k}} \tag{4.40}$$

where ω is the frequency associated with the electron of energy E; i.e., $\omega = E/\hbar$:

$$\begin{aligned}\mathbf{v}_g &= \frac{1}{\hbar}\frac{dE}{d\mathbf{k}} \\ &= \frac{1}{\hbar}\nabla_{\mathbf{k}}E(\mathbf{k})\end{aligned}$$

If we have an electric field $\boldsymbol{E}$ present, the work done on the electron during a time interval δt is

$$\delta E = -e\boldsymbol{E}\cdot\mathbf{v}_g\delta t \tag{4.41}$$

We may also write, in general,

$$\begin{aligned}\delta E &= \left(\frac{dE}{d\mathbf{k}}\right)\delta\mathbf{k} \\ &= \hbar\mathbf{v}_g\cdot\delta\mathbf{k}\end{aligned} \tag{4.42}$$

Comparing the two equations for δE, we get

$$\delta \mathbf{k} = -\frac{e\boldsymbol{E}}{\hbar}\delta t$$

giving us the relation

$$\hbar\frac{d\mathbf{k}}{dt} = -e\boldsymbol{E} \tag{4.43}$$

In general, we may write

$$\boxed{\hbar\frac{d\mathbf{k}}{dt} = \mathbf{F}_{ext}} \tag{4.44}$$

Eqn. 4.44 looks identical to Newton's second law of motion,

$$\frac{d\mathbf{p}}{dt} = \mathbf{F}_{ext}$$

in free space if we associate the quantity $\hbar\mathbf{k}$ with the momentum of the electron in the crystal. The term $\hbar\mathbf{k}$ responds to the *external forces as if it is the momentum of the electron, although, as can be seen by comparing the true Newtons equation of motion, it is clear that* $\hbar\mathbf{k}$ *contains the effects of the internal crystal potentials and is therefore not the true electron momentum.* The quantity $\hbar\mathbf{k}$ is called the crystal momentum. Once the E versus k relation is established, we can, for all practical purposes, forget about the background potential $U(\mathbf{r})$ and treat the electrons as if they are free and obey the effective Newtons equation of motion. This physical picture is summarized in Fig. 4.18.

4.6 APPLICATION EXAMPLE: METALS, INSULATORS, SEMICONDUCTORS, AND SUPERCONDUCTORS

Quantum mechanics in general and quantum statistics in particular have had spectacular success in explaining properties of solid-state materials. Let us consider the question: Why are some materials metals and have very high conductivity, while others are insulators with low conductivity? Note that both metals and insulators have a very high density of electrons (about 10^{22} cm^{-3}). The answer to the question lies in Fermi statistics and in the band theory of solids.

We discussed in Section 4.5 how the spectra of electrons in a crystalline material are described by a series of allowed bands separated by forbidden bandgaps. The question now arises: Which of these allowed states are occupied by electrons and which are unoccupied? Two important situations arise when we examine the electron occupation of allowed bands: In one case we have a situation where an allowed band is completely filled with electrons, while the next allowed band is separated in energy by a gap E_g and is completely empty at 0 K. In a second case, the highest occupied band is only half full (or partially full). These cases are shown in Fig. 4.19.

At this point a very important concept needs to be introduced. *When an allowed band is completely filled with electrons, the electrons in the band cannot conduct any current.* This important concept is central to the special properties of

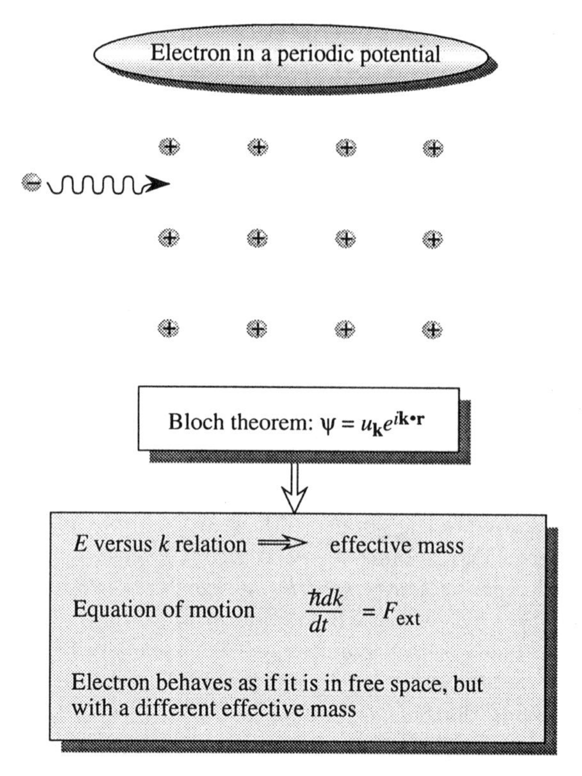

Figure 4.18: A physical description of electrons in a periodic potential. As shown the electrons can be treated as if they are in free space except that their energy-momentum relation is modified because of the potential. Near the bandedges the electrons respond to the outside world as if they have an effective mass m^*. The effective mass can have a *positive or negative value.*

metals and insulators. Being fermions the electrons cannot carry any net current in a filled band since an electron can only move into an empty state. One can imagine a net cancellation of the motion of electrons moving one way and those moving the other. Because of this effect, when we have a material in which a band is completely filled, while the next allowed band is separated in energy and empty, the material has, in principle, infinite resistivity and is called an *insulator* or a *semiconductor.* The material in which a band is only half full with electrons has a very low resistivity and is called a *metal.*

The band that is normally filled with electrons at 0 K in semiconductors is called the valence band, while the upper unfilled band is called the conduction band. The energy difference between the vacuum level and the highest occupied electronic state in a metal is called the metal work function. The energy between the vacuum level and the bottom of the conduction band is called the electron affinity. This is shown schematically in Fig. 4.19.

The metals have a very high conductivity because of the very large number of electrons that can participate in current transport. It is, however, difficult to alter the

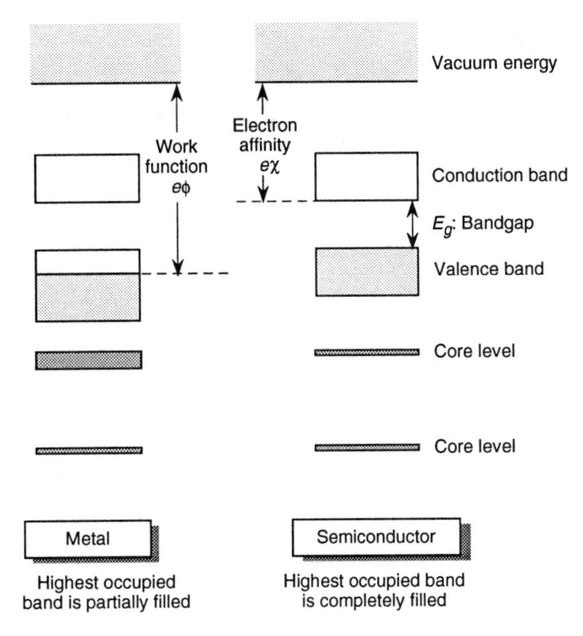

Figure 4.19: A schematic description of electron occupation of the bands in a metal and semiconductor (or insulator). In a metal, the highest occupied band at 0 K is partially filled with electrons. Also shown is the metal work function. In a semiconductor at 0 K, the highest occupied band is completely filled with electrons and the next band is completely empty. The separation between the two bands is the bandgap E_g. The electron affinity and work function are also shown.

conductivity of metals in any simple manner as a result of this. On the other hand, semiconductors have zero conductivity at 0 K and quite low conductivity at finite temperatures, but it is possible to alter their conductivity by orders of magnitude. This is the key reason why semiconductors can be used for active devices.

As noted, semiconductors are defined as materials in which the valence band is full of electrons and the conduction band is empty at 0 K. At finite temperatures some of the electrons leave the valence band and occupy the conduction band. The valence band is then left with some unoccupied states. Let us consider the situation as shown in Fig. 4.20, where an electron with momentum $\mathbf{k}_e$ is missing from the valence band.

When all of the valence band states are occupied, the sum over all wavevector states is zero; i.e.,

$$\sum \mathbf{k}_i = 0 = \sum_{\mathbf{k}_i \neq \mathbf{k}_e} \mathbf{k}_i + \mathbf{k}_e \tag{4.45}$$

This result is just an indication that there are as many positive k states occupied as there are negative ones. Now in the situation where the electron at

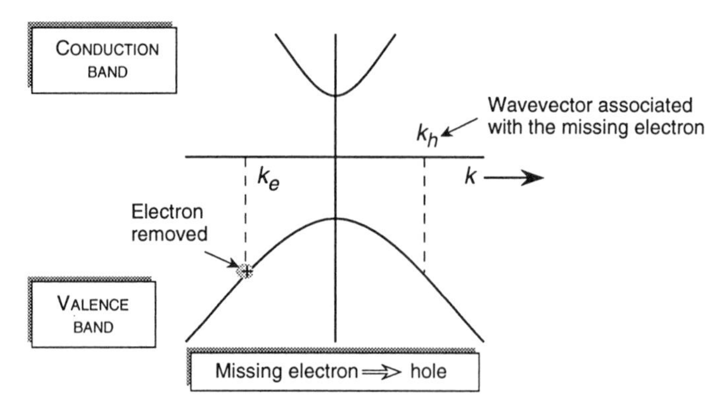

Figure 4.20: Illustration of the wavevector of the missing electron k_e. The wavevector is $-\mathbf{k}_e$, which is associated with the hole.

wavevector $\mathbf{k}_e$ is missing, the total wavevector is

$$\sum_{\mathbf{k}_i \neq \mathbf{k}_e} \mathbf{k}_i = -\mathbf{k}_e \tag{4.46}$$

The missing state is called a hole and the wavevector of the system $-\mathbf{k}_e$ is attributed to it. It is important to note that the electron is missing from the state $\mathbf{k}_e$ and the momentum associated with the hole is at $-\mathbf{k}_e$. The position of the hole is depicted as that of the missing electron. But in reality the hole wavevector $\mathbf{k}_h$ is $-\mathbf{k}_e$, as shown in Fig. 4.20.

$$\mathbf{k}_h = -\mathbf{k}_e \tag{4.47}$$

Note that the hole is a representation for the valence band with a missing electron. As discussed earlier, if the electron is not missing the valence band electrons cannot carry any current. However, if an electron is missing the current flow is allowed. If an electric field is applied, all the electrons move in the direction opposite to the electric field. This results in the unoccupied state moving in the field direction. *The hole thus responds as if it has a positive charge.* It therefore responds to external electric and magnetic fields $\mathbf{E}$ and $\mathbf{B}$, respectively, according to the equation of motion

$$\hbar \frac{d\mathbf{k}_h}{dt} = e\left[\mathbf{E} + \mathbf{v}_h \times \mathbf{B}\right] \tag{4.48}$$

where $\hbar\mathbf{k}_h$ and $\mathbf{v}_h$ are the momentum and velocity of the hole.

Thus the equation of motion of holes is that of particles with a *positive* charge e. The mass of the hole has a positive value, although the electron mass in its valence band is negative. *When we discuss the conduction band properties of semiconductors or insulators we refer to electrons, but when we discuss the valence band properties, we refer to holes. This is because in the valence band only the missing electrons or holes lead to charge transport and current flow.*

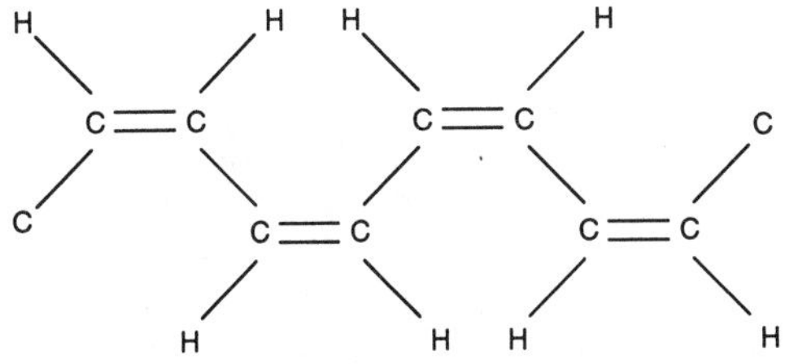

CIS-POLYACETYLENE

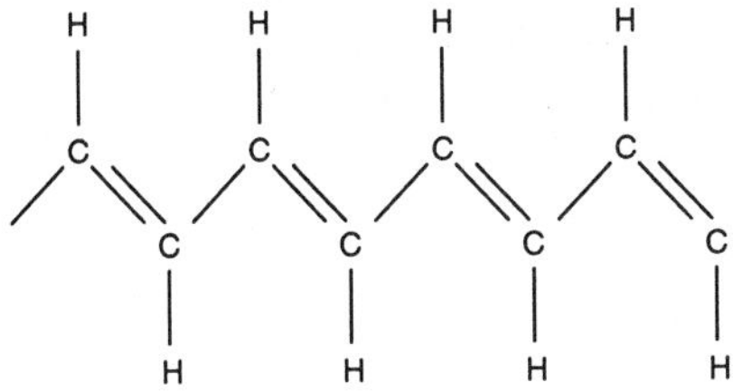

TRANS-POLYACETYLENE

Figure 4.21: Schematic of chains forming the polymer polyacetylene. Two forms of this polymer are shown.

4.6.1 Polymers

At present, most of the information-processing devices are based on semiconductor technology. The most commonly used materials are Si and GaAs. However, there are many other material systems that have semiconducting properties and may be very useful if their technologies can be improved. Polymers, which are a very familiar part of everyday life, have the potential of becoming important materials for electronics and optoelectronics. Most of the current uses of polymers rely on their properties such as chemical inertness and durability. In electronics, polymers are primarily used as insulators. However, new kinds of materials and better understanding is now allowing scientists to develop polymers with properties such as controlled conductivity and light detection and emission. Given the ability of chemists to synthesize large-area polymers, if these properties can be harnessed, polymers could become extremely important in future information-processing systems.

Polymers are formed from long chains of molecules. If there is a good fit between the molecules, the materials can crystallize upon drawing or cooling. Advances in crystallization techniques have led to high-quality polymer crystals which display electronic bands similar to those shown by "traditional" semiconductors. In Fig. 4.21, we show chains of polyacetylene—an important polymer.

At present, the conductivity of polymers is not as easy to control as that of "traditional" semiconductors (Si, GaAs, etc.). As a result, they are not used for high performance logic or memory applications. However, areas where the electronic properties are useful are *i*) transducers, i.e., to convert mechanical stresses to electrical signals. This is based on the piezoelectric behavior of polymers. An important polymer is polyvinylidene fluoride (PVF_2). In such materials there is a built-in polarization due to charge separation. Any stress alters this polarization resulting in an electrical signal, *ii*) devices using their pyroelectric properties, i.e., generation of an electric response with change of temperature. These devices are primarily used as infrared detectors and heat sensors. Since polymers can be fabricated in large areas, these "sheets" can be very effective as burglar and fire alarms; *iii*) devices based on triboelectricity, i.e., transfer of charge between two solids due to their contact and separation. Polymer spheres ($\sim$ 10 μm in diameter) form the black or colored "toner" particles in xerography to help transfer images from an original to a copy; *iv*) photoresponse of polymers can be tailored to make them very useful as photodetectors. Recent work has also show that polymers can also radiate light in visible when a current passes through them. This can have very important applications in display devices.

4.6.2 Normal and Superconducting States

One of the most fascinating observations in solid-state physics is the occurrence of superconductivity. It is found that certain metals lose their resistance when the temperature is lowered below a critical temperature. Recently, a new class of ceramic materials has been discovered, which displays a very high critical temperature (and are thus known as high-T_c materials). While there are still aspects of high-T_c materials which are unexplained, quantum theory has been remarkably successful in explaining metallic superconductivity.

The basis of the theory known as the *Bardeen-Cooper-Schrieffer* (BCS) theory is that in "normal" metals electrons behave as fermions, while in the superconducting state they behave as bosons. Of course, an individual electron is always a fermion, but they can form "pairs," known as *Cooper pairs*, and in this paired state they can act as if they are bosons. This distinction is of central importance in determining whether or not one has superconductivity.

Normally, electrons, being charged particles, repel each other and do not form bound pairs. However, inside a material the electrons interact with the ions on the crystal lattice and create an interaction between each other. We can physically think of this interaction via the simple picture in Fig. 4.22. An electron interacts with the positively charged background ions, creating a local potential disturbance. Another electron can then be attracted by this disturbance. The binding energy of the two electrons is extremely small ($\sim$ 1 meV), and at high temperatures the pairs dissociate and we have the usual unpaired electrons. However, at low temperatures, the electrons can pair up (forming Cooper pairs) and exist in the bound state.

Details of the BCS theory show that the lowest state of the system is one in which Cooper pairs are formed. The lowest state is separated from the next excited state by a gap which is of the order of 1 meV (the gap energy depends upon the nature of the ions), as shown in Fig. 4.23a. In the ground state, since the electron pairs are bosons, a very large number of pairs can occupy the same state. There, the Pauli exclusion principle no longer applies. Thus, to carry current, the electron pairs do not have to move from an occupied state to an unoccupied one. Also, since the excited state (which forms the normal state) is separated by a gap, as long as

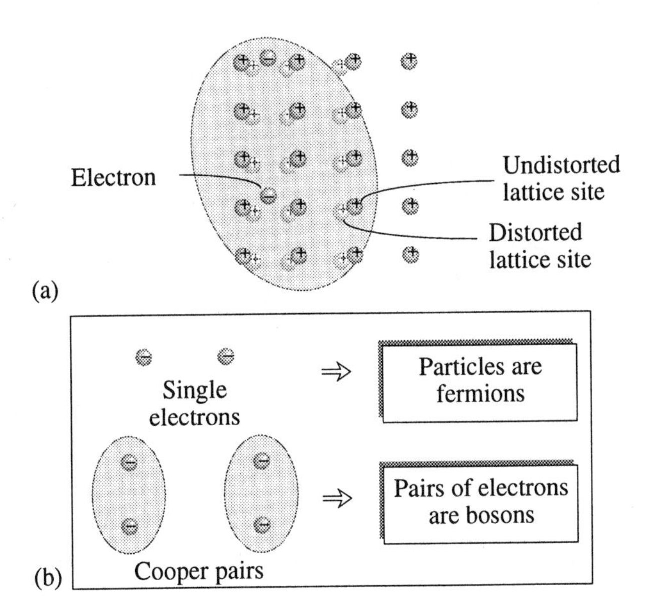

Figure 4.22: (a) A schematic showing how electrons distort the lattice occupied by positive ions to create an attractive potential. A bound pair can be formed in this attractive potential. (b) Single-electron states are antisymmetric under exchange, but if the electrons pair up the overall wavefunction is symmetric under exchange.

the temperature is small, the electrons do not suffer scattering (which is a source of resistance).

At low temperatures, the occupation of the ground state is high, as shown in Fig. 4.23b. As the temperature increases, the occupation decreases, until, at a critical temperature, T_c, there are no Cooper pairs. Thus superconductivity disappears above the critical temperature.

4.7 BANDSTRUCTURES OF SOME MATERIALS

When the Schrödinger equation is solved for electrons in a semiconductor crystal (usually on a computer), one obtains an energy versus effective momentum relation, or an E–k relation. This relation is called the bandstructure of the semiconductor. Every semiconductor has its own unique bandstructure. For the purpose of understanding semiconductor devices, we are interested in knowing the bandstructure in the conduction band and in the valence band. In particular we are interested in what the E–k relation is near the top of the valence band and near the bottom of the conduction band.

4.7.1 Direct and Indirect Semiconductors: Effective Mass

The top of the valence band of most semiconductors occurs at k=0, i.e., at effective momentum equal to zero. A typical bandstructure of a semiconductor near the top

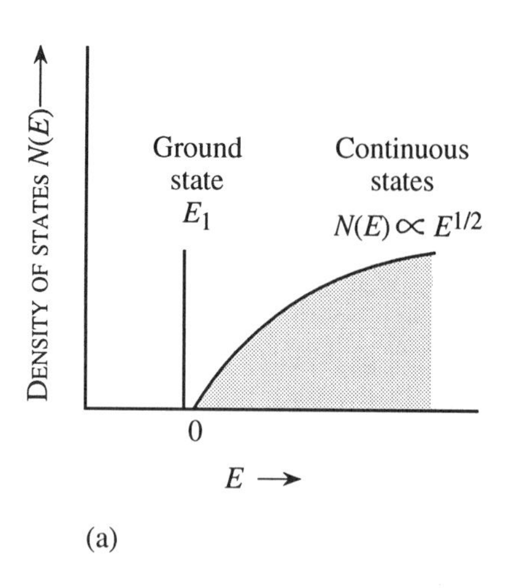

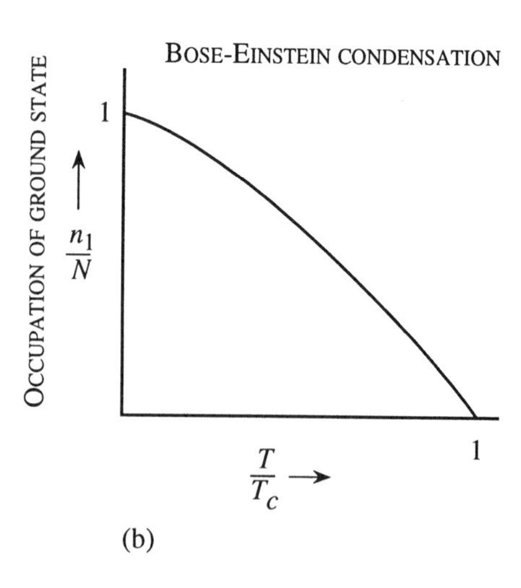

- ABOVE CRITICAL TEMPERATURE: Electrons are in the continuum states
- BELOW CRITICAL TEMPERATURE: Bosons start to condense into the ground state

Figure 4.23: (a) A density of states model to describe a system of bosons. (b) The occupation of the lowest (discrete) state as a function of temperature.

of the valence band is shown in Fig. 4.24. We notice the presence of three bands near the valence bandedge. These curves or bands are labeled I, II, and III in the figure and are called the heavy hole (HH) and the light hole (LH).

The bottom of the conduction band in some semiconductors occurs at k=0. Such semiconductors are called direct bandgap materials. Semiconductors such as GaAs, InP, InGaAs, etc., are direct bandgap semiconductors. In other semiconductors, the bottom of the conduction band does not occur at the $k = 0$ point, but at certain other points. Such semiconductors are called indirect semiconductors. Examples are Si, Ge, AlAs, etc.

An important outcome of the alignment of the bandedges in the valence band and the conduction band is that direct gap materials have a strong interaction with light. Indirect gap materials have a relatively weak interaction with electrons. This is a result of the law of momentum conservation and will be discussed in Chapter 8.

When the bandedges are at $k = 0$, it is possible to represent the bandstructure by a simple relation of the form

$$E(k) = E_c + \frac{\hbar^2 k^2}{2m^*} \tag{4.49}$$

where E_c is the conduction bandedge, and the bandstructure is a simple parabola. The equation for the E–k relation looks very much like that of an electron in free space except that the *free electron mass, m_0 is replaced by a new quantity m^*.*

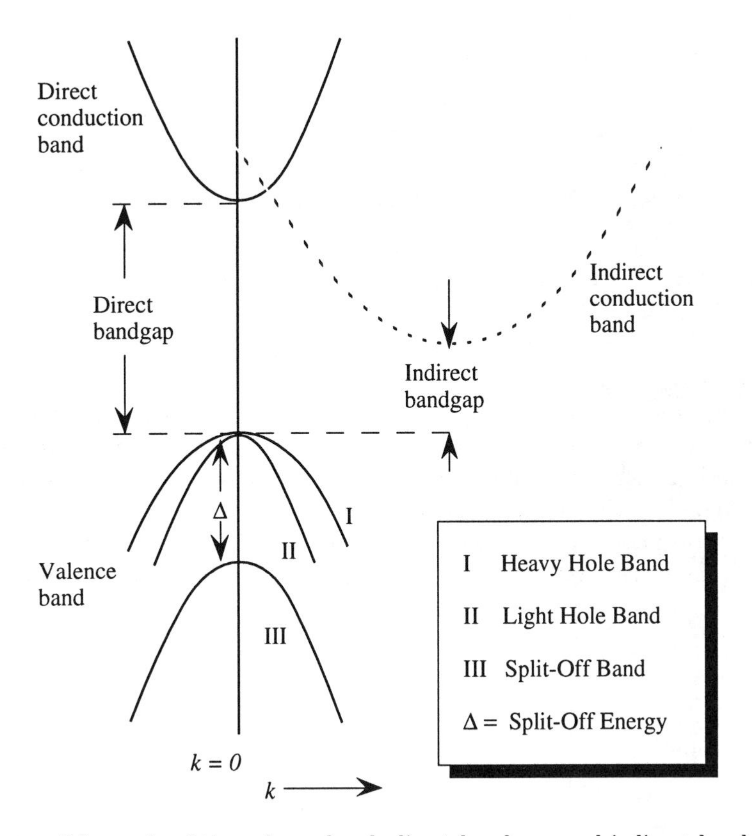

Figure 4.24: Schematic of the valence band, direct bandgap, and indirect bandgap conduction bands. The conduction band of the direct gap semiconductor is shown as the solid line, while the conduction band of the indirect semiconductor is shown as the dashed line. The curves I, II, III in the valence band are called heavy hole, light hole, and split-off hole states, respectively.

Silicon, the most important semiconductor in modern microelectronics, is an indirect semiconductor with a bandstructure, as shown in Fig. 2.25. The bottom of the conduction band does not occur at $k = 0$ but at 6 equivalent minima along the x-, y- and z-axes. The k-values at the minima are: $\frac{2\pi}{a}$ (0.85,0,0), $\frac{2\pi}{a}$ (0,0.85,0), $\frac{2\pi}{a}$ (0,0,0.85), and their inverses $\frac{2\pi}{a}(-0.85,0,0)$, $\frac{2\pi}{a}(0,-0.85,0)$, $\frac{2\pi}{a}(0,0,-0.85)$. The quantity a is the lattice constant (=5.43 Å for Si).

At each of the six k-points given above, the conduction band energy reaches a minimum value and as k moves away from these points, the energy rises. It is as if there are 6 *valleys* in the conduction band. The energy momentum relation in these valleys has the form given below.

- Valleys along the x-axis and $-x$-axis: $k_{0x} = \frac{2\pi}{a}(0.85, 0, 0)$ and $k_{0x} = \frac{2\pi}{a}(-0.85, 0, 0)$:

$$E(k) = E_c + \frac{\hbar^2}{2}\left[\frac{(k_x - k_{0x})^2}{m_l^*} + \frac{k_y^2 + k_z^2}{m_t^*}\right] \tag{4.50}$$

• Valleys along the y-axis and $-y$-axis: $k_{0y} = \frac{2\pi}{a}(0, 0.85, 0)$ and $k_{0y} = \frac{2\pi}{a}(0, -0.85, 0)$:

$$E(k) = E_c + \frac{\hbar^2}{2}\left[\frac{(k_y - k_{0y})^2}{m_l^*} + \frac{k_x^2 + k_z^2}{m_t^*}\right] \tag{4.51}$$

• Valleys along the z-axis and $-z$-axis: $k_{0z} = \frac{2\pi}{a}(0, 0, 0.85)$ and $k_{0z} = \frac{2\pi}{a}(0, 0, -0.85)$:

$$E(k) = E_c + \frac{\hbar^2}{2}\left[\frac{(k_z - k_{0z})^2}{m_l^*} + \frac{k_x^2 + k_y^2}{m_t^*}\right] \tag{4.52}$$

The effective mass m_l^* is called the longitudinal mass and the mass m_t^* is called the transverse mass.

Near the top of the valence band, as noted earlier (see Fig. 4.24), there are two curves. The heavier mass band is called the heavy hole band (I in Fig. 4.24). The second lighter band is called light hole band (II in Fig. 4.24). The masses of the valence band electrons are usually much heavier than those in the conduction band and are also negative. The energy–momentum relation in the valence band can be written as (here the quantities m_{hh}^* and m_{lh}^* are defined as positive values):

Heavy-hole band:

$$E = E_v - \frac{\hbar^2 k^2}{2m_{hh}^*} \tag{4.53}$$

Light-hole band:

$$E = E_v - \frac{\hbar^2 k^2}{2m_{lh}^*} \tag{4.54}$$

Split-off band:

$$E = E_v - \Delta - \frac{\hbar^2 k^2}{2m_{so}^*} \tag{4.55}$$

The effective masses of holes in silicon are $m_{lh}^* = 0.19\ m_0$; $m_{hh}^* = 0.49\ m_0$. The longitudinal and transverse conduction band masses are 0.98 m_0 and 0.19 m_0, respectively. Fig. 4.26 shows the bandstructure of GaAs, another important semiconductor. The conduction band effective masses in GaAs is 0.067 m_0. The heavy and light hole masses are 0.08 m_0 and 0.45 m_0.

In Section 4.2, we discussed the concept of the density of states. We use the same concept here, except we use the appropriate effective mass instead of the free electron mass. If the E–k relation is of the form

$$E = E_c + \frac{\hbar^2 k^2}{2m^*}$$

which is appropriate for the conduction band, the density of states is (for $E > E_c$)

$$N(E) = \frac{\sqrt{2}m^{*3/2}(E - E_c)^{1/2}}{\pi^2 \hbar^3} \tag{4.56}$$

If the E–k relation is of the form

$$E = E_v - \frac{\hbar^2 k^2}{2m^*}$$

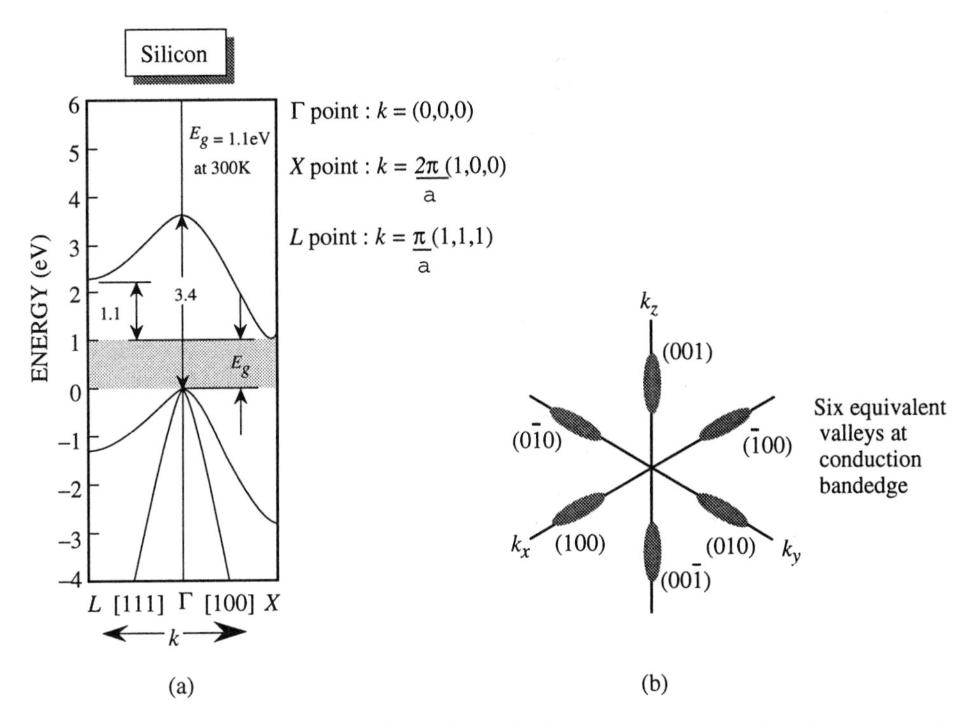

Figure 4.25: (a) Bandstructure of Si. Although the bandstructure of Si is far from ideal, having an indirect bandgap, high hole masses, and small spin-orbit splitting, processing related advantages make Si the premier semiconductor for consumer electronics. (b) Constant energy ellipsoids for Si conduction band. There are six equivalent valleys in Si at the bandedge.

which is relevant for the valence band, the density of states is (for $E < E_v$)

$$N(E) = \frac{\sqrt{2}m^{*3/2}(E_v - E)^{1/2}}{\pi^2\hbar^3} \tag{4.57}$$

We note that in the conduction band of indirect semiconductors, the electron mass is different along different directions. What value of mass should be used to define the density of states? The appropriate mass for a valley turns out to be $(m_1 m_2 m_3)^{1/3}$, where m_1, m_2 and m_3 are the masses along the three principle axes. For Si we get $(m_l m_t^2)^{1/3}$. Accounting for the six valleys in the silicon conduction band, the density of states mass becomes

$$m^*_{dos} = \left[6^{2/3}(m_l^* m_t^{*2})^{1/3}\right] \tag{4.58}$$

For direct gap materials the conduction band mass is isotropic and the density of states mass is just the effective mass.

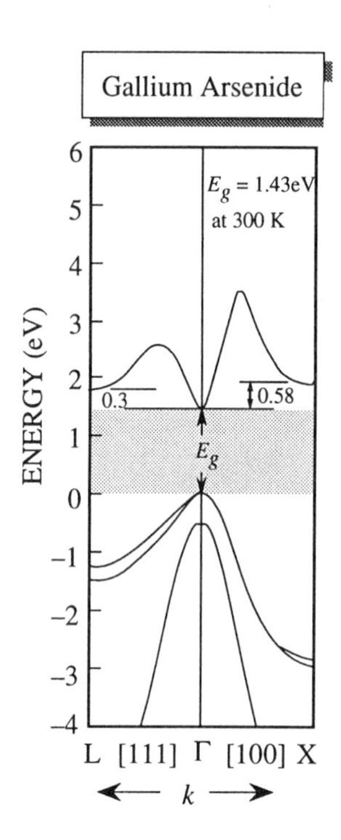

Figure 4.26: Bandstructure of GaAs. The bandgap at 0 K is 1.51 eV and at 300 K it is 1.43 eV. The bottom of the conduction band is at k = (0,0,0).

In the valence band (summing the density from the HH and LH bands),

$$\begin{aligned} N(E) &= \frac{\sqrt{2} m_{hh}^{*3/2} (E_v - E)^{1/2}}{\pi^2 \hbar^3} + \frac{\sqrt{2} m_{lh}^{*3/2} (E_v - E)^{1/2}}{\pi^2 \hbar^3} \\ &= \frac{\sqrt{2} m_{dos}^{*3/2} (E_v - E)^{1/2}}{\pi^2 \hbar^3} \end{aligned} \tag{4.59}$$

where the density of states mass for the valence band is given by

$$m_{dos}^{*3/2} = (m_{hh}^{*3/2} + m_{lh}^{*3/2})^{2/3} \tag{4.60}$$

Table 4.3 provides information on electron and hole masses of several important semiconductors.

4.7.2 Modification of Bandstructure by Alloying

Since essentially all the electronic and optical properties of semiconductor devices are dependent upon the bandstructure, an obvious question that arises is: How can the bandstructure of a material be changed?

In principle, many physical phenomena can modify the electronic bandstructures, but there are two important ones that are widely used for band tailoring. These involve: i) alloying of two or more semiconductors; ii) use of heterostructures

Material	Bandgap (eV)	Relative Dielectric Constant
C	5.5, I	5.57
Si	1.124, I	11.9
Ge	0.664, I	16.2
SiC	2.416, I	9.72
GaAs	1.424, D	13.18
AlAs	2.153, I	10.06
InAs	0.354, D	15.15
GaP	2.272, I	11.11
InP	1.344, D	12.56
InSb	0.230, D	16.8
CdTe	1.475, D	10.2
AlN	6.2, D	9.14
GaN	3.44, D	10.0
ZnSe	2.822, D	9.1
ZnTe	2.394, D	8.7

Material	Electron Mass (m_0)	Hole Mass (m_0)
AlAs	0.1	
AlSb	0.12	$m^*_{dos} = 0.98$
GaN	0.19	$m^*_{dos} = 0.60$
GaP	0.82	$m^*_{dos} = 0.60$
GaAs	0.067	$m^*_{lh} = 0.082$ $m^*_{hh} = 0.45$
GaSb	0.042	$m^*_{dos} = 0.40$
Ge	$m_l = 1.64$ $m_t = 0.082$ $m_{dos} = 0.56$	$m^*_{lh} = 0.044$ $m^*_{hh} = 0.28$
InP	0.073	$m^*_{dos} = 0.64$
InAs	0.027	$m^*_{dos} = 0.4$
InSb	0.13	$m^*_{dos} = 0.4$
Si	$m_l = 0.98$ $m_t = 0.19$ $m_{dos} = 1.08$	$m^*_{lh} = 0.16$ $m^*_{hh} = 0.49$

Table 4.3: Properties of some semiconductors. D and I stand for direct and indirect gaps, respectively. The data are at 300 K. Note that Si has six conduction band valleys, while Ge has four.

to cause quantum confinement or formation of "superlattices." These two concepts are increasingly being used for improved-performance electronic and optical devices, and their importance is expected to become greater with each passing year. In this section we will briefly examine the effect of alloying on the bandstructure. In the next chapter we will see how quantum wells change the electronic properties of materials.

When two semiconductors A and B are mixed via an appropriate growth technique, one has the following properties of the alloy:

a) *The crystalline structure of the lattice*: In most semiconductors the two (or more) components of the alloy have the same crystal structure so that the final alloy also has the same crystalline structure. For the same lattice structure materials the lattice constant obeys the Vegard's law for the alloy A_xB_{1-x}

$$\boxed{a_{alloy} = xa_A + (1-x)a_B} \tag{4.61}$$

where a_A and a_B are the lattice constants of A and B.
b) *Bandstructures of alloys* The bandstructure of alloys is difficult to calculate in principle since alloys are *not perfect crystals even if they have a perfect lattice.* This is because the atoms are placed randomly and not in any periodic manner. A simple approach used for the problem is called the virtual crystal approximation according to which the bandstructure of the alloy A_xB_{1-x} is simply the weighted bandstructure of the individual bandstructures of A and B. Thus the bandstructure is given by

$$E_{alloy}(k) = xE_A(k) + (1-x)E_B(k) \tag{4.62}$$

With this we have, for the mass of an alloy,

$$\frac{1}{m^*_{alloy}} = \frac{x}{m^*_A} + \frac{(1-x)}{m^*_B}$$

The relations given above for the lattice constant and bandgap are strictly valid if there is a good "mixing" in the alloy formation process. Thus, if an alloy A_xB_{1-x} is grown, the probability on an average that an A-type atom is surrounded by a B-type atom should be $(1-x)$ and the probability that a B-type atom has an A-type neighbor should be x. If this is true, the alloy is called a random alloy. If, on the other hand, the probability that an A-type atom is next to a B-type atom is smaller than x, the alloy is clustered or phase separated.

It is essential that an alloy not be phase separated; otherwise the material is not very useful for devices and cannot be used in any reliable device process.

EXAMPLE 4.7 Consider electrons at the bottom of the conduction band in silicon. A force is applied so that the k-vector for the electrons increases in the x-direction by 0.1 Å^{-1}. Calculate the energies of the electrons in each of the six valleys.

The electron wavevector k has been increased in the x-direction. Thus for the valleys (100) and ($\bar{1}$00), the momentum is longitudinal. For the other four valleys, the momentum is transverse. For the (100) and ($\bar{1}$00) valleys, the energy is

$$E = \frac{\hbar^2 \Delta k_l^2}{2m_l^*} + \frac{\hbar^2 \Delta k_t^2}{2m_t^*}$$

Putting $\Delta k_l = 0.1$ Å^{-1} and $\Delta k_t = 0$, we get, using $m_l^* = 0.98\ m_0$

$$E = \frac{(1.05 \times 10^{-34}\ \text{J.s})^2 (0.1 \times 10^{10}\ \text{m}^{-1})^2}{2(0.98 \times 0.91 \times 10^{-30}\ \text{kg})} = 6.16 \times 10^{-21}\ \text{J}$$
$$= 38.5\ \text{meV}$$

For the four valleys perpendicular to the x-direction, $\Delta k_l = 0$, $\Delta k_t = 0.1$ Å^{-1} and we get, using $m_t^* = 0.19\ m_0$,

$$E = \frac{(1.05 \times 10^{-34}\ \text{J.s})^2 (0.1 \times 10^{10}\ \text{m}^{-1})^2}{2(0.19 \times 0.91 \times 10^{-30}\ \text{kg})} = 3.18 \times 10^{-20}\ \text{Js}$$
$$= 198.8\ \text{meV}$$

The electrons in different valleys gain different energies from the change in the x-direction momentum.

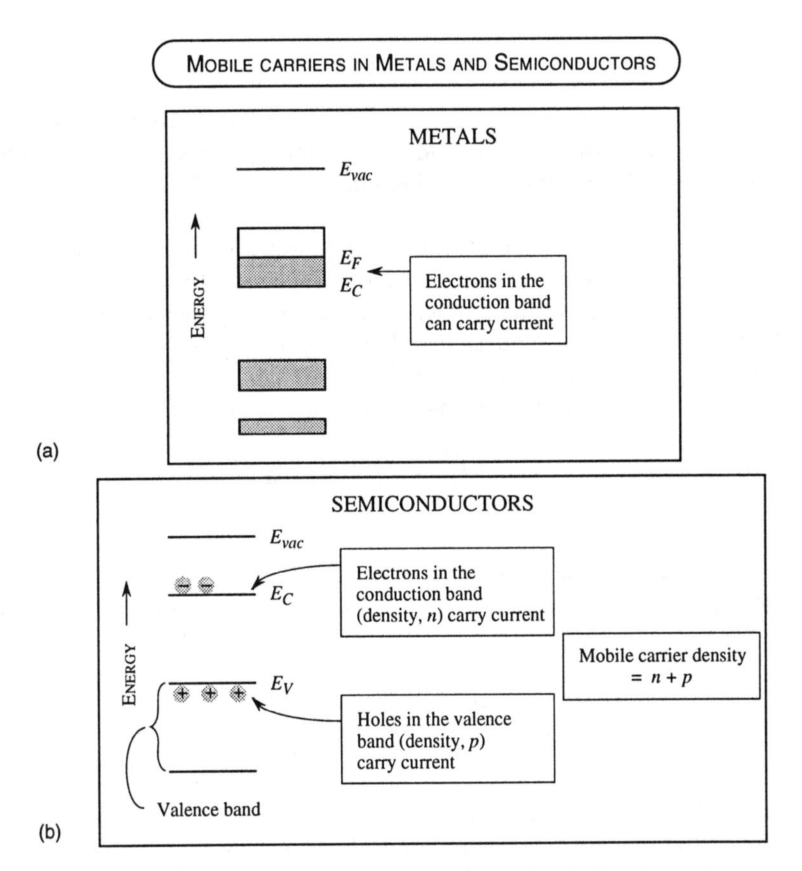

Figure 4.27: (a) A schematic showing mobile carrier (capable of carrying current) in metals; (b) In semiconductor, only electrons in the conduction band holes in the value band can carry current.

4.8 INTRINSIC CARRIER CONCENTRATION

From our discussion on metals and semiconductors we see that in a metal, current flows because of the electrons present in the highest (partially) filled band. This is shown schematically in Fig. 4.27a. The density of such electrons is very high ($\sim 10^{22}$ cm^{-3}). In a semiconductor, on the other hand, no current flows if the valence band is filled with electrons and the conduction band is empty of electrons. However, if somehow empty states or holes are created in the valence band by removing electrons, current can flow through the holes. Similarly, if electrons are placed in the conduction band, these electrons can carry current. This is shown schematically in Fig. 4.27b. If the density of electrons in the conduction band is n and that of the holes in the valence band is p, the total mobile carrier density is $n + p$.

How do holes appear in the valence band and how do electrons occupy the conduction band? In a pure semiconductor, at finite temperatures the thermal energy of the crystal knocks some electrons from the valence band into the conduction band. When this happens, an electron-hole pair is produced. In the pure semiconductor, the electron density is equal to the hole density and is denoted by n_i and p_i, the subscript i standing for intrinsic. It is also possible to create holes in the valence band and electrons in the conduction band by using impurity atoms with special properties. This process is called doping, and we will discuss this in the next chapter.

The intrinsic carrier concentration refers to the electrons (holes) present in the conduction band (valence band) of a pure semiconductor. The intrinsic carrier concentration depends upon the bandgap and temperature as well as the details of the bandedge masses. The conduction and valence band density of states are shown in Fig. 4.28 along with the position of a Fermi level.

The concentration of electrons in the conduction band is

$$n = \int_{E_c}^{\infty} N_e(E) f(E) dE$$

where $N_e(E)$ is the electron density of states in the conduction bandedge and $f(E)$ is the Fermi function. Using the appropriate expressions for N_e and f (the conduction band density of states starts at $E = E_c$ as shown in Fig. 4.28), we get

$$n = \frac{1}{2\pi^2}\left(\frac{2m_e^*}{\hbar^2}\right)^{3/2} \int_{E_c}^{\infty} \frac{(E-E_c)^{1/2} dE}{\exp\left(\frac{E-E_F}{k_B T}\right)+1}$$

If the Fermi level is far from the bandedge, then the unity in the denominator can be neglected. This approximation, called the Boltzmann approximation, is valid when n is small ($\lesssim 10^{16}$ cm^{-3} for most semiconductors), and is usually valid for intrinsic concentrations. Then we get (we denote the conduction band density of states mass by m_e^*)

$$\begin{aligned} n &= \frac{1}{2\pi^2}\left(\frac{2m_e^*}{\hbar^2}\right)^{3/2} \exp\left(\frac{E_F}{k_B T}\right) \int_{E_c}^{\infty} (E-E_c)^{1/2} \exp(-E/k_B T) dE \\ &= 2\left(\frac{m_e^* k_B T}{2\pi\hbar^2}\right)^{3/2} \exp\left[(E_F - E_c)/k_B T\right] \\ &= N_c \exp\left[(E_F - E_c)/k_B T\right] \end{aligned} \tag{4.63}$$

where

$$N_c = 2\left(\frac{m_e^* k_B T}{2\pi\hbar^2}\right)^{3/2} \tag{4.64}$$

N_c is known as the effective density of states at the conduction bandedge. Note that the units of the density of states N_e are eV^{-1} cm^{-3}, while those of the effective density of states N_c are cm^{-3}.

The carrier concentration is known *when E_F is calculated.* To find the intrinsic carrier concentration, this requires finding the hole concentration p as well.

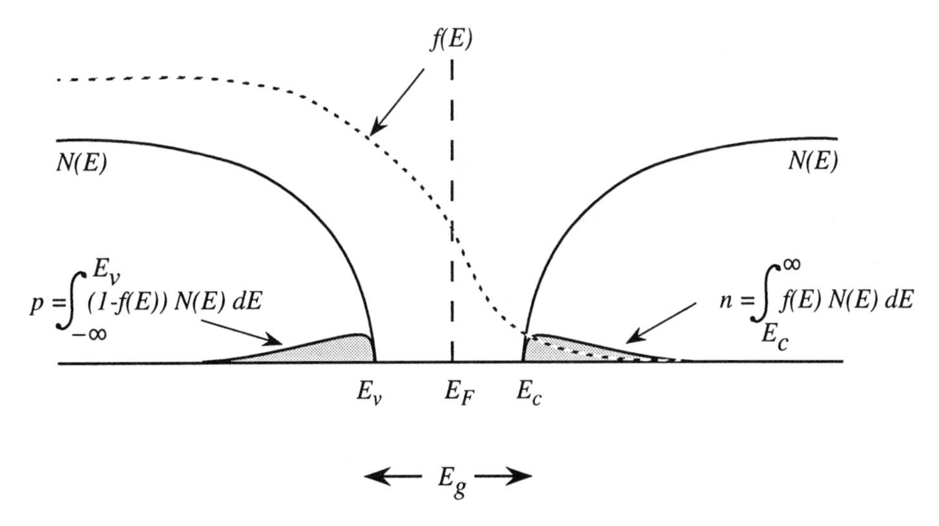

Figure 4.28: A schematic of the density of states of the conduction and valence band. N_e and N_h are the electron and hole density of states. Also shown is the Fermi function giving the occupation probability for the electrons. The resulting electron and hole concentrations are shown. For an intrinsic semiconductor $n = p$, since each electron produced in the conduction band leaves behind a hole in the valence band.

The hole distribution function f_h is given by (remember the hole is the absence of an electron)

$$f_h = 1 - f_e = 1 - \frac{1}{\exp\left(\frac{E-E_F}{k_BT}\right) + 1} = \frac{1}{\exp\left[\frac{(E_F-E)}{k_BT}\right] + 1}$$

$$\cong \exp - \left[\frac{(E_F - E)}{k_BT}\right] \tag{4.65}$$

The approximation is again based on our assumption that $E_F - E \gg k_BT$, which is a good approximation for pure semiconductors. Carrying out the mathematics similar to that for electrons, we find that (denoting the valence band density of states mass by m_h^*)

$$\boxed{\begin{aligned} p &= 2\left(\frac{m_h^* k_BT}{2\pi\hbar^2}\right)^{3/2} \exp\left[(E_v - E_F)/k_BT\right] \\ &= N_v \exp\left[(E_v - E_F)/k_BT\right] \end{aligned}} \tag{4.66}$$

where N_v is the effective density of states for the valence bandedge.

In Section 4.7 we have discussed the proper density of states mass that should be used to calculate N_c and N_v. In Example 4.8 we have explicitly calculated these quantities for Si and GaAs.

In intrinsic semiconductors, the electron concentration is equal to the hole concentration since each electron in the conduction band leaves a hole in the valence band. If we multiply the electron and hole concentrations, we get

$$\boxed{np = 4\left(\frac{k_B T}{2\pi\hbar^2}\right)^3 (m_e^* m_h^*)^{3/2} \exp\left(-E_g/k_B T\right)} \tag{4.67}$$

and since for the intrinsic case $n = n_i = p = p_i$, we have from the square root of the equation above,

$$\boxed{n_i = p_i = 2\left(\frac{k_B T}{2\pi\hbar^2}\right)^{3/2} (m_e^* m_h^*)^{3/4} \exp\left(-E_g/2k_B T\right)} \tag{4.68}$$

If we set $n = p$, we also obtain the Fermi level position measured from the valence bandedge using Eqns. 4.63 and 4.66. We denote the intrinsic Fermi level by E_{Fi}

$$\exp\ (2E_{Fi}/k_B T) = (m_h^*/m_e^*)^{3/2}\ \exp\ ((E_c + E_v)/k_B T)$$

or

$$\boxed{E_{Fi} = \frac{E_c + E_v}{2} + \frac{3}{4} k_B T \ell n\, (m_h^*/m_e^*)} \tag{4.69}$$

Thus the Fermi level of an intrinsic material lies close to the midgap.

We note that the carrier concentration increases exponentially as the bandgap decreases. Results for the intrinsic carrier concentrations for some semiconductors are shown in Fig. 4.29. (The strong temperature dependence and bandgap dependence of intrinsic carrier concentration can be seen from this figure.) In electronic devices where current has to be modulated by some means, the concentration of intrinsic carriers is fixed by the temperature and therefore is detrimental to device performance. Once the intrinsic carrier concentration increases to $\sim 10^{15}$ cm^{-3}, the material becomes unsuitable for electronic devices. A growing interest in high bandgap semiconductors such as diamond (C), SiC, etc., is partly due to the potential applications of these materials for high-temperature devices where, due to their larger gap, the intrinsic carrier concentration remains low up to very hig-temperatures.

We note that the product np calculated above is independent of the Fermi level E_F. This is an expression of the law of mass action. This result is valid not only for the intrinsic case but also when we have dopants, as long as the Fermi level is not too close to the bandedges (i.e., the Boltzmann approximation is valid).

EXAMPLE 4.8 Calculate the effective density of states for the conduction and valence bands of GaAs and Si at 300 K.

Let us start with the GaAs conduction band case which is the simplest one because of the spherical band effective mass. The effective density of states is

$$N_c = 2\left(\frac{m_e^* k_B T}{2\pi\hbar^2}\right)^{3/2}$$

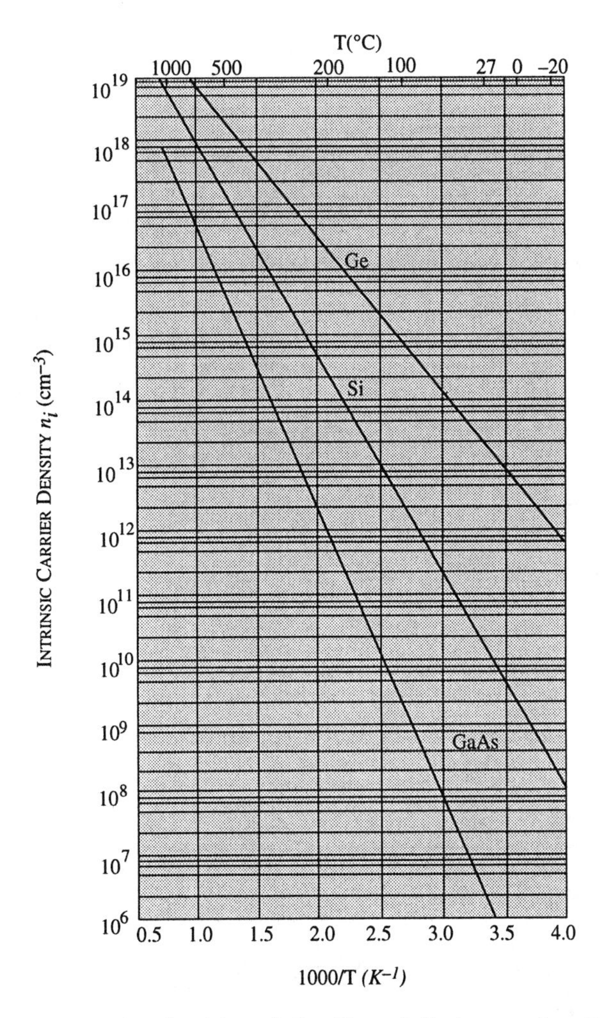

Figure 4.29: Intrinsic carrier densities of Ge, Si, and GaAs as a function of reciprocal temperature.

Note that at 300 K, $k_B T = 26$ meV $= 4 \times 10^{-21}$ J.

$$N_c = 2\left(\frac{0.067 \times 0.91 \times 10^{-30}(\text{kg}) \times 4.16 \times 10^{-21}(\text{J})}{2 \times 3.1416 \times (1.05 \times 10^{-34}(\text{J.s}))^2}\right)^{3/2} \text{m}^{-3}$$

$$= 4.45 \times 10^{23}\text{m}^{-3} = 4.45 \times 10^{17}\text{cm}^{-3}$$

In silicon, the density of states mass is to be used in the effective density of states. This is given by

$$m^*_{dos} = 6^{2/3}(0.98 \times 0.19 \times 0.19)^{1/3} m_0 = 0.32 m_0 \times 6^{2/3}$$

The effective density of states is

$$\begin{aligned} N_c &= 2\left(\frac{m^*_{dos}k_BT}{2\pi\hbar^2}\right)^{3/2} \\ &= 12\left(\frac{0.32\times 0.91\times 10^{-30}(\mathrm{kg})\times 4.16\times 10^{-21}(\mathrm{J})}{2\times 3.1416\times(1.05\times 10^{-34}(\mathrm{J.s}))^2}\right)^{3/2}\ \mathrm{m}^{-3} \\ &= 2.78\times 10^{25}\mathrm{m}^{-3} = 2.78\times 10^{19}\mathrm{cm}^{-3} \end{aligned}$$

One can see the large difference in the effective density between Si and GaAs. In the case of the valence band, one has the heavy hole and light hole bands, both of which contribute to the effective density. The effective density is

$$N_v = 2\left(m_{hh}^{3/2} + m_{\ell h}^{3/2}\right)\left(\frac{k_BT}{2\pi\hbar^2}\right)^{3/2}$$

For GaAs we use $m_{hh} = 0.45m_0, m_{\ell h} = 0.08m_0$, and for Si we use $m_{hh} = 0.49m_0, m_{\ell h} = 0.16m_0$, to get

$$\begin{aligned} N_v(GaAs) &= 7.72\times 10^{18}\mathrm{cm}^{-3} \\ N_v(Si) &= 9.84\times 10^{18}\mathrm{cm}^{-3} \end{aligned}$$

EXAMPLE 4.9 Calculate the position of the intrinsic Fermi level in Si at 300 K.

The density of states effective mass of the combined six valleys of silicon is

$$m^*_{dos} = (6)^{2/3}\left(m^*_\ell\ m_t^2\right)^{1/3} = 1.08m_0$$

The density of states mass for the valence band is $0.56m_0$. The intrinsic Fermi level is given by (referring to the valence bandedge energy as zero)

$$\begin{aligned} E_{Fi} &= \frac{E_g}{2} + \frac{3}{4}k_BT\ell n\left(\frac{m^*_h}{m^*_e}\right) = \frac{E_g}{2} + \frac{3}{4}(0.026)\ell n\left(\frac{0.56}{1.08}\right) \\ &= \frac{E_g}{2} - (0.0128\ \mathrm{eV}) \end{aligned}$$

The Fermi level is then 12.8 meV below the center of the mid-bandgap.

EXAMPLE 4.10 Calculate the intrinsic carrier concentration in InAs at 300 K and 600 K.

The bandgap of InAs is 0.35 eV and the electron mass is $0.027m_0$. The hole density of states mass is $0.4m_0$. The intrinsic concentration at 300 K is

$$\begin{aligned} n_i &= p_i = 2\left(\frac{k_BT}{2\pi\hbar^2}\right)^{3/2}\left(m^*_e\ m^*_h\right)^{3/4}\exp\left(\frac{-E_g}{2k_BT}\right) \\ &= 2\left(\frac{(0.026)(1.6\times 10^{-19})}{2\times 3.1416\times(1.05\times 10^{-34})^2}\right)^{3/2} \\ &\quad \left(0.027\times 0.4\times(0.91\times 10^{-30})^2\right)^{3/4}\ \exp\left(-\frac{0.35}{0.052}\right) \\ &= 1.025\times 10^{21}\mathrm{m}^{-3} = 1.025\times 10^{15}\ \mathrm{cm}^{-3} \end{aligned}$$

The concentration at 600 K becomes

$$n_i(600 \text{ K}) = 2.89 \times 10^{16} \text{ cm}^{-3}$$

4.9 ELECTRONS IN METALS

We have seen from our discussion that in a metal, we have a series of filled bands and a partially filled band called the conduction band. The filled bands are inert as far as electrical and optical properties of metals are concerned. The conduction band of metals can be assumed to be described by the parabolic energy-momentum relation

$$E(k) = E_c + \frac{\hbar^2 h^2}{2m_0} \tag{4.70}$$

Note that we have used an effective mass equal to the free electrons mass. This is a reasonable approximation for metals. The large electron density in the band "screens" out the background potential quite effectively and the electron effective mass is quite close to the free space value.

The electron density in the conduction band of a metal is related to the Fermi level by the relation

$$n = \int_{E_c}^{\infty} \frac{\sqrt{2}m_0^{3/2}}{\pi^2 \hbar^3} \frac{E^{1/2} dE}{\exp\left(\frac{E - E_F}{k_B T}\right) - 1} \tag{4.71}$$

This integral is particularly simple to evaluate as 0 K, since, at this temperature

$$\frac{1}{\exp\left(\frac{E - E_F}{k_B T}\right) + 1} = 1 \text{ if } E \leq E_F$$

$$= 0 \quad \text{otherwise}$$

Thus (choosing the conduction bandedge as the origin)

$$n = \int_0^{E_F} N(E) dE$$

We then have

$$n = \frac{\sqrt{2}m_0^{3/2}}{\pi^2 \hbar^3} \int_0^{E_F} E^{1/2} dE$$

$$= \frac{2\sqrt{2}m_0^{3/2}}{3\pi^2 \hbar^3} E_F^{3/2}$$

or

$$E_F = \frac{\hbar^2}{2m_0} \left(3\pi^2 n\right)^{2/3} \tag{4.72}$$

The expression is applicable to metals such as copper, gold, etc. In Table 4.4 we

ELEMENT	VALENCE	DENSITY (gm/cm^3)	CONDUCTION ELECTRON DENSITY (10^{22} cm^{-3})
Al	3	2.7	18.1
Ag	1	10.5	5.86
Au	1	19.3	5.90
Na	1	0.97	2.65
Fe	2	7.86	17.0
Zn	2	7.14	13.2
Mg	2	1.74	8.61
Ca	2	1.54	4.61
Cu	1	8.96	8.47
Cs	1	1.9	0.91
Sn	4	7.3	14.8

Table 4.4: Properties of some metals. In case of elements that display several values of chemical valence, one of the values has been chosen arbitrarily.

show the conduction band electron densities for several metals. The quantity E_F which is the highest occupied energy state at 0 K is called the Fermi energy. We can define a corresponding wavevector k_F called the Fermi vector and a velocity v_F called the Fermi velocity as

$$\begin{aligned} k_F &= \left(3\pi^2 n\right)^{1/3} \\ v_F &= \left(\frac{\hbar}{m_0}\right)\left(3\pi^2 n\right)^{1/3} \end{aligned} \tag{4.73}$$

It is important to note that even at 0 K, the velocity of the highest occupied state is v_F and not zero, as would be the case if we used classical statistics.

At finite temperatures, the Fermi level is approximately given by

$$E_F(T) = E_F(0)\left[1 - \frac{\pi^2}{12}\frac{(k_B T)^2}{(E_F(0))^2}\right] \tag{4.74}$$

where $E_F(T)$ and $E_F(0)$ are the Fermi levels at temperatures T and 0 K, respectively.

EXAMPLE 4.11 A particular metal has 10^{22} electrons per cubic centimeter. Calculate the Fermi energy and the Fermi velocity (at 0 K).

The Fermi energy is the highest occupied energy state at 0 K and is given by (measured from the conduction bandedge)

$$\begin{aligned} E_F &= \frac{\hbar^2}{2m_0}\left(3\pi^2 n\right)^{2/3} \\ &= \frac{(1.05\times 10^{-34})^2[3\pi^2(10^{28})]^{2/3}}{2(0.91\times 10^{-30})} = 2.75\times 10^{-19}\ \text{J} \\ &= 1.72\ \text{eV} \end{aligned}$$

The Fermi velocity is

$$\begin{aligned} v_F &= \frac{\hbar}{m_0}\left(3\pi^2 n\right)^{1/3} \\ &= \frac{(1.05\times 10^{-34}\ \text{J.s})(3\pi^2\times 10^{28}\ \text{m}^{-3})^{1/3}}{0.91\times 10^{-30}\ \text{kg}} = 7.52\times 10^5\ \text{m/s} \\ &= 7.52\times 10^7\ \text{cm/s} \end{aligned}$$

The highest energy electron thus has a large energy and is moving with a very large speed. In a classical system the electron energy would be $\sim \frac{3}{2}k_BT$ which would be zero at 0 K. The electron velocity will also be zero at 0 K in classical physics.

4.10 CHAPTER SUMMARY

The areas discussed in this chapter are summarized in Tables 4.5–4.8.

4.11 PROBLEMS

Problem 4.1 Plot the density of states for electrons moving in free space with a uniform background potential of −3.0 eV. Solve the problem for 3-, 2- and 1-dimensional space. Plot the results from −4.0 eV to 1.0 eV.

Problem 4.2 Calculate and plot the 3-dimensional density of states for a particle with energy-momentum relation given by

$$E = \frac{\hbar^2 k^2}{m'} - 2.0\ \text{eV}$$

with $m' = -m_0$. Plot the results in the range −4.0 eV to 0 eV.

Problem 4.3 Calculate the form of the density of states for a particle which has the following energy-momentum relation

$$\frac{\hbar^2 k^2}{2m_0} = E(1+\alpha E)$$

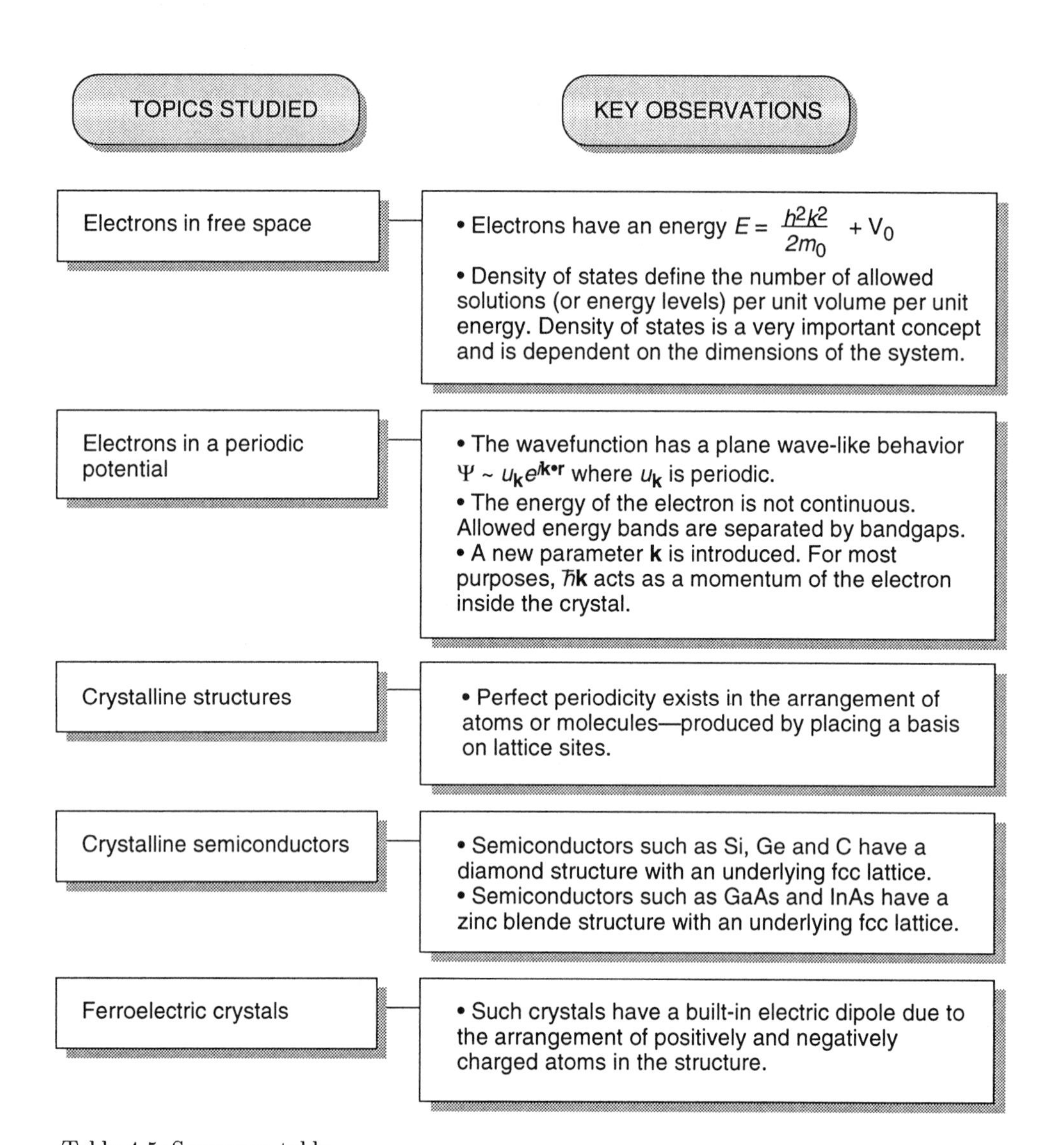

Table 4.5: Summary table.

TOPICS STUDIED	KEY OBSERVATIONS
Kronig-Penney model for bandstructure	A simple model for electronic properties in a periodic structure. The outcome of this model is: • There are a series of allowed energy bands separated by bandgaps. • Electrons cannot have energies in the bandgap region.
Insulators, semiconductors, and metals	• If, at 0 K, the highest occupied band is filled completely and the next band (separated by a bandgap) is completely empty, a semiconductor or insulator results. • If the highest band is partially filled, a metal results.
Valence and conduction bands	• The highest occupied band in a semiconductor at 0 K is called the *valence band.* • A completely filled band cannot carry any current, so the conductivity of a pure semiconductor at 0 K is zero.

Table 4.6: Summary table.

The parameter α is called the non-parabolicity factor.

Problem 4.4 Suppose that identical solid spheres are placed in space so that their centers lie on the atomic points of a crystal and the spheres on the neighboring sites touch each other. Assuming that the spheres have unit density, show that density of such spheres is the following for the various crystal structures:

$$\begin{aligned} \text{fcc} &: \sqrt{2}\pi/6 = 0.74 \\ \text{bcc} &: \sqrt{3}\pi/8 = 0.68 \\ \text{sc} &: \pi/6 = 0.52 \\ \text{diamond} &: \sqrt{3}\pi/16 = 0.34 \end{aligned}$$

Problem 4.5 Calculate the number of cells per unit volume in GaAs ($a = 5.65$ Å). Si has a 4% larger lattice constant. What is the unit cell density for Si? What is the number of atoms per unit volume in each case?

Problem 4.6 A Si wafer is nominally oriented along the (001) direction, but is found to be cut 2° off, toward the (110) axis. This off axis cut produces "steps" on the surface which are 2 monolayers high. What is the lateral spacing between the steps of the 2° off-axis wafer?

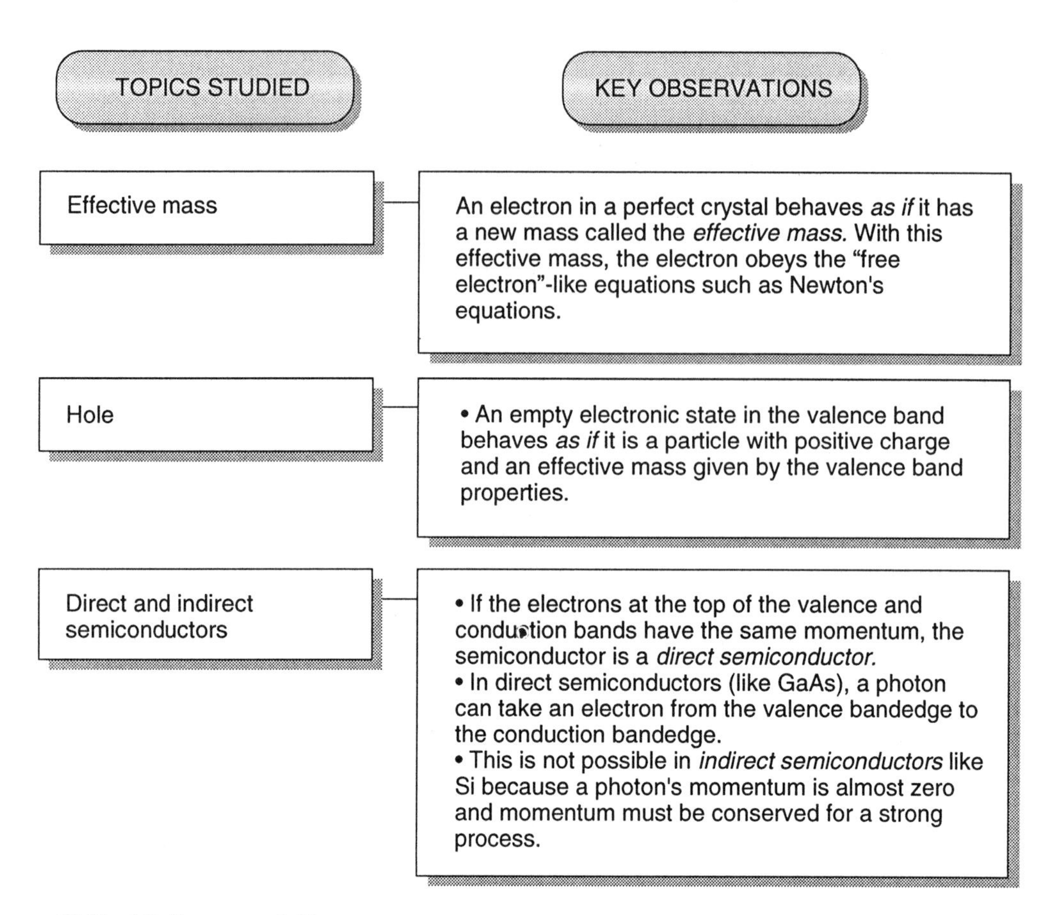

Table 4.7: Summary table.

Problem 4.7 In high-purity Si crystals, defect densities can be reduced to levels of 10^{13} cm^{-3}. On an average, what is the spacing between defects in such crystals? In heavily doped Si, the dopant density can approach 10^{19} cm^{-3}. What is the spacing between defects for such heavily doped semiconductors?

Problem 4.8 Assume that a Ga–As bond in GaAs has a bond energy of 1.0 eV. Calculate the energy per unit area needed to cleave GaAs in the (001) and (110) planes.

Problem 4.9 Aluminum crystallines in the fcc structure with a lattice constant of 4.05 Å. Calculate the number density of Al atoms. The atomic mass of Al is 26.98. Calculate the mass density of aluminum metal.

Problem 4.10 Silver atoms have an atomic mass of 107.87. Calculate the atomic density and mass density of silver which has an fcc structure with a lattice constant of 4.09 Å.

Problem 4.11 A crystal of table salt (NaCl) is radiated with X-rays of wavelength 3 Å. The first Bragg peak is observed at an angle of 32.13°. Calculate the atomic spacing between the planes in NaCl.

Problem 4.12 A piece of Al metal is irradiated with X-rays of wavelength 2.5 Å.

TOPICS STUDIED	KEY OBSERVATIONS
Intrinsic carrier density	In pure semiconductors, there is a certain electron and hole density which is determined by the temperature, bandgap and the carrier masses. This intrinsic density should be small for good device performance.
Mobile carriers in a semiconductor (insulator)	In a semiconductor or insulator only electrons in the conduction band and holes in the valence band are mobile.
Conduction electrons in a metal	The electrons in the highest energy partially filled bands can cary current.

Table 4.8: Summary table.

Calculate the angles for the first two peaks.

Problem 4.13 Consider an electron in a Kronig-Penney model with $a = b = 2$ Å and $U_0 = 3.0$ eV. Calculate the positions (i.e., the starting energy and ending energy) of the lowest allowed band.

Problem 4.14 Consider electrons in a Kronig-Penney model with the following parameters:

$$a = b = 3 \text{ Å}$$
$$U_0 = 10.0 \text{ eV}$$

Calculate the effective mass of electrons near the start of the first allowed band. You can find this by fitting the E–k results to an equation

$$E = E_1 + \frac{\hbar^2 k^2}{2m^*}$$

where E_1 is the start of the first allowed band.

Problem 4.15 Consider the previous problem. Calculate the effective mass in the first allowed band at i) the bottom of the band, ii) at the top of the band, and iii) at the middle of the band.

Problem 4.16 Consider electrons in a Kronig-Penney potential with the following parameters

$$a = 5 \text{ Å}$$

$$b = 1 \text{ Å}$$
$$U_0 = 4 \text{ eV}$$

Calculate the positions of the bandedges of the first two allowed energy bands. Also calculate the effective mass of electrons near the start of the second energy band. Redo the problem if a and b decrease by 0.2 Å.

Problem 4.17 Plot the conduction band and valence band density of states in Si and GaAs from the bandedges to 0.5 eV into the bands. Use the units eV^{-1} cm^{-3}. Use the following data:

$$\begin{aligned}
\text{Si} \;:\; m_1^* &= m_\ell^* = 0.98\ m_0 \\
m_2^* &= m_3^* = m_t^* = 0.19\ m_0 \\
m_{hh}^* &= 0.49\ m_0 \\
m_{\ell h}^* &= 0.16\ m_0 \\
\text{GaAs} \;:\; m_e^* &= m_{dos}^* = 0.067\ m_0 \\
m_{hh}^* &= 0.45\ m_0 \\
m_{\ell h}^* &= 0.08\ m_0
\end{aligned}$$

Problem 4.18 The wavevector of a conduction band electron in GaAs is $k = (0.1, 0.1, 0.0)\ \AA^{-1}$. Calculate the energy of the electron measured from the conduction bandedge.

Problem 4.19 A conduction band electron in silicon is in the (100) valley and has a k-vector of $2\pi/a$ (1.0, 0.1, 0.1). Calculate the energy of the electron measured from the conduction bandedge. Here a is the lattice constant of silicon.

Problem 4.20 Calculate the energies of electrons in GaAs and InAs conduction band with k-vectors (0.01, 0.01, 0.01) Å^{-1}. Refer the energies to the conduction bandedge values.

Problem 4.21 Calculate the lattice constant, bandgap; and electron effective mass of the alloy $\text{In}_x\text{Ga}_{1-x}\text{As}$ as a function of composition from $x = 0$ to $x = 1$.

Problem 4.22 Using Vegard's law for the lattice constant of an alloy (i.e., lattice constant is the weighted average), find the bandgaps of alloys made in InAs, InP, GaAs, GaP which can be lattice matched to InP.

Problem 4.23 For long-haul optical communication, the optical transmission losses in a fiber dictate that the optical beam must have a wavelength of either 1.3 μm or 1.55 μm. Which alloy combinations lattice matched to InP have a bandgap corresponding to these wavelengths?

Problem 4.24 Calculate the composition of $\text{Hg}_x\text{Cd}_{1-x}\text{Te}$ which can be used for a night vision detector with bandgap corresponding to a photon energy of 0.1 eV. Bandgap of CdTe is 1.6 eV *and that of HgTe is* -0.3 eV at low temperatures around 4 K.

Problem 4.25 Calculate the effective density of states at the conduction and valence bands of Si and GaAs at 77 K, 300 K, and 500 K.

Problem 4.26 Estimate the intrinsic carrier concentration of diamond at 700 K. You can assume that the carrier masses are similar to those in Si. Compare the results with those for GaAs and Si. The result illustrates one reason why diamond is useful for high-temperature electronics.

Problem 4.27 Estimate the change in intrinsic carrier concentration per K change in temperature for InAs, Si and GaAs near room temperature.

Problem 4.28 Calculate the position of the intrinsic Fermi level measured from the midgap for GaAs and InAs.

Problem 4.29 Calculate the Fermi energy and Fermi velocity for the following metals: Ag, Au, Ca, Cs, Cu, Na.

Problem 4.30 Calculate the de Broglie wavelength of electrons at the Fermi energy for the following metals: Ag, Cu, Au, Al.

Problem 4.31 Calculate the change in the Fermi level as temperature changes from 0 to 300 K for Al and Cu.

CHAPTER 5

PARTICLES IN ATTRACTIVE POTENTIALS

MAGIC OF TECHNOLOGY

• In a high-tech company specializing in components for optical communications, scientists are developing new lasers that will switch in microamperes of current. Such lasers could be integrated—millions to a chip—to link vast networks of data and voice lines. The scientists are designing materials such that electrons will *feel like they are in a 0-dimensional world.* This dramatically reduces the current needed to operate a laser.
• A company specializing in novel *intelligent* glasses is designing a three-dimensional memory where images could be stored in a three-dimensional matrix. Tiny particles, each only a millionth of a centimeter across, are imbedded in the glass during the glass formation. The glass then develops special properties that allows it to store images at storage densities not possible in any other storage medium.
• As engineers at a semiconductor memory plant shrink their feature sizes to accommodate the next-generation *gigabit scale integration*, they find that the electrons in the transistor behave as if they are in a 2-dimensional world. Their properties are quite different from those of electrons in the 3-dimensional world. By understanding the physics of the 2-dimensional electrons, scientists are able to come up with new design rules for the next-generation chips.

5.1 INTRODUCTION

There are a number of very important problems in quantum mechanics where a particle finds itself in a potential energy profile which is such that the particle is confined to a certain region in space. The wavefunction of the particle is not extended over all space, but is "bound" in space. Examples of such problems include the atomic problem where the electron–proton attraction produces atomic levels where an electron is "bound" to the nucleus. Often, in a crystalline material if a defect is produced, it produces a defect level where the electron is "trapped," i.e., it is spatially confined to the defect region. A very important problem in quantum mechanics is that of a particle moving in a harmonic potential. This problem has widespread applications.

One of the early successes of quantum mechanics was to understand the atomic spectra and then to understand the chemical properties of atoms in the periodic chart. These problems also fall in the category of bound state problem.

This chapter will address a number of important bound state problems and will examine their implications in technology.

5.2 PARTICLE IN A QUANTUM WELL

We will start with a problem of a particle in a simple potential $V(r)$, which is such that it produces "bound states" for the particle when the Schrödinger equation is solved. The bound state corresponds to a wavefunction which is finite over some finite region in space and goes to zero at large distances. In this section we will also

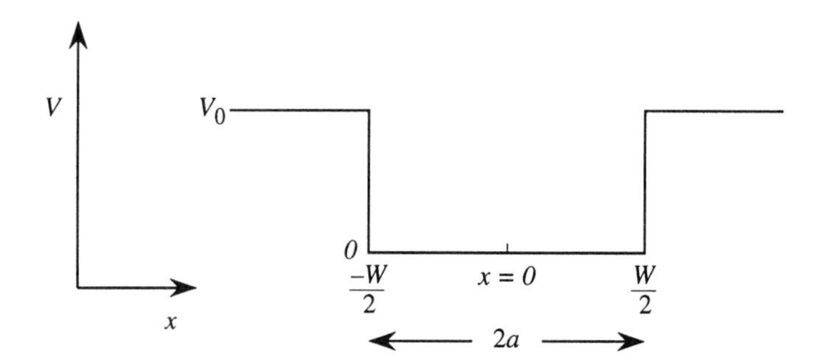

Figure 5.1: A quantum well of width $2a$ and infinite barrier height or barrier height V_0.

assume that the potential has a form

$$V(r) = V(x) + V(y) + V(z) \tag{5.1}$$

so that the wavefunction is separable and of the form

$$\psi(r) = \psi(x)\psi(y)\psi(z) \tag{5.2}$$

The quantum well problem has become of great importance in applied physics and engineering because of the advent of high-precision material growth techniques. Semiconductor devices have greatly benefited from the ability to create a *sub-three-dimensional world* for electrons.

We will discuss the problem of the square potential well, which has acquired a great deal of importance in applied physics due to the use of quantum wells in optoelectronic devices such as semiconductor lasers and modulators. We will also discuss results for the problem of a particle in a triangular quantum well, a problem of importance in electronic devices.

5.2.1 Square Quantum Well

The square quantum well shown in Fig. 5.1 is the simplest potential with important implications for devices. In semiconductor technology a good approximation to such quantum wells is produced when a narrow-bandgap material of width $W = 2a$ is sandwiched between a large-bandgap material. The simplest form of the quantum well is one where the potential is zero in the well and infinite outside. The equation to solve then is (the wavefunction is non-zero only in the well region)

$$-\frac{\hbar^2}{2m}\frac{d^2\psi}{dx^2} = E\psi \tag{5.3}$$

which has the general solutions

$$\begin{aligned} \psi(x) &= B\cos\frac{n\pi x}{2a}, \quad n \text{ odd} \\ &= A\sin\frac{n\pi x}{2a}, \quad n \text{ even} \end{aligned} \tag{5.4}$$

The energy is

$$\boxed{E = \frac{\pi^2 \hbar^2 n^2}{8ma^2}} \tag{5.5}$$

Note that the well size is $2a$.

The normalized particle wavefunctions are

$$\begin{aligned} \psi(x) &= \sqrt{\frac{2}{W}} \cos \frac{n\pi x}{W}, \quad n \text{ odd} \\ &= \sqrt{\frac{2}{W}} \sin \frac{n\pi x}{W}, \quad n \text{ even} \end{aligned} \tag{5.6}$$

If the potential barrier is not infinite, we cannot assume that the wavefunction goes to zero at the boundaries of the well. The equation for the barrier region is

$$\frac{-\hbar^2}{2m} \frac{d^2\psi}{dx^2} + V_0 \psi = E\psi \quad \text{for } |x| \geq a \tag{5.7}$$

where V_0 is the potential step.

The general bound-state ($E < V_0$) solution of the problem can now be taken to be of the form

$$\psi(x) = \begin{cases} Ae^{\beta x}, & x \leq -a \\ B \cos \alpha x + C \sin \alpha x, & -a \leq x \leq a \\ De^{-\beta x}, & x \geq a \end{cases} \tag{5.8}$$

where

$$\begin{aligned} \alpha &= \sqrt{\frac{2mE}{\hbar^2}} \\ \beta &= \sqrt{\frac{2m(V_0 - E)}{\hbar^2}} \end{aligned} \tag{5.9}$$

We now impose the boundary conditions that at $x = \pm a$, ψ and $d\psi/dx$ are continuous. This corresponds to saying that the electron probability and the electron current do not suffer a discontinuity at the boundaries.

Matching the wavefunction and its derivative at the boundaries, we have the conditions

$$\begin{aligned} B \cos \frac{\alpha W}{2} - C \sin \frac{\alpha W}{2} &= Ae^{-\beta W/2} \\ \alpha B \sin \frac{\alpha W}{2} + \alpha C \cos \frac{\alpha W}{2} &= \beta A e^{-\beta W/2} \\ B \cos \frac{\alpha W}{2} + C \sin \frac{\alpha W}{2} &= De^{-\beta W/2} \\ -\alpha B \sin \frac{\alpha W}{2} + \alpha C \cos \frac{\alpha W}{2} &= -\beta D e^{-\beta W/2} \end{aligned} \tag{5.10}$$

From these equations we get two pairs of conditions on the solutions

$$
\begin{aligned}
2B\ \cos\ \frac{\alpha W}{2} &= (A+D)e^{-\beta W/2} \\
2\alpha B\ \sin\ \frac{\alpha W}{2} &= \beta(A+D)e^{-\beta W/2}
\end{aligned}
\tag{5.11}
$$

and

$$
\begin{aligned}
2C\ \sin\ \frac{\alpha W}{2} &= (D-A)e^{-\beta W/2} \\
2\alpha C\ \cos\ \frac{\alpha W}{2} &= -\beta(D-A)e^{-\beta W/2}
\end{aligned}
\tag{5.12}
$$

These pairs give us two separate conditions for the solutions, obtained by dividing one equation by the other within each pair. The conditions for the allowed energy levels are the transcendental equations

$$\boxed{\frac{\alpha W}{2}\ \tan\ \frac{\alpha W}{2} = \frac{\beta W}{2}} \tag{5.13}$$

and

$$\boxed{\frac{\alpha W}{2}\ \cot\ \frac{\alpha W}{2} = -\frac{\beta W}{2}} \tag{5.14}$$

Eqn. 5.13 results in states that have even parity (i.e., with $A = D$), while Eqn. 5.14 gives states with odd parity (with $A = -D$).

Eqns. 5.13 and 5.14 can be solved by numerical techniques. One useful approach is a graphical technique shown in Fig. 5.2a. One starts out with plotting curves in the $\beta W/2 - \alpha W/2$ plane which satisfy Eqns. 5.13 or 5.14. Note that several α-values satisfy the equations for a given value of β. Next, we note that we have the equality (from Eqn. 5.9)

$$\left(\frac{\alpha W}{2}\right)^2 + \left(\frac{\beta W}{2}\right)^2 = \frac{mV_0W^2}{2\hbar^2} \equiv R(d)^2 \tag{5.15}$$

We therefore draw a circle with radius $R(d)$. For given values of V_0 and W, there is one such circle. The intersection of this circle with the first set of curves gives the desired solutions. There may be several solutions for a given well thickness. As the well thickness increases, the number of allowed states also increases, as shown in Fig. 5.2a. To find the highest allowed state for a given well thickness, we note that $(\alpha W/2)\tan(\alpha W/2)$ and $(\alpha W/2)\cot(\alpha W/2)$ intersect the $\alpha W/2$ axis at values of $m\pi/2$. Thus the well width at which a state n is just allowed is given by

$$R(W_c) = \frac{n\pi}{2}$$

The maximum mode number for a given well width is then the integer part of

$$\frac{mV_0W^2}{\pi\hbar^2} \tag{5.16}$$

Typical shapes of the wavefunctions corresponding to the various energy levels are shown in Fig. 5.2b. Note that in the case of the finite barrier potential the wavefunctions penetrate the barrier region and the energy values are lower than the values obtained from the infinite barrier model.

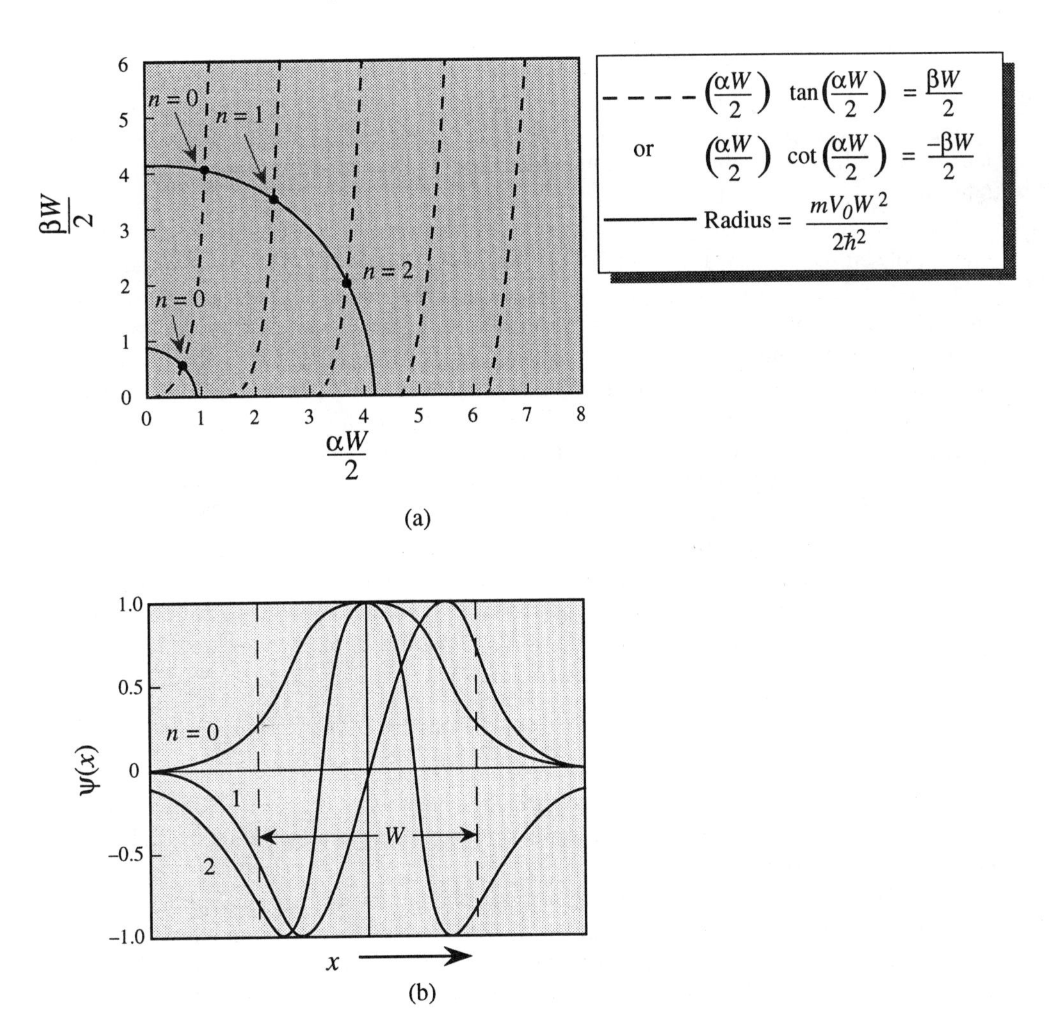

Figure 5.2: (a) The graphical approach to solving for the allowed modes in a finite quantum well. (b) Typical solutions for the particle wavefunctions.

5.2.2 Application Example: Semiconductor Heterostructures

In Section 4.4.7 we have discussed how epitaxial crystal growth techniques can produce atomically abrupt interfaces between two materials. Such techniques can be used to fabricate quantum wells. Such confinement structures are produced primarily in semiconductor heterostructures. Typically when a small bandgap material is sandwiched between a large bandgap material, the electrons see a potential well. Such wells can be grown with great precision (of ~ 3 Å) and serve as important structures for modern transistors and lasers.

In order to apply the results obtained to heterostructures created from semiconductors, we will review some of the properties of semiconductors relevant to the quantum wells fabricated from semiconductors.

As far as the response of the electrons to any forces is concerned, the electrons behave as if they are in free space except that their energy–momentum relation is not the simple parabolic relation of free space. Near the conduction band and valence bandedges, the electron can be described by an effective mass discussed in Chapter 4.

As noted, by using epitaxial growth techniques it is possible to grow a sequence of semiconductor layers so that a narrow-bandgap material is surrounded by a larger-bandgap material. In some semiconductors making up such a heterostructure, the bandgap of the narrow-gap material is completely enclosed by the bandgap of the larger-gap material. An important example is the GaAs and the AlAs system. In Fig. 5.3 we show schematically the change in the potential created when a heterostructure is formed. Quantum wells are formed in the conduction and valence bands. As a result, the electrons in the narrow-gap material are unable to move freely in the crystal growth (confinement) direction. They can still move freely in the plane perpendicular to the growth direction.

An extremely important parameter in the quantum well problem is the bandedge discontinuity produced when two semiconductors are brought together. A part of the bandgap discontinuity ($Eg^A - Eg^B$) of two semiconductors A and B would appear in the conduction band and a part would appear in the valence band.

Note that the important semiconductor system of GaAs and AlAs has a discontinuity ratio given by

$$\frac{\Delta E_c}{\Delta E_v} \cong \frac{60}{40} \text{ to } \frac{65}{35} \tag{5.17}$$

i.e., 60% to 65% of the *direct bandgap difference* between AlAs or the alloy AlGaAs and GaAs is in the conduction band and the remaining direct bandgap difference is in the valence band.

Once the band discontinuity is known, the bandstructure of a quantum well structure can be calculated. The simplest way to do this is to use the effective mass theory in which the electron in each region of the structure is represented by its effective mass. The effect of the quantum well is to impose a background confining potential.

The simple quantum well structure such as shown in Fig. 5.3 is one of the most-studied heterostructures. Its simple square well shape allows us to use the results obtained in this section. If the crystal growth is along the z-axis, the Schrödinger equation for the electron states in the quantum well can be simply written as

$$\left[-\frac{\hbar^2\nabla^2}{2m^*} + V(z)\right]\psi = E\psi \tag{5.18}$$

where m^* is the effective mass of the electron.

The total energy of the electron has two parts. The first part is due to the confinement of the electron in the z-direction and is given by the discussion of the previous subsection. The second part comes from the motion in the x-y plane. If, for example, we assume an infinite barrier potential well, the total energy is (measured from the bandedge)

$$E(n, k_x, k_y) = \frac{\pi^2\hbar^2 n^2}{2m^* W^2} + \frac{\hbar^2 k_x^2}{2m^*} + \frac{\hbar^2 k_y^2}{2m^*};\; n = 1, 2, \ldots \tag{5.19}$$

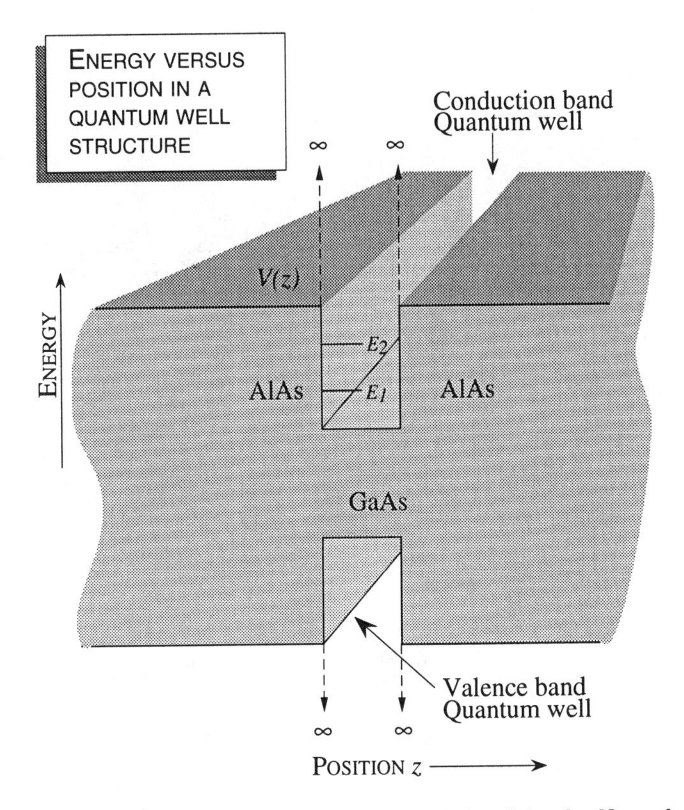

Figure 5.3: Schematic of a quantum well and the sub-band levels. Note that in a semiconductor quantum well, one has a quantum well for the conduction band and one for the valance band. In the infinite barrier model, the barriers are chosen to have an infinite potential, as shown.

For more accurate results, the first term in this equation should be calculated using the numerical or graphical approach mentioned in the previous subsection.

It is important to examine what happens to valence band states in a semiconductor quantum well. At the top of the valence band the effective mass of the electron is negative and there are two bands: a heavy hole band, and a light hole band. Due to the negative mass, the quantization energy pushes the energy levels downwards and there are two sets of energy levels; one for the heavy hole, and one for light hole, as shown in Fig. 5.4.

In Chapter 4, Section 4.2, we have discussed the density of states in two-dimensional systems. Each subband in the quantum well contributes a constant density of states, as shown in Fig. 5.4. Thus we have a step density of states, as shown in the conduction and valence bands.

This discussion has been given for quantum wells where the electron confinement is only along one direction. It is possible (though not as easy) to use crystal growth techniques and/or semiconductor processing to create structures where confinement is along two directions or even in all three dimensions. These structures are called quantum wires and quantum dots respectively. If the shape of the potential

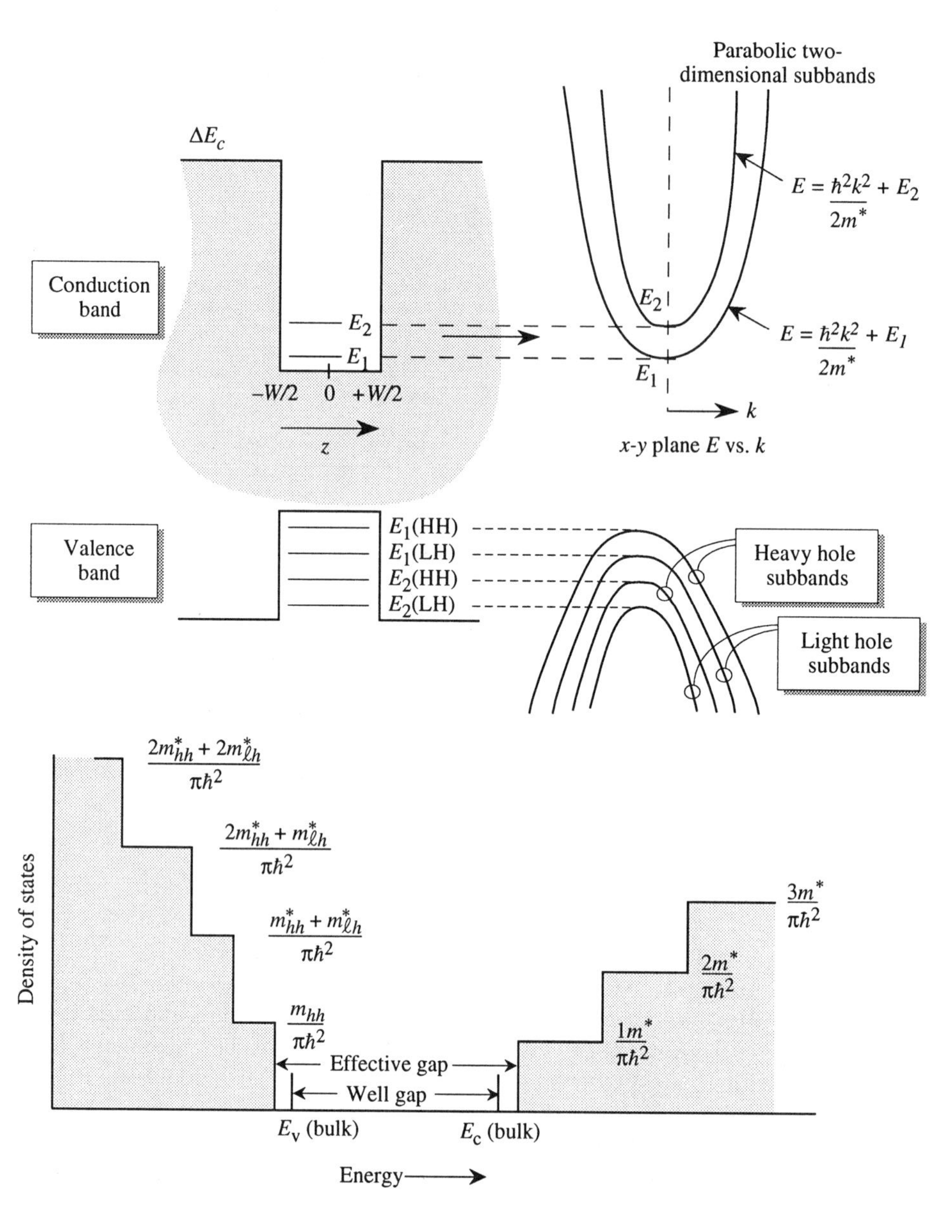

Figure 5.4: Schematic of a quantum well and the subband levels. In the x-y plane the subbands can be represented by parabolas. Subbands are produced in the conduction band and the valence band.

is such that the potential can still be written as

$$V(r) = V(x) + V(y) + V(z)$$

it is possible to simply extend our results since the x-, y- and z-direction problems are separable. Thus for the quantum wire along the y-axis, assuming an infinite confining barrier, we get (assuming an infinite barrier)

$$E(\ell, n, k_y) = \frac{\pi^2 \hbar^2 \ell^2}{2m^* W_x^2} + \frac{\pi^2 \hbar^2 n^2}{2m^* W_z^2} + \frac{\hbar^2 k_y^2}{2m^*} \tag{5.20}$$

where W_x and W_z are the well sizes, along the x- and z-directions. For the quantum dot (ℓ, n, p are positive integers)

$$E(\ell, n, p) = \frac{\pi^2 \hbar^2 \ell^2}{2m^* W_x^2} + \frac{\pi^2 \hbar^2 n^2}{2m^* W_y^2} + \frac{\pi^2 \hbar^2 p^2}{2m^* W_z^2} \tag{5.21}$$

In real structures one has to obtain the energy levels numerically or graphically, as discussed earlier.

The density of states in three-dimensional and sub-three-dimensional systems has been discussed in Chapter 4, Section 4.2. Each subband produces a density of states and the total density of states for any energy is the sum of density of states due to all subbands at or below that energy. The density of states in three-dimensional, quasi-two-dimensional, quasi-one-dimensional, and quasi-zero-dimensional structures is shown in Fig. 5.5. The term "quasi" is included to remind us that the structure has a finite width in the confining direction.

If the only occupied level for a structure is the ground state subband, the system displays low-dimensional effects. If higher levels are occupied, the system loses its low-dimensional characteristics.

5.2.3 Particle in a Triangular Quantum Well

In many electronic devices used for information processing, a quantum well with a "triangular" shape is produced. An example is the metal oxide field effect transistor or MOSFET, shown in Fig. 5.6. The potential for electrons may be written in the form

$$\begin{aligned} V(x) &= \infty \qquad x < 0 \\ &= Cx \qquad x > 0 \end{aligned} \tag{5.22}$$

Such a potential is created for electrons in devices where electrons are confined by a homogeneous field. Let us take the electric field $\tilde{E}$ in the direction of x. The potential energy of the particle is of the form

$$V(x) = Fx + \text{constant} \tag{5.23}$$

where F is the force on the particle (say, an electron) and has a value $e\tilde{E}$.

The triangular well problem can be shown to have the following solutions for the energy levels:

$$E_n = \left(\frac{\hbar^2}{2m^*}\right)^{1/3} \left(\frac{3}{2}\pi F\right)^{2/3} \left(n - \frac{1}{4}\right)^{2/3}, \quad n = 1, 2, \ldots \tag{5.24}$$

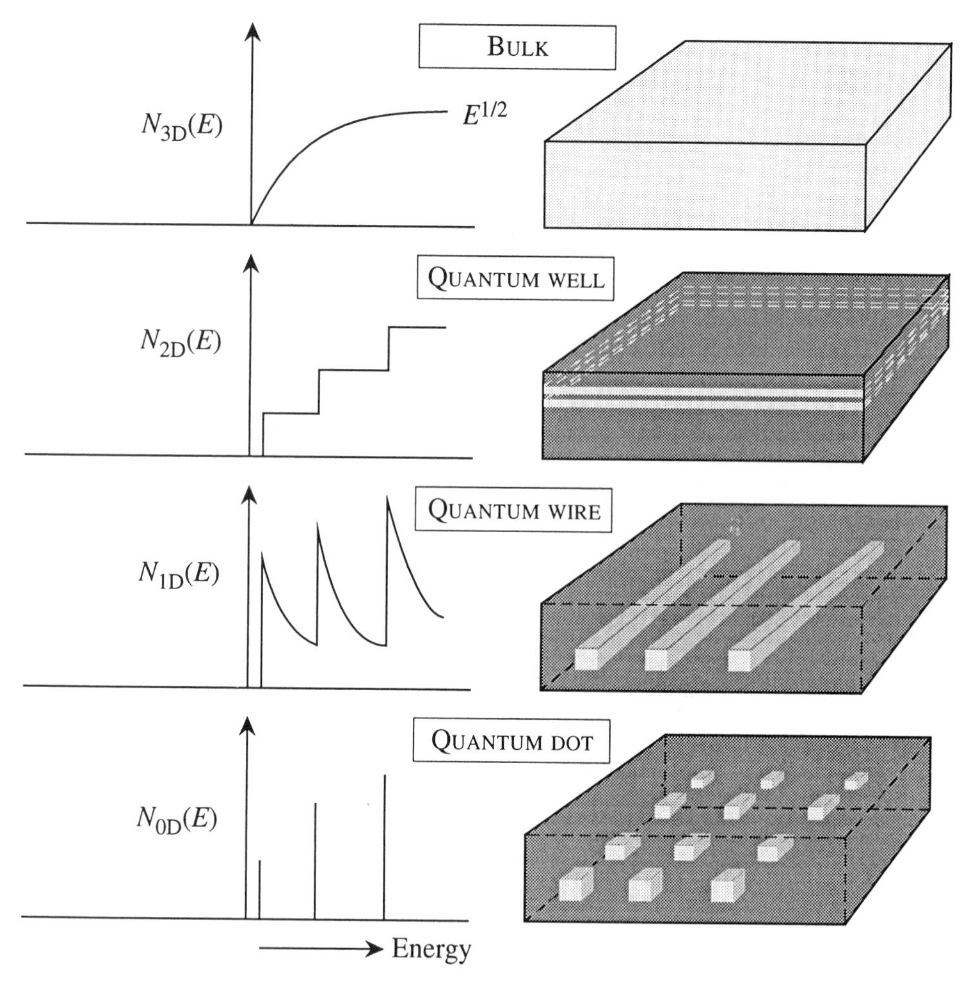

Figure 5.5: A schematic of the density of states in a 3D, quasi-2D, quasi-1D, and quasi-0D system.

5.2.4 Application Example: Confined Levels in Semiconductor Transistors

In many semiconductor devices electrons are confined in almost triangular quantum wells. Examples of such devices are Si MOSFETs used for digital applications and GaAs/AlGaAs MODFETs (modulation-doped-field-effect transistors) used for high-speed applications.

As shown in Fig. 5.6, the device consists of an insulator–semiconductor junction. Electrons are injected at the interface on the semiconductor side by a controlling electrode (the gate). The free charge causes a bending of the semiconductor band to produce an approximately triangular quantum well, as shown. The triangular quantum well is defined by an electric field E_s, which is related to the

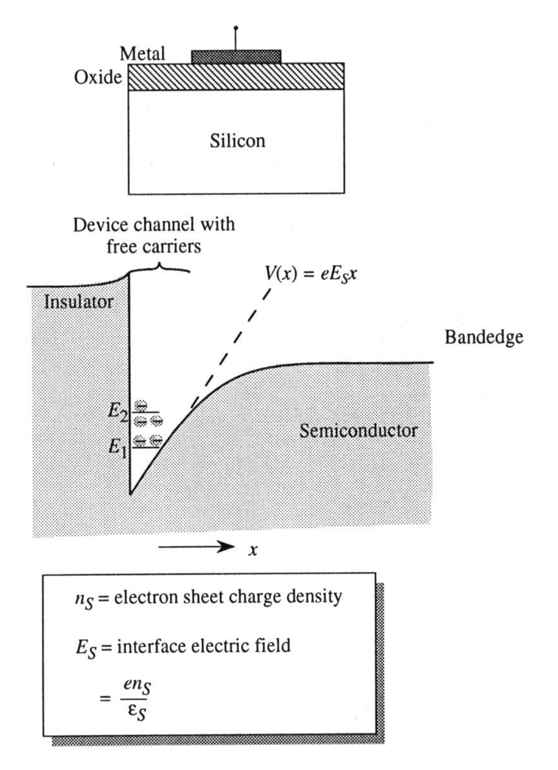

Figure 5.6: A schematic of a metal oxide semiconductor field effect transistor (MOSFET). Electrons at the interface of the oxide and semiconductor produce an approximately triangular quantum well.

areal charge density by Gauss's law

$$E_s = \frac{en_s}{\epsilon_s} = \tilde{E} \tag{5.25}$$

As a result of the confinement, quantized energy levels are formed in the triangular well. Approximate positions of these levels can be obtained from the results derived in the previous section. The electrons are free to move in the plane perpendicular to the interface. The device works by using a bias on the gate to modulate the charge density n_s, and thus the current flowing in the channel.

EXAMPLE 5.1 Calculate the "pressure" exerted on the walls of a rectangular potential box by a particle inside it. The sides of the box are a, b, and c.

Let us consider the pressure on the face with area $b \times c$, i.e., the face normal to the x-axis. The force is given by (H is the energy function)

$$F = \frac{-\partial H}{\partial a}$$

Let us differentiate the equation $(H - E_n)\psi_n = 0$ with respect to a. This gives

$$\left(\frac{\partial H}{\partial a} - \frac{\partial E_n}{\partial a}\right)\psi_n = -(H - E_n)\frac{\partial \psi_n}{\partial a}$$

Multiplying both sides by ψ_n^* and integrating over x, we see that the right-hand side is zero, and

$$F = \frac{-\partial H}{\partial a} = \frac{-\partial E_n}{\partial a}$$

Using the infinite barrier potential result

$$E_n = \frac{\pi^2 \hbar^2}{2m}\left(\frac{n_1^2}{a^2} + \frac{n_2^2}{b^2} + \frac{n_3^2}{c^2}\right) \quad (n_1, n_2, n_3 = 1, 2, \ldots)$$

we get

$$F = \frac{\pi^2 \hbar^2 n_1^2}{ma^3}$$

The pressure is obtained by dividing this by the area bc.

EXAMPLE 5.2 In the GaAs/AlGaAs heterostructure, 60% of the bandgap discontinuity is in the conduction band of the narrow-gap material. Calculate the conduction band and valence band quantum well potentials for GaAs/$Al_{0.3}Ga_{0.7}As$. Can one use the infinite barrier model for obtaining the confined levels? The bandgap difference between GaAs and $Al_{0.3}Ga_{0.7}As$ is 0.374 eV. For the GaAs/AlGaAs system 60% of the bandgap difference is in the conduction band. The discontinuity of the quantum well formed in the conduction band is (this is the barrier height)

$$\Delta E_c = 0.374 \times 0.6 = 0.224 \text{ eV}$$

The barrier height for the valence band is

$$\Delta E_v = 0.15 \text{ eV}$$

These barrier heights are not infinite, so our simple model is only an approximation to the real problem. However, the use of the infinite barrier problem is reasonable once the well size is ~150 Å for this system.

EXAMPLE 5.3 Using a simple infinite barrier approximation, calculate the "effective bandgap" of a 100 Å GaAs/AlAs quantum well. If there is a one-monolayer fluctuation in the well size, how much will the effective bandgap change? This example gives an idea of how stringent the control has to be to exploit heterostructures. We will solve this problem using the infinite barrier model.

The confinement of the electron ($m^* = 0.067\ m_0$) pushes the effective conduction band up, and the confinement of electrons at the valence band push the effective edge down ($m_v^* = -0.4\ m_0$). The change in the electron ground state is (using $n = 1$) 55.77 meV. The shift (downward) in the valence band is 9.34 meV. The net shift is 65.11 meV. The effective bandgap is thus 65.11 meV larger than the bulk GaAs bandgap.

If the well size changes by one monolayer (e.g., goes from 100 Å to 102.86 Å), the change in the electron level is

$$\begin{aligned}\Delta E_e &= E_e\left[1 - \frac{(100)^2}{(102.86)^2}\right] = E_e \times 0.055 \\ &= 3.06 \text{ meV}\end{aligned}$$

The hole energy changes by

$$\Delta E_{hh} = E_{hh} \times 0.055 = 0.51 \text{ meV}$$

Thus the bandgap changes by 3.56 meV for a one-monolayer variation. For many applications, particularly in optics, this change is too large to be acceptable.

EXAMPLE 5.4 GaAs and Si are used as well regions to form two sets of quantum wells. Estimate the maximum widths of the quantum wells that will display two-dimensional effects for electrons at (i) 77 K and at (ii) 300 K.

Use the infinite barrier model for your results.

In this problem we have to find the condition under which the well size is such that the first subband is occupied but the higher subbands are not. This requires that the separation between the first and second subband should at least be equal to k_BT.

For GaAs we use the effective mass of 0.067 m_0 and the direction of the quantum well is unimportant since the mass is isotropic. For Si we need to choose a direction in which the highest mass for any valley is as small as possible. If we choose a direction [001] the maximum mass is m_l. We choose a direction [111]. *The mass for all valleys is then* $(m_t^2 m_l)^{1/3} = 0.33 m_0$.

Using the infinite barrier model the separation between the first two subbands is

$$E_2 - E_1 = \frac{3\pi^2 \hbar^2}{2m^* W^2}$$

Equating this to k_BT we gate for the maximum well width

$$W_{max} = \left(\frac{3\pi^2 \hbar^2}{2m^* k_B T} \right)^{1/2}$$

We get the following results:

$$\begin{aligned}
&T = 77 \text{ K} \\
&\quad \text{GaAs}: \quad W_{max} = 499.4 \text{ Å} \\
&\quad \text{Si}: \quad W_{max} = 225 \text{ Å} \\
&T = 300 \text{ K} \\
&\quad \text{GaAs}: \quad W_{max} = 253 \text{ Å} \\
&\quad \text{Si}: \quad W_{max} = 114 \text{ Å}
\end{aligned}$$

If the Si quantum well was grown along [001] direction we would need to use the longitudinal mass $\sim 1.0 m_0$ which would give

$$77 \text{ K}: \; W_{max} = 129 \text{ Å}; \quad 300 \text{ K}: \; W_{max} = 65.5 \text{ Å}$$

We can see that we can relax the width constraints by going to the [111] direction in Si.

Note that 2D effects become apparent in GaAs at wider wells in GaAs than in Si. Also they are easier to observe at lower temperatures.

EXAMPLE 5.5 A GaAs/AlGaAs MODFET has 10^{12} cm^{-2} electrons in the GaAs channel. Calculate the first two energy levels produced by the confining potential using the triangular well approximation. The effective mass of the electrons in GaAs is 0.067 m_0.

The energy levels are given by

$$E_n = \left(\frac{\hbar^2}{2m^*}\right)^{1/3} \left(\frac{3}{2}\pi e E_s\right)^{2/3} \left(n - \frac{1}{4}\right)^{2/3}$$

This gives (E_s is in V/m)

$$\begin{aligned} E_1 &= 1.98 \times 10^{-6} \ E_s^{2/3} \ \text{eV} \\ E_2 &= 3.48 \times 10^{-6} \ E_s^{2/3} \ \text{eV} \end{aligned}$$

In the case given ($E_s = en_s/\epsilon_s$)

$$\begin{aligned} E_s &= \frac{(1.6 \times 10^{-19} \ \text{C})(10^{16} \ \text{m}^{-2})}{13.2 \times 8.85 \times 10^{-12} \ \text{F/m}} \\ &= 1.37 \times 10^7 \ \text{V/m} \end{aligned}$$

This gives

$$E_1 = 11.4 \ \text{meV}; \quad E_2 = 20 \ \text{meV}$$

5.3 HARMONIC OSCILLATOR PROBLEM

A problem in quantum mechanics with far-reaching consequences is the problem of a point mass attracted to a fixed center with a force proportional to the distance of the particle from the center. This problem, called the harmonic oscillator problem, is well known in classical physics and is used for understanding problems such as vibrations of a mass on a spring, vibrations of atoms in a solid, studying pendulum motion, etc. In quantum mechanics also, the harmonic oscillator problem is quite important. Many important quantum problems can be expressed in terms of the harmonic oscillator problem.

The potential energy for the harmonic oscillator is of the form shown in Fig. 5.7 and is given by

$$V(x) = \frac{1}{2}Kx^2 \tag{5.26}$$

The energy operator of the harmonic oscillator is

$$H = \frac{-\hbar^2}{2m}\nabla^2 + \frac{1}{2}Kx^2 \tag{5.27}$$

where K is the force constant describing the attractive force ($= -Kx$). If ψ represents the wavefunction of the time-independent Schrödinger equation, we have for a one-dimensional case

$$\frac{-\hbar^2}{2m}\frac{d^2\psi}{dx^2} + \frac{Kx^2\psi}{2} = E\psi \tag{5.28}$$

This equation can be recast so that it becomes a well-known differential equation with solutions known as Hermite polynomials, H_n. We will not discuss the mathematical details of how this equation is solved.

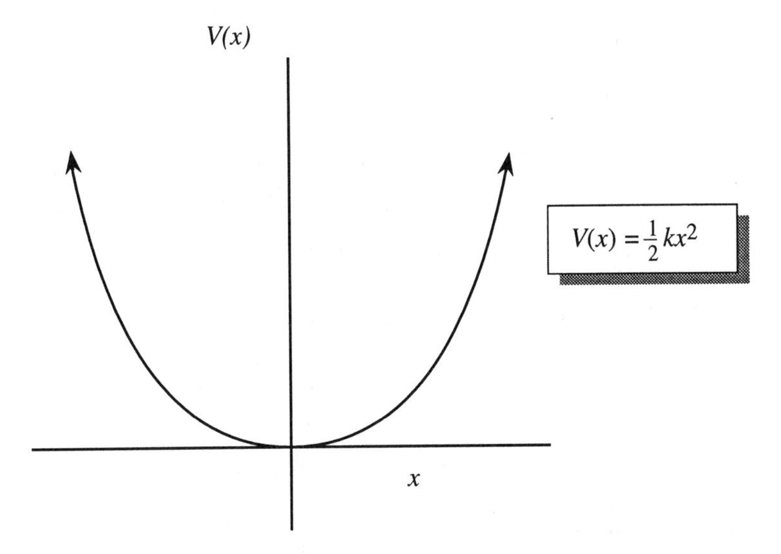

Figure 5.7: The potential energy profile for a harmonic oscillator.

The general solutions are

$$\psi_n(x) = N_n H_n(\alpha x) e^{-\frac{1}{2}\alpha^2 x^2} \tag{5.29}$$

where

$$\alpha^4 = \frac{mK}{\hbar^2}$$

The factor N_n is chosen so that

$$\int_{-\infty}^{\infty} | \psi_n(x) |^2 \, dx = \frac{| N_n |^2}{\alpha} \int_{-\infty}^{\infty} H_n^2(\xi) e^{-\xi^2} \, d\xi = 1$$

The function in the integral is integrable, and it is seen that the normalization factor is

$$N_n = \left(\frac{\alpha}{\pi^{1/2} 2^n n!} \right)^{1/2} \tag{5.30}$$

The first few Hermite polynomials are

$$\begin{aligned} H_0(\xi) &= 1 \\ H_1(\xi) &= 2\xi \\ H_2(\xi) &= 4\xi^2 - 2 \end{aligned} \tag{5.31}$$

In Fig. 5.8 we show some of the lower states of the harmonic oscillator problem.

The eigenenergy corresponding to the wavefunction $\psi_n(x)$ is given by the relation

$$\boxed{E_n = \left(n + \frac{1}{2} \right) \hbar \omega_c \qquad n = 0, 1, 2, \ldots} \tag{5.32}$$

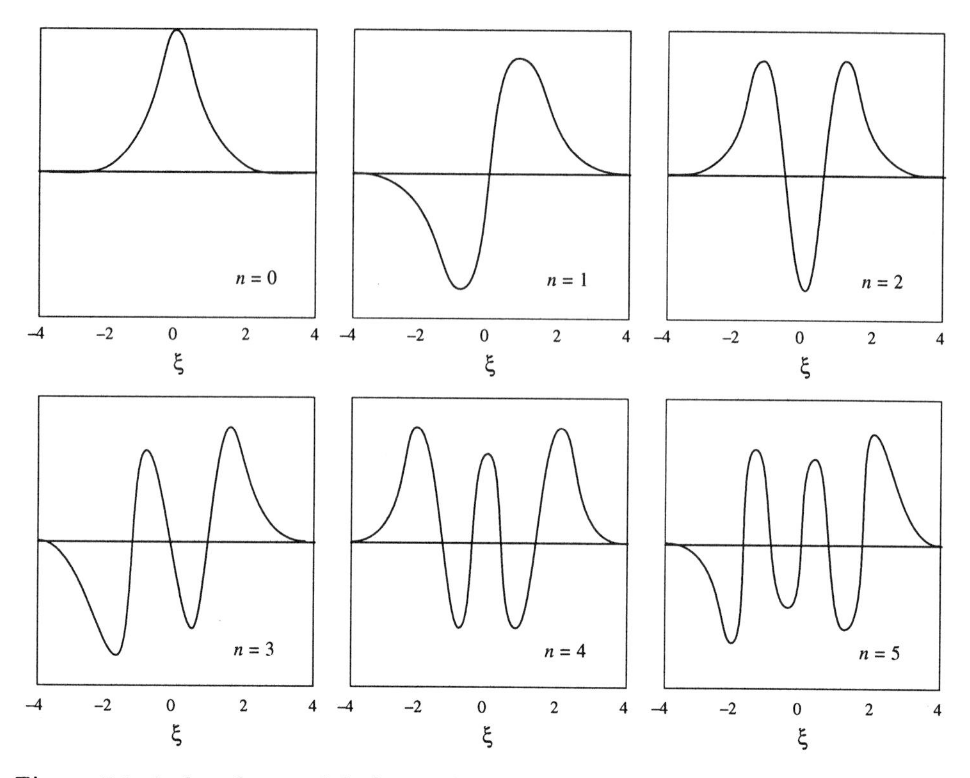

Figure 5.8: A plot of some of the harmonic oscillator wavefunctions corresponding to the lower energy states.

Here the quantity ω_c is given by the classical frequency of the harmonic oscillator

$$\omega_c = \left(\frac{K}{m}\right)^{1/2} \tag{5.33}$$

In a classical harmonic oscillator, the energy of the oscillator is continuous and can be increased by increasing the amplitude of vibration. In the quantum treatment, the energy increases in steps of $\hbar\omega_c$. The number n in Eqn. 5.32 is called the *occupation number* and tells us how many "quanta" of energy are in the oscillator. An important point is that the lowest energy of the quantum oscillator is not zero but $\frac{1}{2}\hbar\omega_c$. This lowest energy is called the *zero point energy* or, depending upon the context, in some problems, the *vacuum energy*.

5.3.1 Application Example: Second Quantization of Wavefields

In Chapter 3 we have discussed how classical waves are represented by "particles" or quanta in quantum mechanics—a process called second quantization. The harmonic oscillator problem discussed above has very important implications for such wave-particles.

A wave equation representing a classical wave has the same form as that of a harmonic oscillator. If ω is the frequency of the wave, the time-dependent wave equation (for a pure mode) is

$$\frac{d^2u}{dt^2} + \omega^2 u = 0 \tag{5.34}$$

where u is the amplitude of vibration. The quantization of the "oscillator" gives energy eigenvalues

$$E(\omega) = (n + 1/2)\hbar\omega \tag{5.35}$$

Thus the mode can have energies which are discrete and change by one quantum of energy, $\hbar\omega$. As noted above, in classical physics the energy is given by the amplitude and is continuous. However, in quantum mechanics it is quantized. For most classical vibrations, the quantum number n, known as the *occupation number*, is very large (see Example 5.7) and the energy seems continuous.

Photons

An important wavefield is the electromagnetic field where as noted in Chapter 1, the time-averaged value of power in a mode of frequency ω is

$$< S >_{\text{time}} = 2v\epsilon\omega^2 \mid A_0 \mid^2 \tag{5.36}$$

where v is the velocity of light in the media of dielectric constant ϵ. A_0 is the amplitude of the vector potential describing the electromagnetic radiation.

The energy density is

$$\left|\frac{S}{v}\right| = 2\epsilon\omega^2 \left|A_0\right|^2 \tag{5.37}$$

In quantum mechanics if n_{ph} is the occupation number for the mode ω, the energy density is

$$\frac{n_{ph}\hbar\omega}{V} \tag{5.38}$$

so that the classical amplitude $|A_0|$ and photon number are related by

$$\left|A_0\right|^2 = \frac{\hbar n_{ph}}{2\epsilon\omega V} \tag{5.39}$$

Phonons

Another important classical vibration is that of atoms in a crystalline structure. These lattice vibrations are an important source of scattering of electrons as they move in the crystal (we will discuss this problem in Chapter 8). Let us consider a diatomic lattice (two atoms per basis) as shown in Fig. 5.9. The atoms are at an equilibrium position around which they vibrate. There is a restoring force (let us assume this force is between the nearest neighbors only). We assume that the atoms have masses M_1 and M_2.

If u_s and v_s represent the displacements of the two kinds of atoms of the unit cell s (see Fig. 5.9), we get the following equations of motion for the atoms in

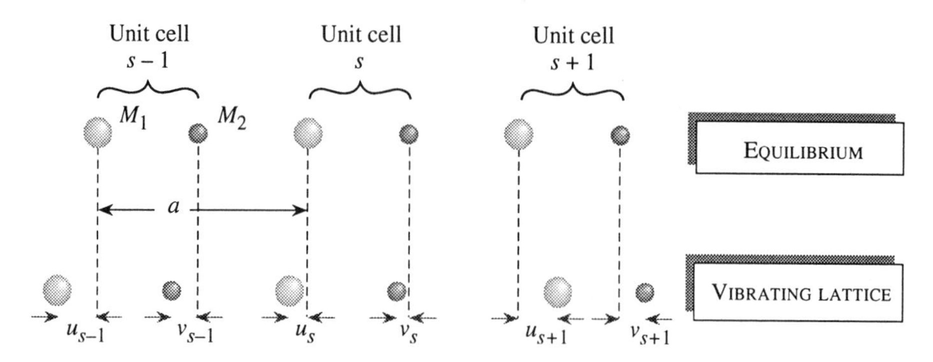

Figure 5.9: Vibrations in a crystal with two atoms per unit cell with masses M_1, M_2 connected by force constant C between adjacent planes.

the unit cell s:

$$M_1 \frac{d^2 u_s}{dt^2} = C(v_s + v_{s-1} - 2u_s) \tag{5.40}$$

$$M_2 \frac{d^2 v_s}{dt^2} = C(u_{s+1} + u_s - 2v_s) \tag{5.41}$$

We look for solutions of the traveling wave form, but with different amplitudes u and v on alternating planes

$$\begin{aligned} u_s &= u \exp(iska) \exp(-i\omega t) \\ v_s &= v \exp(iska) \exp(-i\omega t) \end{aligned} \tag{5.42}$$

We note that a is the distance between nearest identical planes and not nearest planes, i.e., it is the minimum distance of periodicity in the crystal as shown in Fig. 5.9. Eqn. 5.42, when substituted in Eqns. 5.40 and 5.41 gives

$$-\omega^2 M_1 u = Cv\,[1 + \exp(-ika)] - 2Cu \tag{5.43}$$

$$-\omega^2 M_2 v = Cu\,[\exp(-ika) + 1] - 2Cv \tag{5.44}$$

These are coupled eigenvalue equations which can be solved by the matrix method. The equations can be written as the matrix vector product

$$\begin{vmatrix} -\omega^2 M_1 + 2C & -C\,[1 + \exp(-ika)] \\ -C\,[\exp(-ika) + 1] & -\omega^2 M_2 + 2C \end{vmatrix} \begin{vmatrix} u \\ v \end{vmatrix} = 0$$

Equating the determinant to zero, we get

$$\left| 2C - M_1\omega^2 \quad - C[1 + \exp(-ika)] - C[1 + \exp(ika)] \; 2C - M_2\omega^2 \right| = 0 \tag{5.45}$$

or

$$M_1 M_2 \omega^4 - 2C(M_1 + M_2)\omega^2 + 2C^2(1 - \cos ka) = 0 \tag{5.46}$$

This gives the solution

$$\omega^2 = \frac{2C(M_1 + M_2) \pm [4C^2(M_1 + M_2)^2 - 8C^2(1 - \cos ka)M_1 M_2]^{1/2}}{2M_1 M_2} \tag{5.47}$$

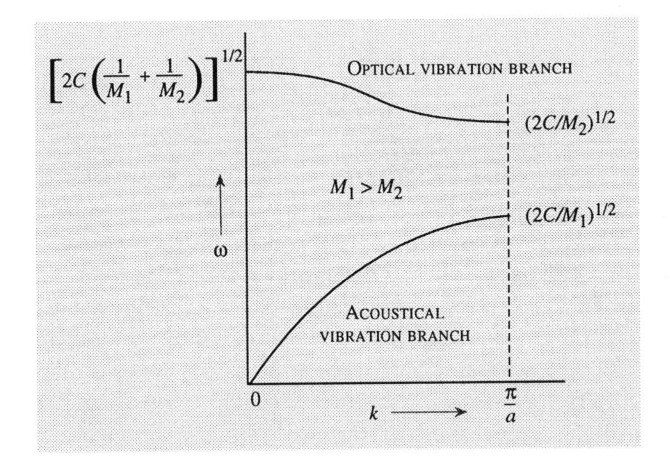

Figure 5.10: Optical and acoustical branches of the dispersion relation for a diatomic linear lattice

It is useful to examine the results at two limiting cases. For small k, we get the two solutions

$$\omega^2 \approx 2C\left(\frac{1}{M_1}+\frac{1}{M_2}\right) \tag{5.48}$$

and

$$\omega^2 \approx \frac{C/2}{M_1+M_2}k^2a^2 \tag{5.49}$$

Near $k = \pi/a$ we get (beyond this value the solutions repeat)

$$\begin{aligned}\omega^2 &= 2C/M_2\\ \omega^2 &= 2C/M_1\end{aligned} \tag{5.50}$$

The general dependence of ω on k is shown in Fig. 5.10. Two branches of lattice vibrations can be observed in the results. The lower branch, which is called the acoustic branch, has the property that as for the monatomic lattice, ω goes to zero as k goes to zero. The upper branch, called the optical branch, has a finite ω even at $k = 0$.

The acoustical branch represents the propagation of sound waves in the crystal. The sound velocity is

$$v_s = \frac{d\omega}{dk} = \sqrt{\frac{C}{M_{av}}}\,a \tag{5.51}$$

where M_{av} is the average mass of the two atoms.

It is important to examine the eigenfunctions (i.e., u_s), for the optical branch and the acoustic branch of the dispersion relation. For $k = 0$, for the optical branch, we have, after substituting

$$\omega^2 = 2C\left(\frac{1}{M_1}+\frac{1}{M_2}\right) \tag{5.52}$$

in the equation of motion (say, Eqn. 5.43)

$$u = \frac{-M_2}{M_1} v \tag{5.53}$$

The two atoms vibrate against each other, but their center of mass is fixed. If we examine the acoustic branch, we get $u = v$ in the long wavelength limit. In Fig. 5.11a we show the different nature of vibration of the acoustic and optical mode.

In ionic materials like GaAs, when optical vibration takes place polarization fields are set up that vibrate as well. These fields are important for longitudinal vibration, but not for translational vibration. As a result, in longitudinal vibrations there is an additional restoring force due to the long-range polarization. In Fig. 5.11b we show the lattice vibration frequency wavevector relation for GaAs. Notice that the longitudinal optical mode frequency is higher than that of the transverse mode frequency.

In quantum mechanics the description of the vibrations is given not by the amplitude of vibration, but by "phonon number." This number is given (in equilibrium) by the Bose-Einstein expression

$$n(\omega) = \frac{1}{e^{\hbar\omega/k_BT} - 1} \tag{5.54}$$

As temperature increases, the phonon number increases as is expected intuitively. Classically this means that the atoms have a larger amplitude of vibration.

EXAMPLE 5.6 The optical phonon energy at $k = 0$ for GaAs is 36 meV and for AlAs is 50 meV. Calculate the force constants for the two materials, assuming a simple linear chain model. What are the sound velocities in these materials? The force constant for the diatomic chain model is given by the relation

$$C = \frac{\mu\omega_0^2}{2}$$

where μ is the reduced mass and ω_0 is the optical phonon frequency. For GaAs the value of the reduced mass is 36.1 atomic mass units. This gives (ω_0 corresponding to a phonon energy of 36 meV is 5.76×10^{13} rad/s)

$$\begin{aligned} C &= \frac{36.1 \times 1.67 \times 10^{-24} \text{ gm})(5.76 \times 10^{13})^2(\text{rad/s})^2}{2} \\ &= 1.0 \times 10^5 \text{ dynes/cm} \end{aligned}$$

To calculate the sound velocity we need to decide what should be used for the periodicity distance a. This depends on the direction of propagation. We will use [110] direction along which the periodicity distance is ~ 4.0 Å. The sound velocity is

$$\begin{aligned} v_s &= \sqrt{\frac{C}{M_{av}}} a = \sqrt{\frac{10^5 \text{ dynes/cm}}{72.3 \times 1.67 \times 10^{-24} \text{ g}}} \times \left(4 \times 10^{-8} \text{ cm}\right) \\ &= 1.15 \times 10^6 \text{ cm/s} \end{aligned}$$

A similar calculation for AlAs gives for the force constant a value of 1.06×10^5 dynes/cm and a sound velocity of 1.37×10^6 cm/s.

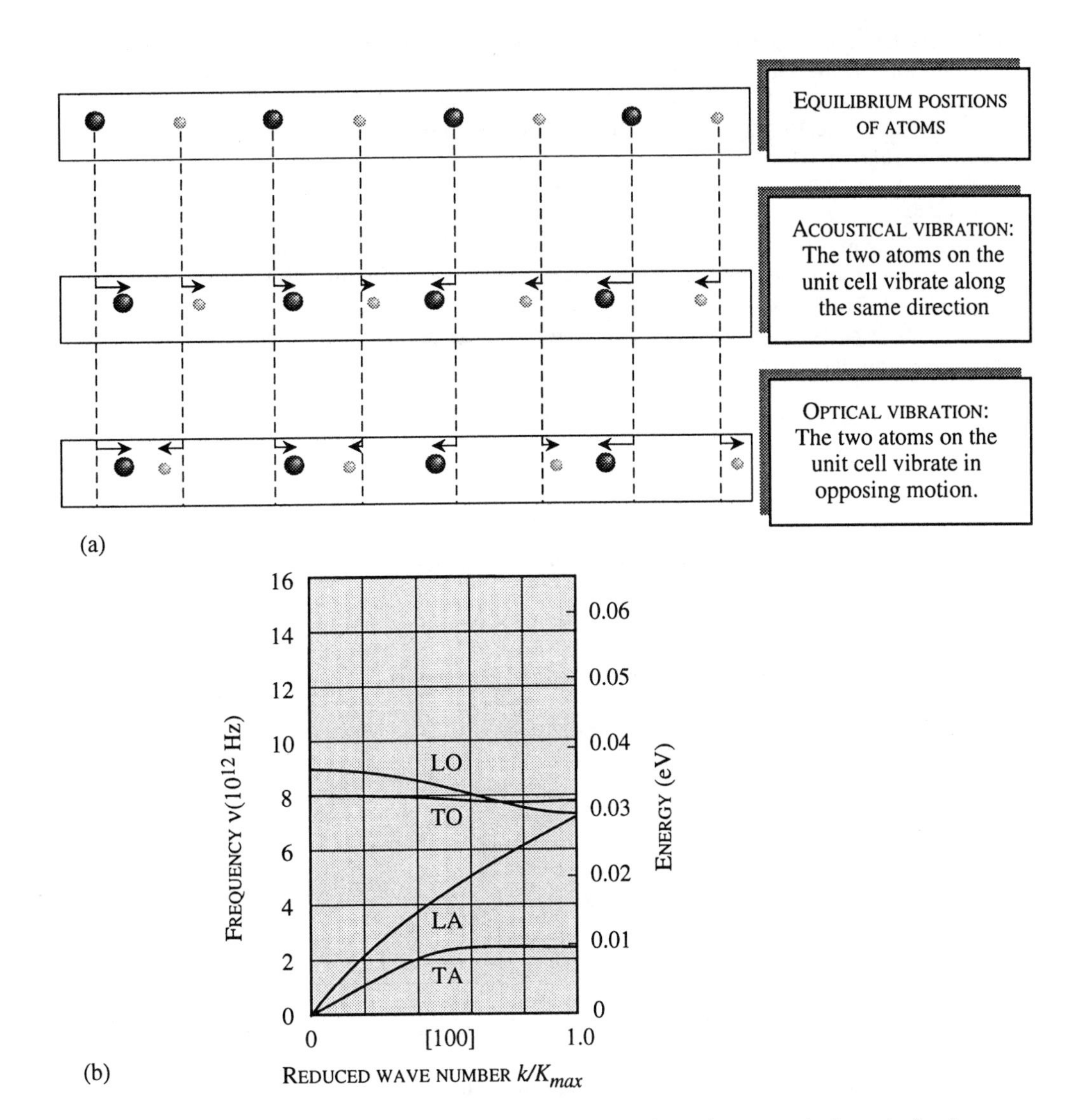

Figure 5.11: (a) The difference between as acoustical mode an optical mode is shown. (b) Phonon dispersion relation in GaAs. The longitudinal (LO, LA) and transverse (TO, TA) optical and acoustical modes are shown.

Note that in the calculation above, the periodicity distance a has been chosen to be 4 Å. This distance changes for different directions. Thus the value used and the results obtained are approximate.

EXAMPLE 5.7 Consider a classical pendulum with mass M and length l. Establish the relation between the quanta number n and the classical amplitude by equating the quantum energy $(n + 1/2)\hbar\omega$ to the classical energy of the pendulum.

Calculate the "quanta number" for a pendulum with mass 1.0 gm, length 1 m, and amplitude of 0.1 cm.

What is the amplitude (using classical energy) of the pendulum when the quanta number is 100?

This problem allows one to estimate the amplitude of vibration of an oscillator in terms of the phonon occupation number. Remember that quantum mechanics gives results similar to classical mechanics when the quantum number is large.

In a classical pendulum with length l, mass M and amplitude a, the vibration frequency and energy are given by

$$E = \frac{1}{2}\frac{Mga^2}{l}$$

$$\omega = \sqrt{\frac{g}{l}}$$

In the quantum oscillator the frequency has the same value. The energy is

$$E = (n + 1/2)\hbar\omega$$

For the pendulum of the problem, the frequency is

$$\omega = 3.13 \text{ s}^{-1}$$

The energy is

$$E = \frac{10^{-3} \times 9.8 \times 10^{-6}}{2 \times 1.0} = 4.9 \times 10^{-9} \text{ J}$$

This gives for the quanta number

$$n + 1/2 = \frac{4.9 \times 10^{-9} \text{ J}}{1.05 \times 10^{-34} \text{ J.s} \times 3.13 \text{ s}^{-1}} = 1.49 \times 10^{25}$$

This is an enormous number and is so large because the pendulum being considered is *very classical in nature.*

If the quanta number is 100, we get from the above equations

$$a^2 = \frac{(n + 1/2)\hbar\omega(2l)}{mg}$$

which gives

$$a = 2.6 \times 10^{-13} \text{ cm}$$

This means that the pendulum will behave as a quantum system only when its amplitude is infinitesimally small. Of course, for atomic oscillators the system shows quantum behavior when the amplitude is a fraction of an Angstrom.

5.4 ONE-ELECTRON ATOM AND THE HYDROGEN ATOM PROBLEM

The hydrogen atom problem represents one of the few problems in quantum mechanics that can be solved exactly. In atoms with more than one electron, the presence of other electrons causes one to resort to numerical techniques for the solutions. However, since the hydrogen atom problem embodies many of the important quantum effects of other atomic systems, its solutions are very instructive. The hydrogen atom problem is also very useful in addressing the problem of doping in semiconductors as well as the exciton problem. Thus this problem has a great impact on problems of great technological importance.

The hydrogen atom problem involves an electron and a proton interacting with each other via the Coulombic interaction, as shown in Fig. 5.12. The Schrödinger equation is

$$\begin{aligned}&\left[\frac{-\hbar^2}{2m_0}\left(\frac{\partial^2}{\partial x_e^2}+\frac{\partial^2}{\partial y_e^2}+\frac{\partial^2}{\partial z_e^2}\right)-\frac{\hbar^2}{2m_p}\right.\\&\left.\left(\frac{\partial^2}{\partial x_p^2}\frac{\partial^2}{\partial y_p^2}+\frac{\partial^2}{\partial z_p^2}\right)+V\left(x_e,y_e,z_e,x_p,y_p,z_p\right)\right]\\&\psi\left(x_e,y_e,z_e,x_p,y_p,z_p\right)=E_{Tot}\psi\left(x_e,y_e,z_e,x_p,y_p,z_p\right)\end{aligned} \tag{5.55}$$

where the subscripts e and p refer to the electron and the proton. Note that the mass of the electron is denoted by m_0. It is well known in classical physics that if the potential energy depends only upon the relative coordinates, i.e., $V = V\left(x_e - x_p, y_e - y_p, z_e - z_p\right)$, the problem can be separated into two one-body problems, as shown schematically in Fig. 5.12. A similar separation is possible in quantum mechanics. This can be done by using the relative coordinates and the center-of-mass coordinates defined by

$$\begin{aligned}x &= x_e - x_p\\ MX &= m_e x_e + m_p x_p\end{aligned} \tag{5.56}$$

etc. Here $M = m_0 + m_p$ is the mass of the electron and the proton. With the new coordinates, Eqn. 5.55 can be rewritten as

$$\left[\frac{-\hbar^2}{2M}\left(\frac{\partial^2}{\partial X^2}+\frac{\partial^2}{\partial Y^2}+\frac{\partial^2}{\partial Z^2}\right)-\frac{\hbar^2}{2\mu}\left(\frac{\partial^2}{\partial x^2}+\frac{\partial^2}{\partial y^2}+\frac{\partial^2}{\partial z^2}\right)+V(x,y,z)\right]\psi=E_{Tot}\psi \tag{5.57}$$

where μ is the reduced mass of the electron–proton system, i.e.,

$$\mu = \frac{m_0 m_p}{m_0 + m_p} \tag{5.58}$$

Since the potential energy depends only upon the relative coordinate, we can make the separation

$$\Psi = \psi(x,y,z)U(X,Y,Z) \tag{5.59}$$

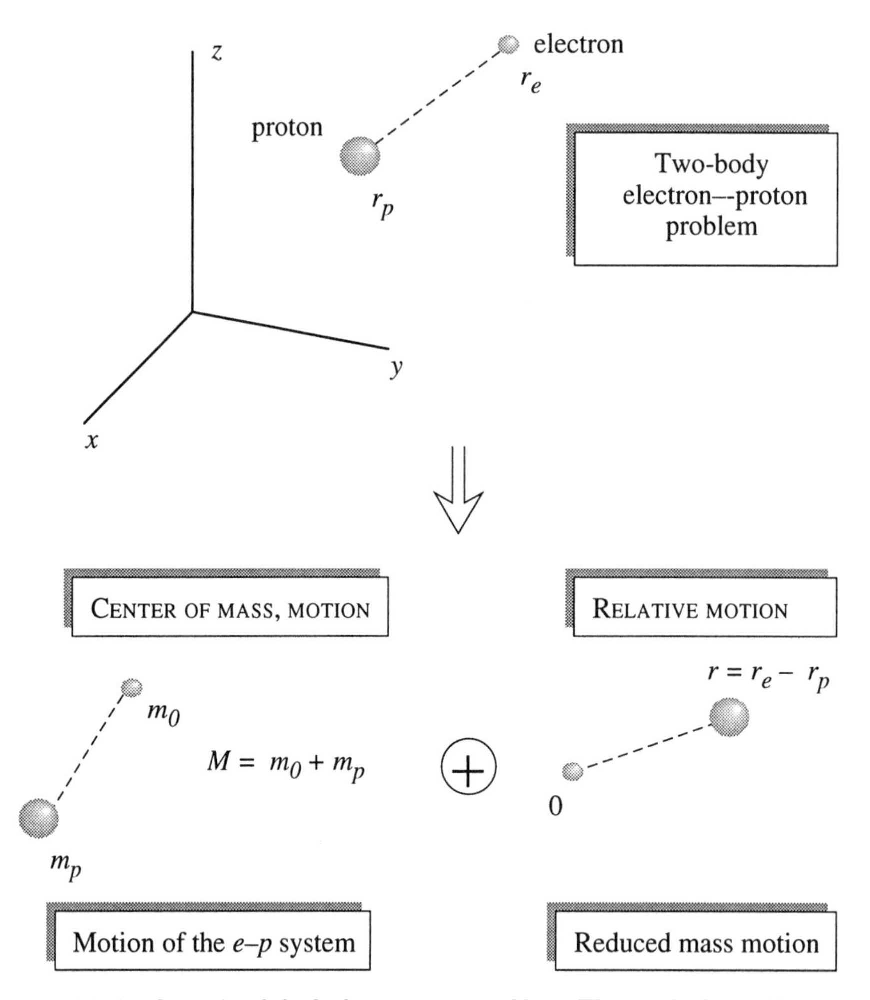

Figure 5.12: A schematic of the hydrogen atom problem. The two-body problem can be represented by the center-of-mass problem and a one-body problem.

and get the two equations

$$\left[\frac{-\hbar^2}{2\mu}\nabla^2 + V(r)\right]\psi(r) = E\psi(r) \tag{5.60}$$

$$\frac{-\hbar^2}{2M}\nabla^2 U = E'U \tag{5.61}$$

and the total energy is

$$E_{Tot} = E + E' \tag{5.62}$$

The solution to Eqn. 5.61 is straightforward and simply represents the "free" motion of the atom

$$E' = \frac{\hbar^2 K^2}{2M} \tag{5.63}$$

We are not interested in the motion of the atom and will assume that the atom is at rest.

We will now discuss the one-body problem described by Eqn. 5.60, with

$$V(r) = -\frac{e^2}{4\pi\epsilon_0 r} \tag{5.64}$$

The time-independent wave equation can be written in the spherical coordinate system as

$$\frac{-\hbar^2}{2\mu}\left[\frac{1}{r^2}\frac{\partial}{\partial r}\left(r^2\frac{\partial}{\partial r}\right) + \frac{1}{r^2\sin\theta}\frac{\partial}{\partial\theta}\left(\sin\theta\frac{\partial}{\partial\theta}\right) + \frac{1}{r^2\sin^2\theta}\frac{\partial^2}{\partial\phi^2}\right]\psi + V(r)\psi = E\psi \tag{5.65}$$

where ψ is the wavefunction and E is the energy of the system. Let us now separate the radial and angular parts of the solution by the substitution

$$\psi(r,\theta,\phi) = R(r)F(\theta)G(\phi)$$

When this problem is solved, the eigenfunctions and eigenvalues that result are described by three quantum numbers (the quantum numbers are 3 because this is a three-dimensional problem). The eigenfunction is given by

$$\psi_{n\ell m}(r,\theta,\phi) = R_{n\ell}(r)F_{\ell m}(\theta)G_m(\phi) \tag{5.66}$$

The symbols n, ℓ, m are the three quantum numbers describing the solution. The three quantum numbers have the following allowed values:

$$\begin{aligned} \text{principle number}, n &: \text{ Takes values } 1, 2, 3, \ldots \\ \text{angular momentum number}, \ell &: \text{ Takes values } 0, 1, 2, \ldots n-1 \\ \text{magnetic number}, m &: \text{ Takes values } -\ell, -\ell+1, \ldots \ell \end{aligned} \tag{5.67}$$

The principle quantum number specifies the energy of the allowed electronic levels. The energy eigenvalues are given by

$$\boxed{E_n = -\frac{\mu e^4}{2\left(4\pi\epsilon_0\right)^2\hbar^2 n^2}} \tag{5.68}$$

Note that the energy levels obtained here have the same values as those obtained by applying Bohr's quantization rules in Chapter 2, Section 2.2.4. The spectrum is shown schematically in Fig. 5.13. Due to the much larger mass of the nucleus as compared to the mass of the electron, the reduced mass μ is essentially the same as the electron mass m_0. The ground state of the hydrogen atom is given by

$$\psi_{100} = \frac{1}{\sqrt{\pi a_0^3}}e^{-r/a_0} \tag{5.69}$$

The parameter a_0 appearing in the functions is called *Bohr radius* and is given by

$$a_0 = \frac{4\pi\epsilon_0\hbar^2}{m_0 e^2} = 0.53 \text{ Å} \tag{5.70}$$

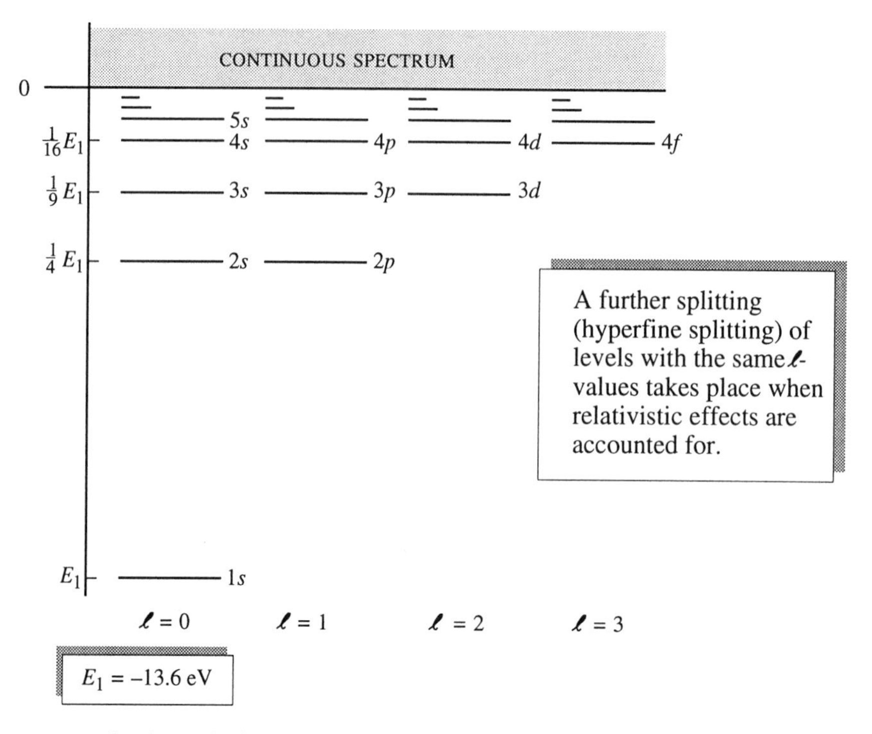

Figure 5.13: A schematic description of the energy levels of electrons in a hydrogen atom.

n	ℓ	m	$R(r)$	$F(\theta)$	$G(\phi)$
1	0	0	$\frac{2}{a_0^{3/2}} e^{-r/a_0}$	$\frac{1}{\sqrt{2}}$	$\frac{1}{\sqrt{2\pi}}$
2	0	0	$\frac{1}{(2a_0)^{3/2}}\left(2-\frac{r}{a_0}\right)e^{-r/2a_0}$	$\frac{1}{\sqrt{2}}$	$\frac{1}{\sqrt{2\pi}}$
2	1	0	$\frac{1}{\sqrt{3}(2a_0)^{3/2}}\frac{r}{a_0}e^{-r/2a_0}$	$\sqrt{\frac{3}{2}}\cos\theta$	$\frac{1}{\sqrt{2\pi}}$
2	1	± 1	$\frac{1}{\sqrt{3}(2a_0)^{3/2}}\frac{r}{a_0}e^{-r/2a_0}$	$\frac{\sqrt{3}}{2}\sin\theta$	$\frac{1}{\sqrt{2\pi}}e^{\pm i\phi}$

Table 5.1: Some low-lying hydrogen atom wavefunctions.

It roughly represents the spread of the ground state. Table 5.1 gives the functional forms of some of the low-lying wavefunctions.

Before ending the discussion on the hydrogen atom problem, it is useful to point out that states with $\ell = 0, 1, 2, 3, \ldots$ are called $s, p, d, f, \ldots$ in atomic physics notation. Such a notation is used for not only the hydrogen atom case, but also for all atomic spectra. The form of these functions, as well as the electron probability function, is shown in Fig. 5.14 for some low-lying states.

5.4.1 Application Example: Doping of Semiconductors

In Chapter 4 we discussed how allowed bands are filled with electrons in metals, semiconductors, and insulators. When a band is completely filled with electrons, the electrons cannot carry any current since to carry current the electrons must move from an occupied state to an unoccupied state. Even at finite temperatures, in a semiconductor, the electron density in the upper conduction band (the band that is completely empty at 0 K) and in the valence band (the band that is totally filled at 0 K) is quite small. To alter the conductivity of the semiconductor, dopants are introduced.

When a dopant (impurity) atom is introduced in a crystal, the perfect periodicity of the crystal is destroyed. At a particular atomic site the background potential of the host lattice is replaced by the potential of the impurity. If the background potential changes due to the impurity is weak and long-ranged (i.e., extends over several tens of atomic sites), a tremendous simplification can be introduced. In this simplification the properties of the dopant atom can be described by a simple hydrogen-like model where the *electron mass is simply the effective mass at the bandedge.* This approximation is called the effective mass approximation for impurities.

There are two kinds of dopants—donors which can donate an electron to the conduction band and acceptors which can accept an electron from the valence band and thus create a hole. To solve the donor (or acceptor) problem, we consider a donor atom on a crystal lattice site. The donor atom could be a pentavalent atom in Si or a Si atom on a Ga site in GaAs. Focusing on the pentavalent atom in Si, four of the valence electrons of the donor atom behave as they would in a Si atom; the remaining fifth electron now sees a positively charged ion to which it is attracted, as shown in Fig. 5.15. The ion has a charge of unity and the attraction is simply Coulombic suppressed by the dielectric constant of the material. The problem is now that of the hydrogen atom case, except that the electron mass is the effective mass at the bandedge. The attractive potential is

$$U(r) = \frac{-e^2}{4\pi\epsilon r} \tag{5.71}$$

where ϵ is the dielectric constant of the semiconductor, i.e., the product of ϵ_0 and the relative dielectric constant. We get the effective mass equation for the donor level which has an energy for E_d of

$$\left[\frac{-\hbar^2}{2m_e^*}\nabla^2 - \frac{e^2}{4\pi\epsilon r}\right] F_c(r) = (E_d - E_c)F_c(r) \tag{5.72}$$

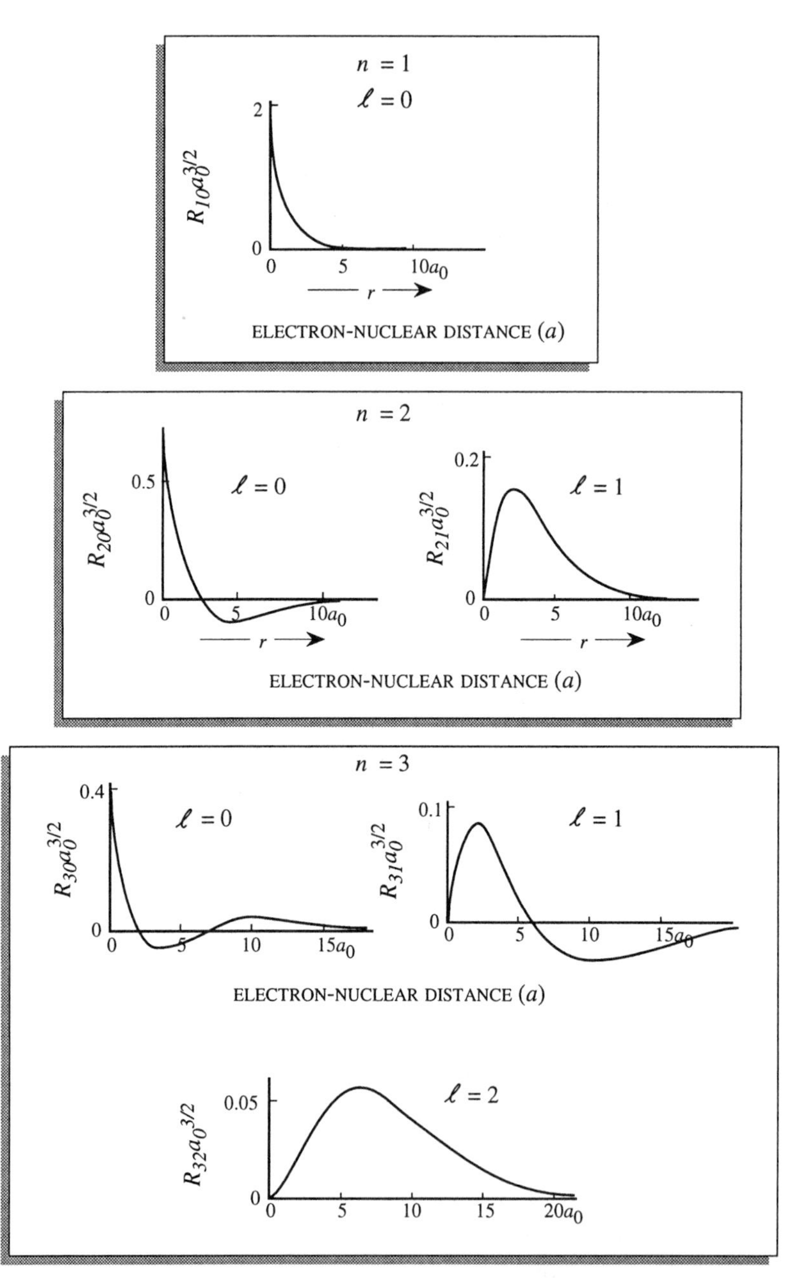

Figure 5.14: A plot of the probability density function as a function of the angle θ for the s, p, d electrons.

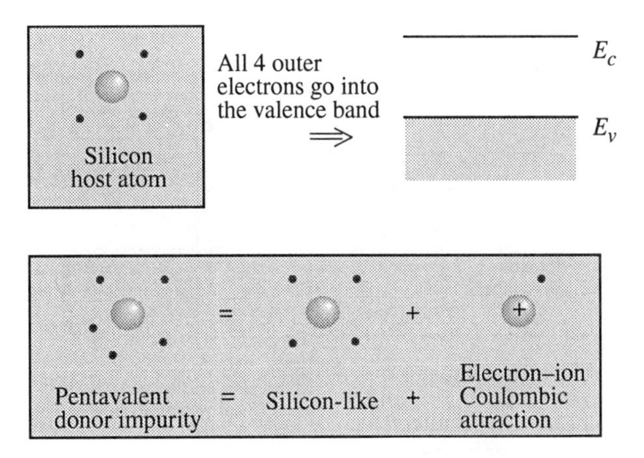

Figure 5.15: A schematic showing the approach one takes to understand donors in semiconductors. The donor problem is treated as the host atom problem together with a Coulombic interaction term. The silicon atom has four "free" electrons per atom. All four electrons occupy the valence band at 0 K. The dopant has five electrons out of which four are contributed to the valence band, while the fifth one can be used for increasing electrons in the conduction band.

where m_e^* is the conduction bandedge mass and $E_d - E_c$ is the impurity energy with respect to the conduction bandedge E_c levels.

This equation is now essentially the same as that of an electron in the hydrogen atom problem. The only difference is that the electron mass is m^* and the Coulombic potential is reduced by ϵ_0/ϵ.

The energy solutions for this problem are

$$E_d = E_c - \frac{e^4 m_e^*}{2(4\pi\epsilon)^2\hbar^2}\frac{1}{n^2}, \quad n = 1, 2, ... \tag{5.73}$$

A series of energy levels are produced, with the ground state energy level being at

$$\begin{aligned} E_d &= E_c - \frac{e^4 m_e^*}{2(4\pi\epsilon)^2\hbar^2} \\ &= E_c - 13.6\left(\frac{m^*}{m_o}\right)\left(\frac{\epsilon_o}{\epsilon}\right)^2 \text{ eV} \end{aligned} \tag{5.74}$$

Note that in the hydrogen atom problem the electron levels are measured from the vacuum energy level which is taken as $E = 0$. *In the donor problem, the energy level is measured from the bandedge.* Fig. 5.16 shows the energy level associated with a donor impurity.

The wavefunction of the ground state is as in the hydrogen atom problem:

$$F_c(r) = \frac{1}{\sqrt{\pi a^3}} e^{-r/a} \tag{5.75}$$

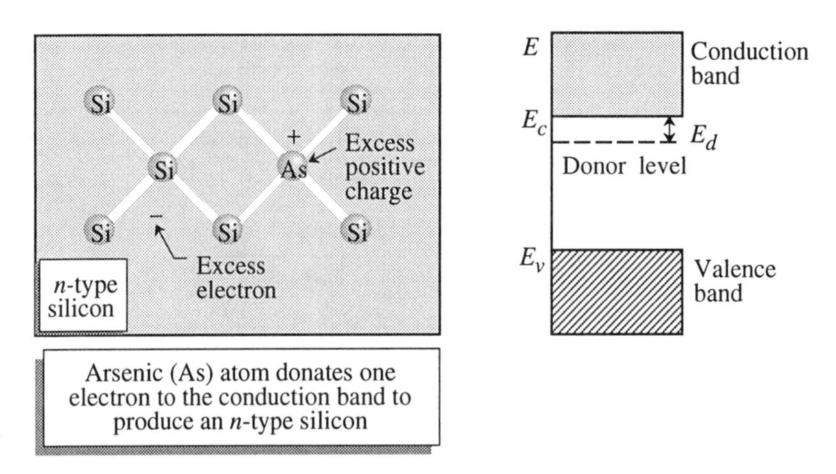

Figure 5.16: A schematic of doping of Si with arsenic. A donor level is produced below the conduction bandedge.

where a is the donor Bohr radius and is given by

$$a = \frac{(4\pi\epsilon)\hbar^2}{m_e^* e^2} = 0.53 \left(\frac{\epsilon/\epsilon_0}{m_e^*/m_0} \right) \text{Å} \tag{5.76}$$

For most semiconductors the donor energies are a few meVs below the conduction bandedge and the Bohr radius is $\sim$100 Å.

Another important class of intentional impurities is the acceptors. Just as donors are defect levels which are neutral when an electron occupies the defect level and positively charged when unoccupied, the acceptors are neutral when empty and negatively charged when occupied by an electron. The acceptor levels are produced when impurities which have a similar core potential as the atoms in the host lattice, but have one less electron in the outermost shell, are introduced in the crystal. Thus group III elements can form acceptors in Si or Ge, while Si could be an acceptor if it replaces As in GaAs.

The acceptor impurity potential could now be considered to be equivalent to a host atom potential together with the Coulombic potential of a negatively charged particle. The "hole" (i.e., the absence of an electron in the valence band) can then bind to the acceptor potential. The effective mass equation can again be used since only the top of the valence band contributes to the acceptor level. The valence band problem is considerably more complex and requires the solution of multiband effective mass theory. However, the acceptor level can be reasonably predicted by using the heavy hole mass. Due to the heavier hole masses, the Bohr radius for the acceptor levels is usually a factor of 2 to 3 smaller than that for donors.

Population of Dopant Levels

We have discussed above how the presence of a dopant impurity creates a bound level E_d (or E_a) near the conduction (or valence) bandedge. If the extra electron associated with the donor occupies the donor level, it does not contribute to the mobile carrier density. The purpose of doping is to create a mobile electron or hole. When the electron associated with a donor (or a hole associated with an acceptor) is in the conduction (or valence) band, the dopant is said to be ionized. To calculate the ionization probability, we refer back to our discussion of Chapter 3, Section 3.6. There we have calculated the occupation probability of discrete levels.

Consider a semiconductor containing both donors and acceptors. At finite temperatures, the electrons will be redistributed, but their numbers will be conserved and will satisfy the following equality resulting from charge neutrality:

$$(n - n_i) + n_d = N_d \tag{5.77}$$

$$(p - p_i) + p_a = N_a \tag{5.78}$$

which gives

$$n + n_d = N_d - N_a + p + p_a \tag{5.79}$$

where

n = total free electrons in the conduction band

n_d = electrons bound to the donors

p = total free holes in the valence band

p_a = holes bound to the acceptors

The number density of electrons attached to the donors has been derived in Section 3.6 and is given by

$$\frac{n_d}{N_d} = \frac{1}{\frac{1}{2}\exp\left(\frac{E_d - E_F}{k_B T}\right) + 1} \tag{5.80}$$

The factor $\frac{1}{2}$ essentially arises from the fact that there are two states an electron can occupy at a donor site corresponding to the two spin states.

The probability of a hole being trapped to an acceptor level is given by

$$\frac{p_a}{N_a} = \frac{1}{\frac{1}{4}\exp\left(\frac{E_F - E_a}{k_B T}\right) + 1} \tag{5.81}$$

The factor of $\frac{1}{4}$ comes about because of the presence of the two bands, light hole, heavy hole, and the two spin-states.

To find the fraction of donors or acceptors that are ionized, one has to use a computer program in which the position of the Fermi level is adjusted so that Eqn. 5.79 is satisfied. Once E_F is known, one can calculate the electron or hole densities in the conduction and valence bands. For doped systems, it is useful to use the Joyce-Dixon approximation which gives the relation between the Fermi level and the free carrier concentration. This approximation is more accurate than the Boltzmann approximation. According to the Joyce-Dixon approximation, we have

$$E_F = E_c + k_B T\left[\ell n\,\frac{n}{N_c} + \frac{1}{\sqrt{8}}\,\frac{n}{N_c}\right] = E_v - k_B T\left[\ell n\,\frac{p}{N_v} + \frac{1}{\sqrt{8}}\,\frac{p}{N_v}\right] \tag{5.82}$$

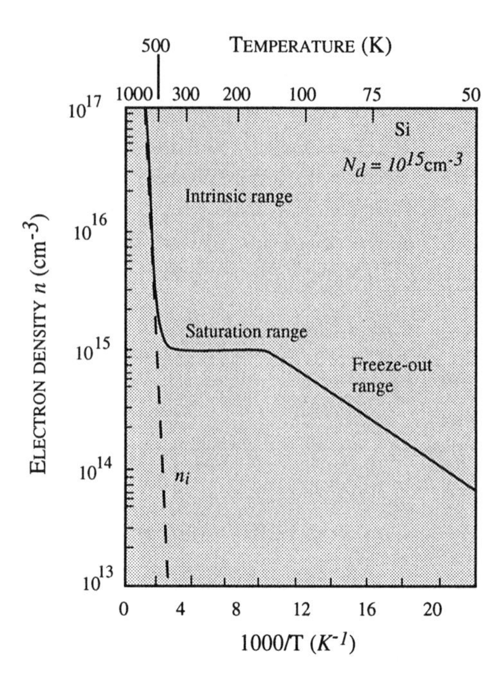

Figure 5.17: Electron density as a function of temperature for a Si sample with donor impurity concentration of 10^{15} cm^{-3}.

This relation can be used to obtain the Fermi level if n is specified. Or else, it can be used to obtain n if E_F is known by solving for n iteratively. *If the term* $(n/\sqrt{8}\ N_c)$ *is ignored, the result corresponds to the Boltzmann approximation.*

In general, if we have a doped semiconductor and we examine its mobile carrier density dependence upon temperature, there are three regimes. Let us consider an n-type semiconductor. At low temperatures, the electrons coming from the donors sit on the impurity levels E_d. Thus there is no contribution to the mobile carrier density from the dopants. This regime is called the *carrier freezeout* regime. At higher temperatures, the dopants ionize until most of them are ionized. Then over a temperature regime, the mobile carrier is essentially equal to the dopant density and independent of temperature. This is the saturation regime and semiconductor devices are operated in this regime. At very high temperatures, the intrinsic carrier density overwhelms the dopant density and the material acts as an intrinsic material. The three regimes are shown in Fig. 5.17.

5.4.2 Application Example: Excitons in Semiconductors

An important problem in semiconductors is that of an electron–hole pair. An electron and a hole can form a bound state via their Coulombic interaction. Such a system is called an exciton. In a semiconductor (or an insulator) the electronic spectra are such that the valence band is completely filled with electrons at 0 K and the conduction band is empty. The two bands are separated by the bandgap. This gives a "single-electron energy-momentum" relation of the form shown in Fig.

5.18a. In the single-electron picture, the properties of an electron are described by assuming that the valence band electrons produce a mean field.

Now consider an electron being removed from the valence band (the ground state of the problem is the filled valence band state) and excited to a higher energy state. An important question now arises: What is the lowest excited state energy of the system? If we ignore the interaction of the electron that is removed from the valence band with the hole in the valence band produced by the absence of the electrons, the lowest energy needed to excite the system is the bandgap energy E_g. However, the electron and hole interact with each other, since the hole responds as if it is a positively charged particle, as shown in Fig. 5.18b. The electron–hole system can be represented by the hydrogen-like model. The exciton problem can be written as

$$\left[-\frac{\hbar^2}{2m_e^*}\nabla_e^2 - \frac{\hbar^2}{2m_h^*}\nabla_h^2 - \frac{e^2}{4\pi\epsilon\left|\mathbf{r}_e - \mathbf{r}_h\right|}\right]\psi_{\text{ex}} = E\psi_{\text{ex}} \tag{5.83}$$

Here m_e^* and m_h^* are the electron and hole effective masses and $|\mathbf{r}_e - \mathbf{r}_h|$ is the difference in coordinates defining the Coulombic interaction between the electron and the hole. The problem is now the standard two-body problem like the electron–proton problem of the H atom. The relative coordinate problem has the form (m_r^* is the reduced mass of the electron–hole pair)

$$\left(\frac{\hbar^2 k^2}{2m_r^*} - \frac{e^2}{4\pi\epsilon\left|\boldsymbol{r}\right|}\right)F(\mathbf{r}) = EF(\mathbf{r}) \tag{5.84}$$

This is the usual hydrogen atom problem and $F(\mathbf{r})$ can be obtained from the mathematics of that problem. The general exciton solution is now (writing $\mathbf{K}_{\text{ex}} = \mathbf{K}$)

$$\psi_{n\mathbf{K}_{\text{ex}}} = e^{i\mathbf{K}_{\text{ex}}\cdot\mathbf{R}}F_n(\mathbf{r})\phi_c(\mathbf{r})\phi_v(\mathbf{r}_h) \tag{5.85}$$

where ϕ_c and ϕ_v represent the electron and hole bandedge states. The excitonic energy levels are then

$$E_{n\mathbf{K}_{\text{ex}}} = E_n + \frac{\hbar^2}{2(m_e^* + m_h^*)}K_{\text{ex}}^2 \tag{5.86}$$

with E_n being the eigenvalues of the hydrogen atom-like problem (the energy is referenced to the bandgap energy E_g)

$$E_n = -\frac{m_r^* e^4}{2(4\pi\epsilon)^2\hbar^2}\frac{1}{n^2} \tag{5.87}$$

and the second term in Eqn. 5.86 represents the kinetic energy of the center of mass of the electron–hole pair.

The exciton energy is thus slightly lower than the bandgap energy of the semiconductor. Excitonic states can be observed in optical absorption spectra and are exploited for many optoelectronic devices.

EXAMPLE 5.8 Calculate the probabilty that an electron in the ground state of hydrogen atoms is within one Bohr radius from the origin. The probability of distribution is

$$P(r)dr = r^2\left|R_{1,0}(r)\right|^2 dr = r^2\frac{4}{a_0^3}e^{-2r/a_0}dr$$

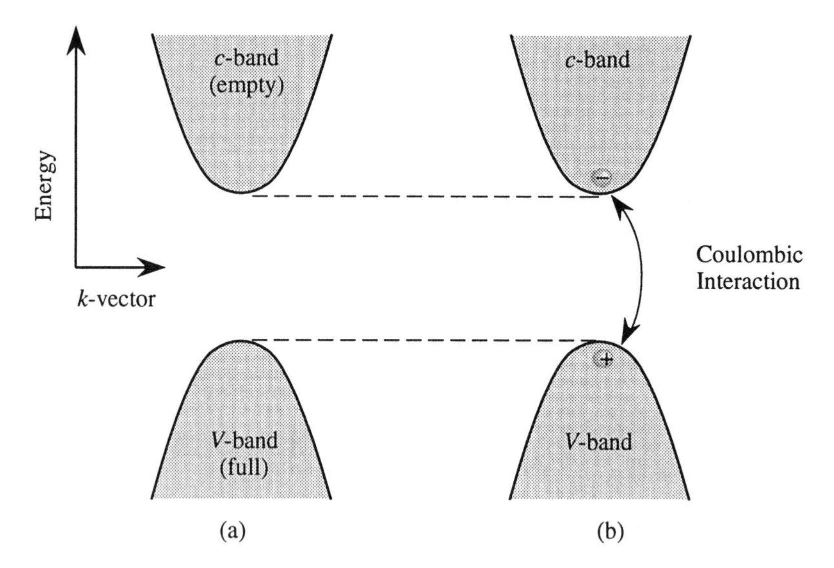

Figure 5.18: (a) The bandstructure in the independent electron picture; and (b) the Coulombic interaction between the electron and hole, which would modify the band picture.

The total probability of finding the electron between $r = 0$ and $r = a_0$ is

$$P = \int_0^{a_0} P(r)dr = \frac{4}{a_0^3} \int_0^{a_0} r^2 e^{2r/a_0} dr$$

Letting $x = 2r/a_0$, we rewrite this as

$$P = \frac{1}{2} \int_0^2 x^2 e^{-x} dx$$

Evaluating the integral gives

$$P = 0.32$$

That is, 32 percent of the time the electron is closer than a Bohr radius to the nucleus.

EXAMPLE 5.9 Show that the most probable distance from the origin of an electron in the $n = 2$, $\ell = 1$ state is $4a_0$.

In the $n = s$, $\ell = 1$ level, the radial probability density is

$$\begin{aligned} P(r) &= r^2 \left|R_{2,1}(r)\right|^2 \\ &= r^2 \frac{1}{24a_0^3} \frac{r^2}{a_0^2} e^{-r/a_0} \end{aligned}$$

We wish to find where this function has its maximum. We take the first derivative of $P(r)$ and set it equal to zero:

$$\frac{dP(r)}{dr} = \frac{1}{24a_0^5} \frac{d}{dr} \left(r^4 e^{-r/a_0}\right) = 0$$

The solution that yields a maximum is $r = 4a_0$.

EXAMPLE 5.10 Calculate the donor and acceptor level energies in GaAs and Si. The shallow level energies are in general given by

$$E_d = E_c - 13.6 \text{ eV} \times \frac{m^*/m_0}{(\epsilon/\epsilon_0)^2}$$

The conduction band effective mass in GaAs is $0.067m_0$ and $\epsilon = 13.2\epsilon_0$, and we get for the donor level

$$E_d(\text{GaAs}) = E_c - 5.5 \text{ meV}$$

The problem in silicon is a bit more complicated since the effective mass is not so simple. For donors we need to use the conductivity mass which is given by

$$m_\sigma^* = \frac{3m_0}{\left(\frac{1}{m_\ell^*} + \frac{2}{m_t^*}\right)}$$

where m_ℓ^* and m_t^* are the longitudinal and transverse masses for the silicon conduction band. Using $m_\ell^* = 0.98$ and $m_t^* = 0.2m_0$, we get

$$m_\sigma^* = 0.26m_0$$

Using $\epsilon = 11.9\epsilon_0$, we get

$$E_d(\text{Si}) = 25 \text{ meV}$$

The acceptor problem is much more complicated due to the degeneracy of the heavy hole and light hole bands. However, a reasonable approximation is obtained by using the heavy hole mass ($\sim 0.45m_0$ for GaAs, $\sim 0.5m_0$ for Si):

$$\begin{aligned} E_a(\text{Si}) &= 48 \text{ meV} \\ E_a(\text{GaAs}) &\cong 36 \text{ meV} \end{aligned}$$

This is the energy above the valence bandedge. However, it must be noted that the use of the heavy hole mass is not strictly valid.

5.5 FROM THE HYDROGEN ATOM TO THE PERIODIC TABLE: A QUALITATIVE REVIEW

Decades before the advent of quantum mechanics, chemists have known that different elements have different chemical reactivities. An important success of quantum mechanics has been to explain the chemical reactivities of elements in the periodic chart. The explanation lies in understanding the atomic spectra and the use of Pauli exclusion principle. In this section, we will discuss a qualitative summary of the interpretation of the periodic table and the chemical properties of elements.

To make the many electron atom problem tractable, formalisms based on the "mean field" approach have been developed. In this approach, a given electron sees the potential energy due to the nucleus and the mean spherically symmetrical field due to the spatial distribution of the remaining electrons. For the purpose of computing the electron–electron interaction, the charge density of the electrons is taken to be equal to the electron charge e times the electron density $P(r, \theta, \phi)$

averaged over a sphere with radius r. To appreciate the nature of the mean potential energy $V_i(r)$, we note the following:

(i) For the innermost electrons, $V_i(r)$ is approximately given by the potential of the nucleus together with a constant term, since the electron is inside an electron cloud created by the outer electrons.

(ii) For the outer electrons, *the potential energy $V_i(r)$ is approximately that of a nucleus with a charge equal to one.*

Using this conceptual picture and the general behavior of the hydrogen-like functions, we are now ready to discuss the periodic table. The chemical properties of elements are controlled by the ground state of the atom. We will discuss the ground state of various elements using the Pauli exclusion principle; i.e., each allowed state can only have two electrons (one with spin up and the other with spin down). *If the ground state is such that the outermost occupied level cannot accept any more electrons (say, from another atom), then the atom is chemically inactive. On the other hand, if the ground state outer electronic state is unfilled, the atom is chemically active.*

The L-Shell: $n = 1$

Hydrogen (H: $Z = 1$). The ground state of the H-atom is the $n = 1, \ell = m = 0$ state, or the $1s$-state. The ionization energy of the electron in this state is 13.6 eV. Helium (He: $Z = 2$). In this case, the two electrons can be in the lowest $1s$-state. If He had only one electron, the lowest energy state would be four times that of the H atom, i.e., -54.4 eV. However, because of the electron–electron interaction, the energy is -24.6 eV. Since the $1s$-state cannot take any more electrons, the He atom is chemically inert.

The M-Shell: $n = 2$

Lithium (Li: $Z = 3$). Two of the three electrons will go into the $n = 1, \ell = 0$ state. The third atom will go into the $n = 2$ state. However, it could go into the $2s$ ($\ell = 0$) or $2p$ ($\ell = 1$) state. These states are degenerate for the H atom. But they are not degenerate for the Li atom due to the presence of the inner two electrons.

The $2s$-state has a high probability of being near the origin where the potential is essentially that of a nucleus with three protons. The $2p$-state has high probability far from the origin where the potential is that of an ion with a single charge. As a result, the $2s$-state has a lower energy (the electron has a stronger binding energy) than the $2p$-state. This argument can be used to predict the ordering of other higher state levels as well.

As a result of the spatial form of the s-state, the third electron in Li goes in the $2s$-state in its ground state. The ionization energy of this electron is only 5.4 eV and the atom is chemically quite active, since the $2s$-state is only half filled.

As we go from Li toward Ne, we gradually fill up the various states in the L-shell. As Ne is reached, the L-shell is completely filled. As a result, Ne is an inert gas.

Electron configurations of some elements							
Z	Element	W_1 (eV)	Electron State Occupation: K-Shell	L-Shell	M-Shell		N-Shell
11	Na *sodium*	5.1			1		
12	Mg *magnesium*	7.6			2		
13	Al *aluminum*	6.0			2 1		
14	Si *silicon*	8.1	—FILLED—		2 2		
15	P *phosphorus*	10.5			2 3		
16	S *sulfur*	10.4	(2)	(8)	2 4		
17	Cl *chlorine*	13.0			2 5		
18	A *argon*	15.8			2 6		
19	K *potassium*	4.3					1
20	Ca *calcium*	6.1					2
21	Sc *scandium*	6.5				1	2
22	Ti *titanium*	6.8				2	2
23	V *vanadium*	6.7		FILLED		3	2
24	Cr *chromium*	6.8				5	1
25	Mn *manganese*	7.4	(2)	(8)	(8)	5	2
26	Fe *iron*	7.9				6	2
27	Co *cobalt*	7.9				7	2
28	Ni *nickel*	7.6				8	2
29	Cu *copper*	7.7				10	1
30	Zn *zinc*	9.4				10	2

$3d$ — $4s$

$3d^5$ — $4s^1$

$3d^{10}$ — $4s^1$

Figure 5.19: The electronic configuration of some elements of the periodic table. Notice the ordering of the $3d$- and $4s$-states based on occupation.

M-Shell and Higher Shells: ($n = 3$ and higher)
The M-shell contains subshells $3s, 3p$, and $3d$, as shown in Fig. 5.19. However, an interesting effect occurs if we consider the ordering of the $4s$-state and the $3d$-state. As noted, since the $3d$-states are spread away from the nucleus, it is possible that an s state with a *higher* principle quantum number may have a *lower* energy. This does occur when we compare the energies of $3d$ and $4s$. *The energy of the* $4s$*-state is lower than that of the* $3d$*-state.* Thus, when we consider potassium (K) with $Z = 19$, the order of the electron states occupied is $1s^2, 2s^2, 2p^6, 3s^2, 3p^6, 4s^1$. As we move above K and the $4s$-states get filled, the $3d$-states start filling.

As we start filling the $3d$-state with electrons, the repulsion of these electrons raises the energy of the $4s$-state. Thus for Cr, instead of the configuration $3d^4 4s^2$, we get $3d^5 4s^1$. When another electron is added to form Mg, the $4s$-state fills again.

As we continue filling up the electronic states, we finally fill the $4s$-and $4p$-states for krypton (Kr). Krypton has a very weak chemical activity because the p-shell is filled.

Based on the qualitative discussion provided here, we can interpret some of the important properties of elements in the periodic chart:
(i) Elements with similar electron occupation of the outermost subshells of the atom have quite similar chemical properties.
(ii) Except for the ($1s^2$), the noble gases are produced for cases where the outer p-shell is completely filled. When the p-shell is filled, there is a significant energy gap between the next d (or s)-state and, as a result, these elements have little incentive to create any transformation in their electronic configuration.

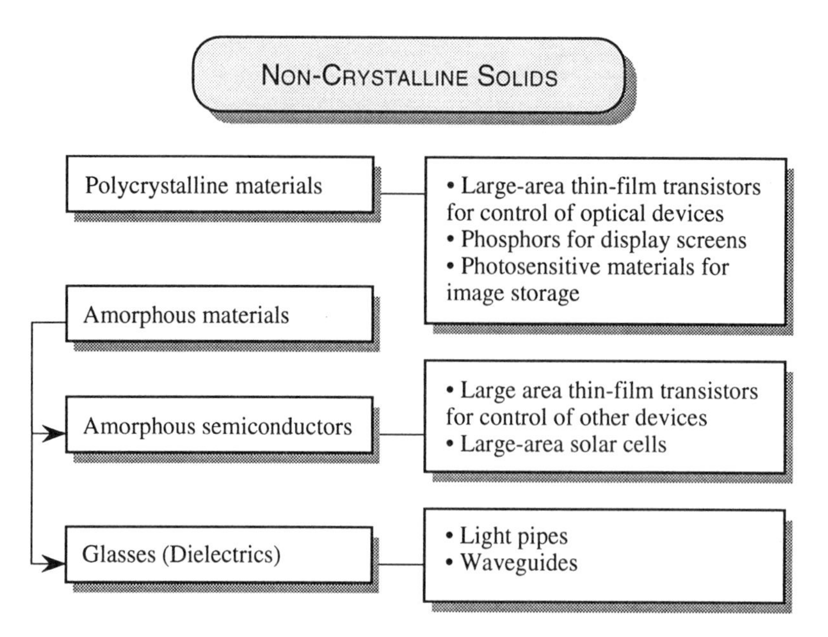

Figure 5.20: Categories of non-crystalline materials and some of their applications.

(iii) The elements with the strongest chemical activity are those which have one electron less than or one electron more than the noble gases. This is because the energy of a completely filled subshell is much lower than if it is lacking one electron. Also, if there is one excess electron, it is bound very weakly to the nucleus. Thus the atom tends to either accept one electron from another atom or donate its electron.

5.6 NON-PERIODIC MATERIALS: STRUCTURAL ISSUES

In Chapter 4 we have discussed the electronic properties of crystalline materials where atoms or molecules are arranged in a perfectly periodic arrangement. We have seen that an outcome of the periodicity is that the electronic spectrum is made of a series of allowed energy bands separated by forbidden gaps. Most of the current information processing technologies are dependent upon crystalline materials. Devices such as computer memories, microprocessors, optical communication lasers and detectors, etc., are all based on crystalline or *almost crystalline* materials. However, it is expensive to make high-quality crystals. Also, it is difficult to grow crystals with a large cross-sectional area. For many applications, low-cost and large-area issues are important. Non-crystalline materials are important for such applications which include large-area display technologies, solar energy devices, "intelligent" window glasses, etc.

Non-crystalline materials can be characterized into two categories, as shown in Fig. 5.20. These are *i)* polycrystalline materials and *ii)* amorphous materials. The important amorphous materials include amorphous semiconductors and glasses.

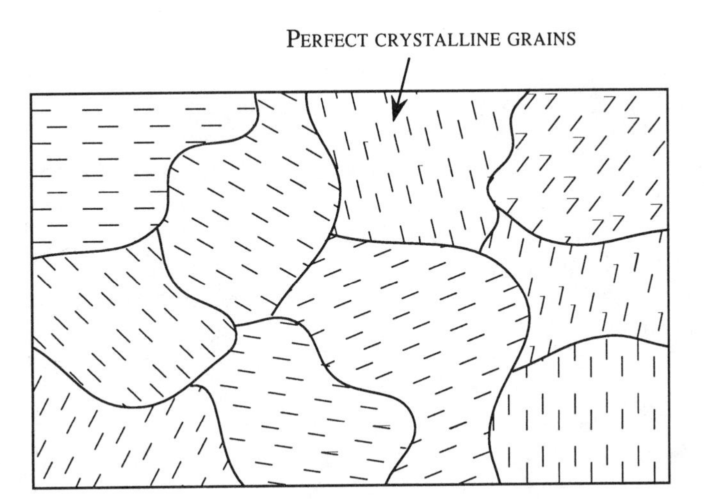

Figure 5.21: A schematic description of a polycrystalline material. Atoms are arranged periodically in a grain, but there is no order between the various grains. The grain boundaries represent regions where defects produced by broken or unfilled bonds are present.

Polycrystalline materials are widely used in electronic and optoelectronic technologies. Polycrystalline structures are produced when a material is deposited on a substrate which does not have a similar crystal structure. For example, if a metal film is deposited on a semiconductor, the film grows in a polycrystalline form. Also, if silicon is deposited on a glass substrate, it grows in a polycrystalline form. It is essential to mention $PbZrO_3$–$PbTiO_3$–La_2O_3 (PLZT), an extremely important polycrystalline material that is finding extensive use in optoelectronics. This material is a ceramic oxide with ferroelectric properties. While single crystal electro-optic materials like potassium dihydrogen phosphate (KDP), $BaTiO_3$, and $Gd(M_0O_4)_3$ are important materials, their applications are limited by cost, size, and susceptibility to moisture (especially for KDP). In contrast, polycrystalline ceramics are not subject to these limitations. The fabrication technology of PLZT is now highly developed and this ceramic is used for a variety of electro-optic devices.

Polycrystalline films are described by their average grain size as shown in Fig. 5.21. Within a grain the atoms are arranged as in crystal, i.e., with perfect order. However, each grain is surrounded by a grain boundary which is a region with a high density of defects. The defects arise due to broken or unfulfilled bonds between atoms. Different grains in the polycrystal have essentially no order between their constituent atoms. Depending upon the growth process and the differences between the substrate and the deposited film, the grain size of a polycrystal can range from 0.1 μm to 10 μm or more.

In amorphous materials (sometimes also called glasses) the order among atoms is even lower than that in polycrystalline materials. The most important amorphous materials are glasses based on SiO_x (with different dopants) and amorphous semiconductors such as amorphous silicon (a-Si).

The amorphous materials are characterized by good short-range order, but poor long-range order. Thus the nearest-neighbor and even second-neighbor coordination is quite good in amorphous materials. However, the arrangement of atoms which are third nearest neighbors (or further away) is random. The amorphous material may also have a high density of broken bonds.

The most important amorphous material in optics is, of course, "glass," which is used in all kinds of optical elements, such as lenses, prisms, etc., as well as in optical fibers. Glass is made from some of the most abundant elements on the earth's crust, such as, oxygen (which forms 62% of earth's crust) and silicon (which forms 21% of the crust). In silica-based glass, silicon and oxygen atoms form a lattice network which is not crystalline, but has a good nearest-neighbor ordering.

Amorphous Si (a-Si) is the most important amorphous semiconductor material for electronic applications due to its importance in solar cell technology and display technology. We note that, as in crystalline silicon, in a-Si, the Si atoms are four-fold coordinated, i.e., they have four nearest neighbors. However, some of the atoms have broken or dangling bonds. A high density of dangling bonds can render the material useless electronically.

5.7 NON-PERIODIC MATERIALS: ELECTRONIC STRUCTURE

In Chapter 4 we have seen that the electronic spectrum of periodic structures consists of well-defined bandgaps and allowed energy bands. To understand the effect of the disorder in non-periodic materials we can invoke our understanding of the problem of a particle in a quantum well. In Fig. 5.22 we show a schematic of the structural arrangement of atoms in a crystalline and polycrystalline material. We then show a schematic of an allowed bandedge in space. In these perfect crystals, the bandedge has no spatial variation as discussed in the previous chapter. In a polycrystalline material, the bandedge is uniform until one reaches a grain boundary. The grain boundary creates a sharp discontinuity in the potential energy profile. This may be viewed as a deep potential well in which electronic states are confined in the well. Such energy states and energy levels are called *trap states* and signify that the electron is physically trapped near the defective region.

In a crystalline material, the electron states are spread over the entire crystal and are "free" states, as shown in Fig. 5.22a. In a polycrystalline material, many of the electron states are localized in space near the grain boundaries. This difference has a profound impact on transport of electrons in these materials.

The amorphous material is represented by a randomly varying potential profile. As we have seen from the quantum well problem, some of these random fluctuations can have sufficient depth and width to have a bound state solution. Electrons in these states will be "localized" within the potential well. As a result of the random potential fluctuations, the bandedges of amorphous materials are characterized by "bandtail" states, which penetrate into the region which would have been a bandgap in the perfect crystal. This is shown schematically in Fig. 5.23.

In an amorphous material there is an energy position which separates localized state from extended states. In the extended states, the electron is able to move across the material, while in the localized state, it is confined to a small region in

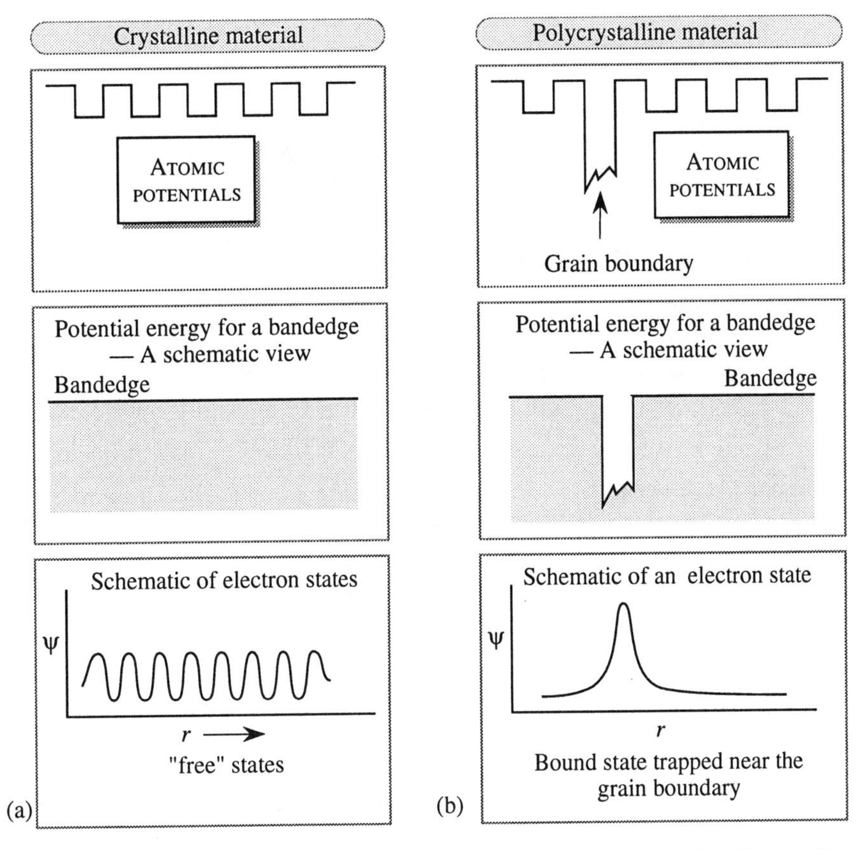

Figure 5.22: (a) A schematic of the structural and electronic properties of crystals; and (b) a schematic of the structural and electronic properites of a polycrystalline material.

space. The energy separating the two kinds of states is called a mobility edge, and is shown in Fig. 5.23.

5.7.1 Defects in Crystalline Materials and Trap States

In our discussion of crystalline materials we have explored the properties of the perfect crystalline structure. In real materials, crystalline structures invariably have some defects that are introduced due to either thermodynamic considerations (nothing in life is perfect!) or the presence of impurities during the crystal growth process. In general, defects in crystalline semiconductors can be characterized as i) point defects; ii) line defects; iii) planar defects, and iv) volume defects. The defects typically cause a potential well in which electrons can be trapped. We will give a brief overview of some important defects.

Point Defects

A point defect is a highly localized defect that affects the periodicity of the crystal only in one or a few unit cells. There are a variety of point defects, as shown in Fig. 5.24. An important point in defects is the vacancy that is produced when an atom

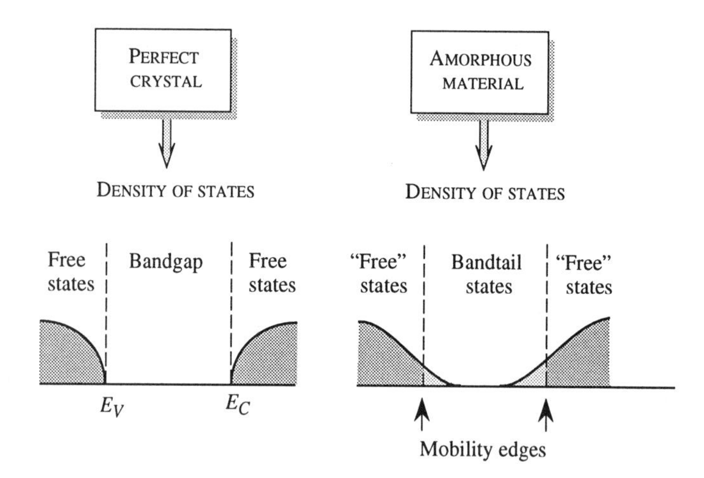

Figure 5.23: Density of states and the influence of disorder. The shaded region represents the region where the electronic states are localized in space.

is missing from a lattice point.

Other point defects are interstitials, in which an atom is sitting in a site that is in between the lattice points as shown in Fig. 5.24, and impurity atoms which involve a wrong chemical species in the lattice. In some cases the defect may involve several sites forming a defect complex.

Line Defects or Dislocations

In contrast to point defects, line defects (called dislocations) involve a large number of atomic sites that can be connected by a line. Dislocations are produced if, for example, an extra half-plane of atoms is inserted (or taken out) of the crystal . Such dislocations are called *edge dislocations*. Dislocations can also be created if there is

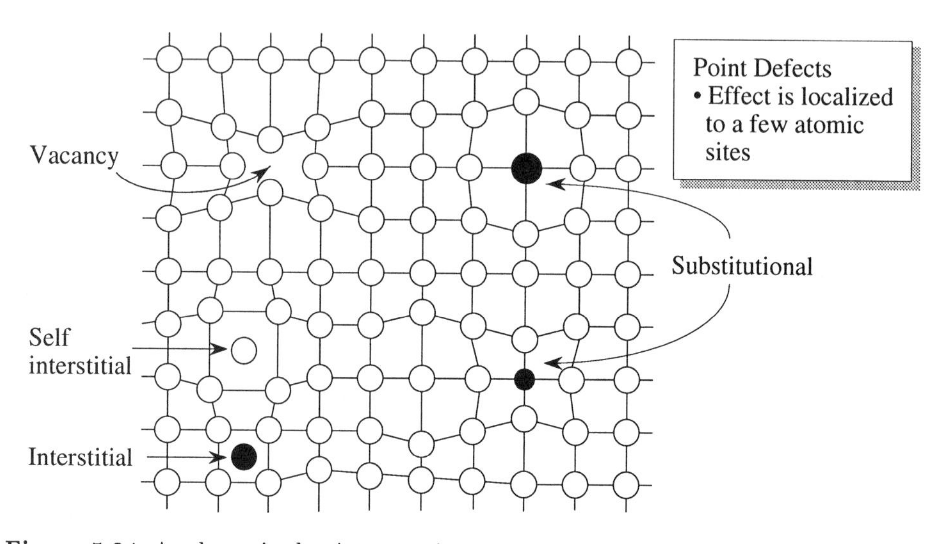

Figure 5.24: A schematic showing some important point defects in a crystal.

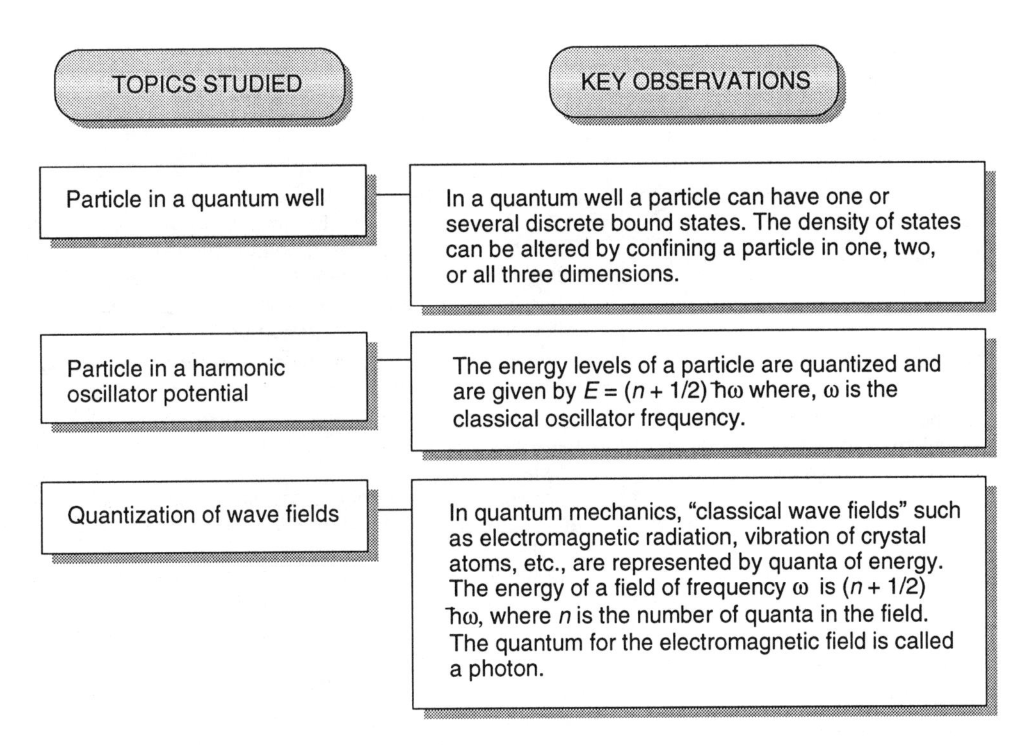

TOPICS STUDIED	KEY OBSERVATIONS
Particle in a quantum well	In a quantum well a particle can have one or several discrete bound states. The density of states can be altered by confining a particle in one, two, or all three dimensions.
Particle in a harmonic oscillator potential	The energy levels of a particle are quantized and are given by $E = (n + 1/2)\hbar\omega$ where, ω is the classical oscillator frequency.
Quantization of wave fields	In quantum mechanics, "classical wave fields" such as electromagnetic radiation, vibration of crystal atoms, etc., are represented by quanta of energy. The energy of a field of frequency ω is $(n + 1/2)$ $\hbar\omega$, where n is the number of quanta in the field. The quantum for the electromagnetic field is called a photon.

Table 5.2: Summary table.

a slip in the crystal so that part of the crystal bonds are broken and reconnected with atoms after the slip.

Dislocations can be a serious problem, especially in the growth of strained heterostructures, i.e., structures where the lattice constant between the overlayer and the substrate is different. In optoelectronic devices, dislocations can ruin the device performance and render the device useless. Thus the control of dislocations is of great importance.

5.8 CHAPTER SUMMARY

The important issues discussed in this chapter are summarized in Tables 5.2–5.4.

5.9 PROBLEMS

Problem 5.1 At room temperature the bandgap of GaAs is 1.43 eV. Assuming an infinite barrier approximation, what is the well size of a GaAs/AlAs quantum well which can produce an effective bandgap of 1.50 eV?
Problem 5.2 If a 100 Å cubic dot of GaAs is embedded in AlAs, what is the approximate bandgap of the quantum dot? Assume an infinite barrier model. Use an effective mass of 0.067 m_0 and 0.4 m_0 for the electron and heavy holes, respectively.
Problem 5.3 In the $In_{0.53}Ga_{0.47}As/InP$ system, 40% of the bandgap discontinuity

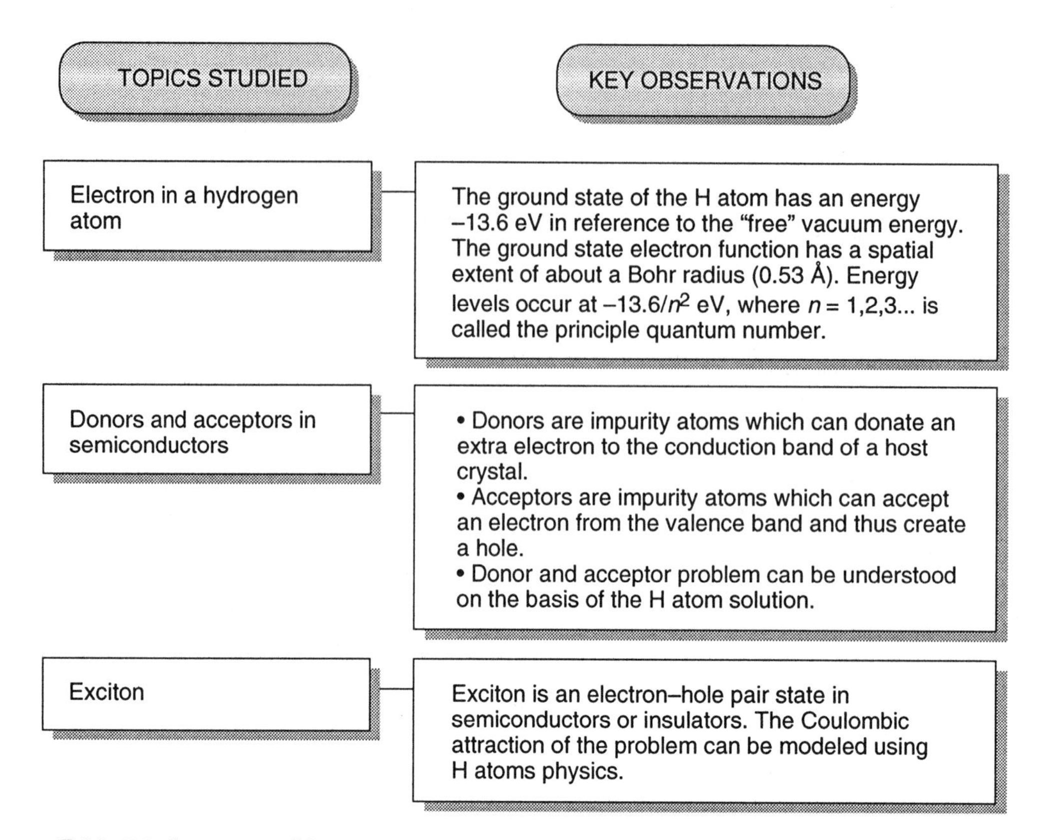

Table 5.3: Summary table.

is in the conduction band. Calculate the conduction and valence band discontinuities. Calculate the effective bandgap of a 100 Å quantum well. Use the infinite potential approximation and the finite potential approximation and compare the results.

Problem 5.4 Calculate the first and second subband energy levels for the conduction band in a GaAs/$Al_{0.3}Ga_{0.7}As$ quantum well as a function of well size. Assume that the barrier height is 0.18 eV.

Problem 5.5 Calculate the width of a GaAs/AlGaAs quantum well structure in which the effective bandgap is 1.6 eV. The effective bandgap is given by

$$E_g^{eff} = E_g(\text{GaAs}) + E_1^e + E_1^h$$

where E_g (GaAs) is the bandgap of GaAs (= 1.5 eV) and E_1^e and E_1^h are the ground state energies in the conduction and valence band quantum wells. Assume that $m_e^* = 0.067\ m_0, m_{hh}^* = 0.45\ m_0$. The barrier heights for the conduction and valence band well is 0.2 eV and 0.13 eV, respectively.

Problem 5.6 Estimate the minimum energy of a proton trapped inside a nucleus of diameter 10^{-14} m. Also find the minimum energy of an electron trapped in the same nucleus.

Problem 5.7 Consider a pendulum of length 10 cm. Regard this as a quantum

TOPICS STUDIED	KEY OBSERVATIONS
Understanding of the periodic chart and the chemical activity of atoms	• The one-electron-atom problem can be extended to study the properties of many-electron-atoms by using a mean field approach and the Pauli exclusion principle. • The chemical activity of different elements is understood on the basis of the filling of the outermost electronic levels.
Polycrystalline materials	• Perfect short- and long-range order exists over the grain size which can be several microns. • The structure loses order as one goes from one grain to another.
Amorphous or glassy materials	• Good short-range order exists, but there is no long-range order. • The materials also have broken or dangling bonds and impurities.
Defects in semiconductors	Defects such as vacancies, impurity atoms, etc., produce levels in the bandgap of the semiconductor. The electron levels are spatially localized, unlike the plane-wave functions describing the electrons in the conduction and valence bands.
Bandstructure of amorphous semiconductors	Amorphous semiconductors have no well-defined bandgap. However, there is an electrical bandgap defined by the conduction band mobility edge and the valence band mobility edge. The mobility edge separates electron states that are spatially localized and those that are extended throughout the material.

Table 5.4: Summary table.

oscillator and calculate the energy difference between successive levels. Is it possible to observe these quantized levels?

Problem 5.8 Consider a carbon atom of atomic mass 12 vibrating in a material. Assume that the oscillator can be described by a force constant of 0.8 Nm^{-1}. The oscillator is in the $n = 10$ state. Calculate the energy of the oscillator. If this was a classical oscillator with the same energy, what would be the amplitude of vibration?

Problem 5.9 The ground state of an oscillator is 10 meV. How much energy is needed to excite this oscillator to the first excited state? If this oscillator is in thermal equilibrium at 300 K, calculate the energy of the system.

Problem 5.10 The optical phonon energies of GaAs and Si are 36 meV and 52 meV, respectively, at $k = 0$. What is the occupation probability of these optical phonons at 77 K and 300 K?

Problem 5.11 The optical phonon energy for Si at $k = 0$ is 55 meV. Calculate the force constants for the material, assuming a simple linear chain model. Based on these results, what is the sound velocity in the semiconductor?

Problem 5.12 Estimate the kinetic energy of an electron confined to a size 1 Å (twice the Bohr radius) using the uncertainty principle. How does this compare to the kinetic energy of an electron in the hydrogen atom?

Problem 5.13 In the $n = 2$ state of hydrogen, find the electron's velocity, kinetic energy, and potential energy.

Problem 5.14 Consider an electron in the $n = 5$ state of the hydrogen atom. Calculate the wavelengths of photons emitted as the electron relaxes to the ground state. Assume that only those transitions in which n changes by unity can be made.

Problem 5.15 Calculate the ionization energy of: (a) the $n = 3$ level of hydrogen. (b) The $n = 2$ level of He^+ (singly ionized helium). (c) The $n = 4$ level of Li^{++} (doubly ionized lithium). You can use the hydrogen atom model for He^+ and Li^{++} with appropriate changes in the nuclear charge.

Problem 5.16 The lifetimes of the levels in a hydrogen atom are of the order of 10^{-8} s. Find the energy uncertainty of the first excited state.

Problem 5.17 Calculate the separation between the ground state and first excited states of (a) a n-type dopant and (b) a p-type dopant in GaAs. The electron mass is 0.067 m_0 and hole mass is 0.45 m_0.

Problem 5.18 Find the binding energy of an exciton in GaAs. Also find the energy separation of the $n = 3$ and $n = 4$ excited states.

Problem 5.19 Calculate the exciton binding energy in InAs ($m_r^* = 0.02$) m_0 and InP ($m_r^* = 0.07$).

Problem 5.20 Calculate the doping density at which the ground state wavefunction of n-type dopants in GaAs start to overlap each other on an average. What is the p-type doping density at which a similar effect occurs?

Problem 5.21 Assume that a particular defect in silicon can be represented by a three-dimensional quantum well of depth 1.5 eV (with reference to the conduction bandedge). Calculate the position of the ground state of the trap level if the defect dimensions are 5 Å× 5 Å× 5 Å. The electron effective mass is 0.26 m_0.

Problem 5.22 A defect level in silicon produces a level at 0.5 eV below the conduction band. Estimate the potential depth of the defect if the defect dimensions is 5 Å× 5 Å×5 Å. The electron mass is 0.25 m_0.

CHAPTER 6

TUNNELING OF PARTICLES

CHAPTER AT A GLANCE

MAGIC OF TECHNOLOGY

• A biologist is developing a model to understand how cockroaches move through the sewer system. For some reason the scientist does not want to spend the next few months with these critters. Technology comes to the rescue. The world's most adaptable creatures are injected with a *dye*, which sends out signals that can be tracked from the streets above.
• Children playing in a lush forest somewhere in Africa come across a strange jawbone of an animal. They have seen nothing like this before and start to use it as a pretend weapon. A passing paleontologist sees them at play and asks to borrow the bone. A few days later a worldwide press release announces: *JAWBONE OF A 20 MILLION YEAR OLD DINOSAUR DISCOVERED.*
• A high-density semiconductor memory manufacturing plant has developed a new *oxidation process* that will be used for gigabit memory chips. However, the engineers find that the yield of the process is extremely low. A team of researchers starts to look at the precise arrangement of silicon and oxygen atoms on the surface using a *scanning tunneling microscope*. They discover that whenever the starting silicon atoms are *bunched up* in a certain way the process fails. This causes the engineers to be more precise in preparing their wafers and the yield shoots up.
• A high-precision machine tool company is developing advanced materials for ball bearings that don't wear out so soon. Special ball bearings are prepared with a *built-in diagnostic tool.* As microscopic wear occurs in the bearings during use; a signal is sent out giving the exact level of degradation.
• Scientists at a research institute are following a promising lead to attack cancerous cells individually. They have developed a *homing* molecule which goes and binds itself to the offending cells. On top of this molecule the researchers manage to place an *armed* atom with a *zap-gun*! Once the homing molecule attaches to the cancerous cell, the zap-gun kills the cell.

In other institutes, scientists are developing *tracer* techniques to study complex biological reactions, blood flow through damaged arteries, and thyroid disorders. Similar techniques are being exploited by chemical companies, agriculture scientists, anthropologists, etc.

6.1 INTRODUCTION

In Chapter 1 we discussed how, in understanding optical phenomena, certain features can be explained without referring to the wave nature of light (i.e., by geometrical optics), while certain other features require one to call upon the wave nature of light. Interference and diffraction effects are examples of phenomena that require us to use the wave nature of light. Another important phenomenon which can only be understood through wave optics is the "tunneling" of light through regions which are "forbidden" in a geometrical optics description. An example is the well-known phenomenon of *total internal reflection*. When light impinges upon the interface between two media with refractive indices n_1 and n_2, it suffers total internal reflection when

$$\theta_i \geq \sin^{-1} \frac{n_1}{n_2}$$

where θ_i is the angle of incidence. According to the geometrical optics description, there is no penetration of light into the rarer medium. However, it is known that, in reality, the light does propagate into the rarer medium, even under total internal reflection conditions. If, for example, we do an experiment where the rarer medium is made to be very thin (comparable to the wavelength of light) and the outside medium is replaced by a denser medium, light can be extracted. The penetration of light into "forbidden" regions is used for a variety of optical devices such as prism couplers, directional couplers, switches, etc.

In quantum mechanics a particle is described through a wave equation, and the particle wave should be able to penetrate classically forbidden regions. We have already seen a manifestation of this in the previous chapter. In the finite problem of a square well, for example, we have seen that for bound state energies, the particle wavefunction in the barrier region is non-zero even though

$$E < V_0$$

where E is the particle energy and V_0 is the barrier height of the well.

In this chapter we will study the general phenomenon of particle tunneling through regions in space where the potential energy is greater than the particle energy. The tunneling problem is extremely important in applied physics. Fig. 6.1 gives an overview of some of the important applications of tunneling in modern technologies.

6.2 GENERAL TUNNELING PROBLEM

The general problem of particle tunneling can be loosely classified into two categories, both of which involve the propagation of a particle with energy E through a region of potential energy $V(r)$, where in some regions, the particle energy is smaller than the potential energy. This region is classically forbidden to the particle, but, as discussed in the Introduction, when the particle is described through a wave description, tunneling can occur. In the first category of problems we have a situation, as shown in Fig. 6.2a, where there is a region of space where the particle can be represented by a "free" state. The particle can be represented by a momentum in this free space region where the electron energy E is greater than the background potential energy, which may be considered to be uniform. In this example, the tunneling problem involves the particle coming from the left and striking the potential barrier, with the particle having a finite probability of tunneling through and a finite probability of being reflected back.

In the second category of tunneling problems, the particle is initially confined to a "quantum well" region in a "quasi-bound" state, as shown in Fig. 6.2b. In the quasi-bound state, the particle is primarily confined to the quantum well, but has a finite probability of tunneling out of the well and escaping. The key difference between the two cases is that in the first problem the wavefunction corresponding to the initial state is essentially unbound, whereas in the second case it is primarily confined to the quantum well region.

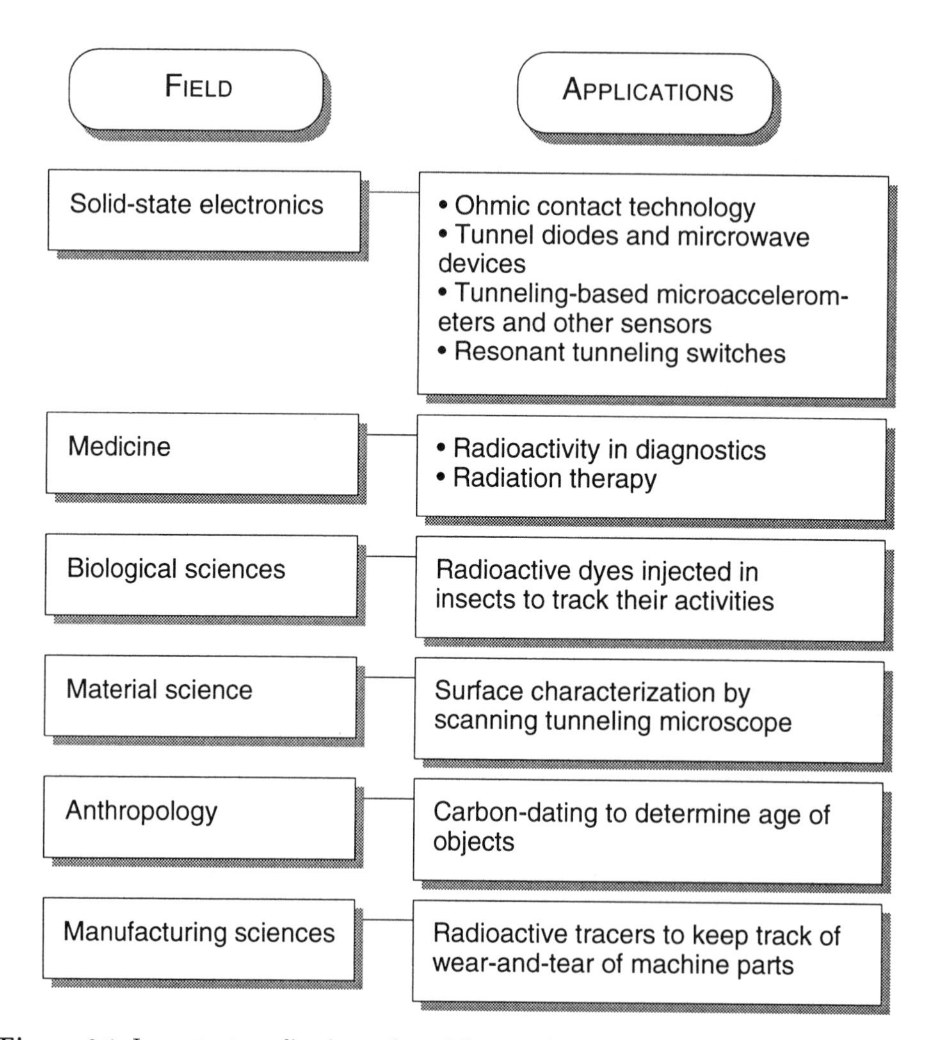

Figure 6.1: Important applications of particle tunneling in various fields.

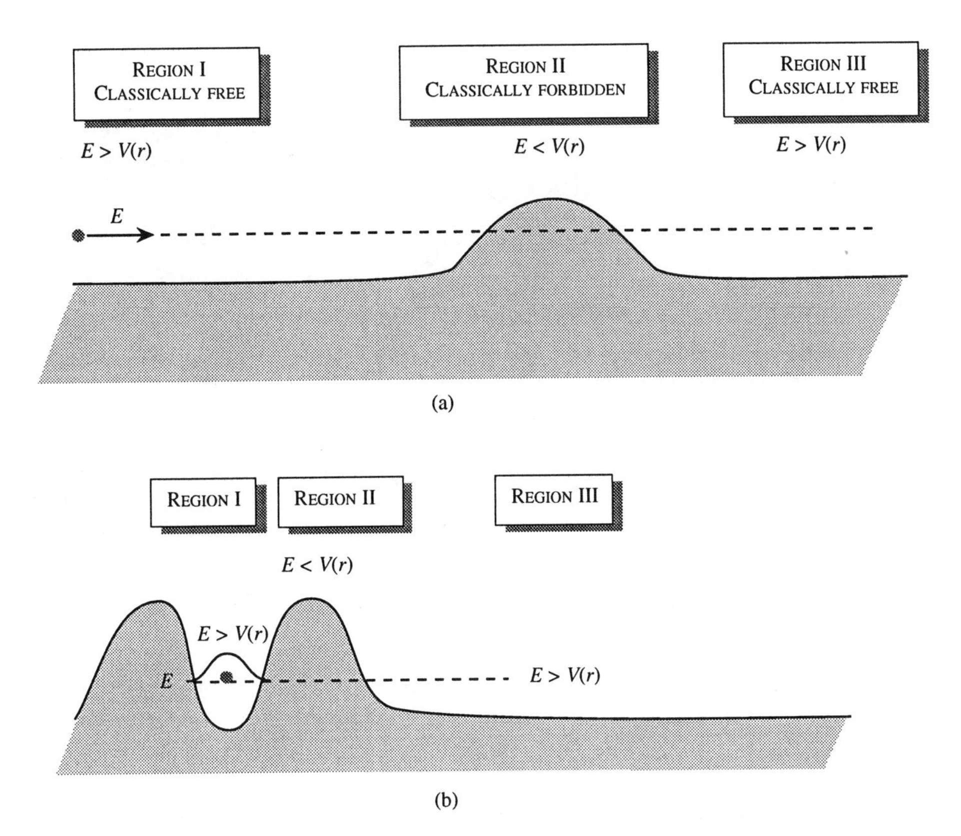

Figure 6.2: A schematic view of two categories of tunneling problems. (a) A particle starts from the unbound state on the left in region I where it is free to move classically (i.e., $E \geq V(r)$) and tunnels through a classically "forbidden" region (II) into a classically allowed region (III). In (b) the particle is initially in "bound region" and tunnels through a "forbidden" region (II) into a classically allowed region (III).

6.3 TIME-INDEPENDENT APPROACH TO TUNNELING

The time-independent approach to tunneling is widely used and provides analytical results for some simple potential barriers. We will develop this approach for some important potential barriers. This approach, shown schematically in Fig. 6.3, usually proceeds as follows (i) the solutions of the Schrödinger equation are written in different regions of space, e.g., inside and outside the potential barrier (ii) the solutions in different regions are then connected at boundaries by using the continuity of ψ and the particle current (iii) the resulting solution is solved to provide information on reflection and transmission properties of the particle. We will illustrate this approach by starting with the simple "square barrier" problem shown in Fig. 6.4.

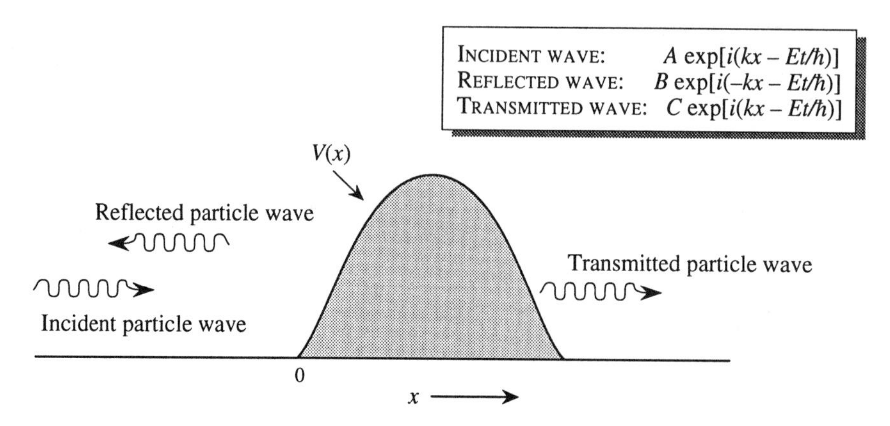

Figure 6.3: A generic description of the tunneling problem. Particles are incident from the left with an energy E. A certain fraction of the particle current is reflected and the rest is transmitted.

6.3.1 Tunneling through a Square Potential Barrier

Referring to Fig. 6.4, we are interested in describing a particle approaching a potential with height V_0 and width a. We have divided the space into three regions. Say a particle has an energy E and impinges from region I onto the barrier. We are interested in the fraction of the particle wave that tunnels through into region III and the fraction of the particle wave which gets reflected. To represent this problem, we choose our wave solutions in the following manner: For $x < 0$, we want the function to represent a particle in a region where $V(x) = 0$ moving to the right (the incident particle) as well as to the left (the reflected particle). The functions $\exp i(kx - \omega t)$ and $\exp -i(kx - \omega t)$ have such forms, respectively. In region $x > a$, the solution should only represent a particle moving to the right (the transmitted particle). Note that for most real-life applications, the particle of interest is the electron.

We start with the case where the energy E of the particle is larger than the barrier height V_0. In classical mechanics, in a case like this, the particle will simply move across the barrier with no reflection. In other words, the transmission probability would be unity. As we will see, this is not the case in the quantum treatment.

The general solutions of the Schrödinger equation in the three regions considered independently are:

$$\text{Region I:}\quad \psi_1(x) = Ae^{ikx} + Be^{-ikx};\quad k^2 = \frac{2mE}{\hbar^2} \tag{6.1}$$

$$\text{Region II:}\quad \psi_2(x) = Ce^{i\alpha x} + De^{-i\alpha x};\quad \alpha^2 = \frac{2m}{\hbar^2}(E - V_0) \tag{6.2}$$

$$\text{Region III:}\quad \psi_3(x) = Fe^{ikx};\quad k^2 = \frac{2mE}{\hbar^2} \tag{6.3}$$

In region I, the wavefunction represents a wave proceeding toward the barrier (Ae^{ikx}) and away from the barrier (Be^{-ikx}), representing the incident and reflected

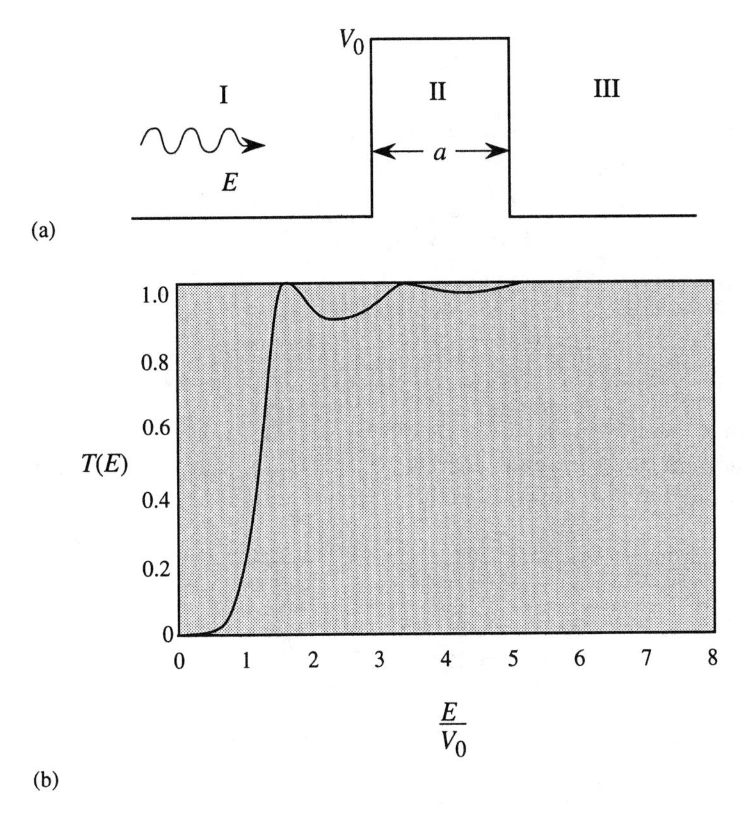

Figure 6.4: (a) Tunneling of a particle through a potential barrier. (b) The transmission coefficients for an electron as a function of energy for $V_0 = 1.0$ eV, $a = 7.77$ Å. The results are typical of transmittance through a barrier. Notice that even for $E > V_0$, the probability goes through values less than unity; i.e., some reflection occurs.

waves. In region III we only consider the wave going away from the barrier, representing the wave that is tunneling out. Note that the solution in region II represents a decaying or growing exponential if $E \leq V_0$, and an oscillatory function otherwise.

We now use the fact that the wavefunction and its derivative are continuous at the boundaries. This gives us the following conditions:

$$\begin{aligned}
&\text{at } x = 0: \quad \begin{cases} A + B = C + D \\ k(A - B) = \alpha(C - D) \end{cases} \\
&\text{at } x = a: \quad \begin{cases} Ce^{i\alpha a} + De^{-i\alpha a} = Fe^{ika} \\ \alpha\left(Ce^{i\alpha a} - De^{-i\alpha a}\right) = kFe^{ika} \end{cases}
\end{aligned} \tag{6.4}$$

Eliminating C and D from these equations, we obtain the following results:

$$\frac{B}{A} = \frac{\left(k^2 - \alpha^2\right)\left(1 - e^{2i\alpha a}\right)}{(k + \alpha)^2 - (k - \alpha)^2 e^{2i\alpha a}}$$

$$\frac{F}{A} = \frac{4k\alpha e^{i(\alpha-k)a}}{(k+\alpha)^2 - (k-\alpha)^2 e^{2i\alpha a}} \tag{6.5}$$

This gives us the following values for the tunneling probability and reflection probability:

$$T = \left|\frac{F}{A}\right|^2 = \frac{4E(E-V_0)}{V_0^2 \sin^2 \alpha a + 4E(E-V_0)} \tag{6.6}$$

and the reflection coefficient is

$$R = \left|\frac{B}{A}\right|^2 = \frac{V_0^2 \sin^2 \alpha a}{V_0^2 \sin^2 \alpha a + 4E(E-V_0)} \tag{6.7}$$

In a classical case, there should be no reflection when $E > V_0$. However, in the quantum problem, even particles with $E > V_0$ can be reflected back with a non-zero probability. If $E = V_0$, we have

$$\begin{aligned} T(E = V_0) &= \left(1 + \frac{ma^2 V_0}{2\hbar^2}\right)^{-1} \\ R(E = V_0) &= \left(1 + \frac{2\hbar^2}{ma^2 V_0}\right)^{-1} \end{aligned} \tag{6.8}$$

Also, we see that the barrier becomes completely transparent ($R = 0, T = 1$) when

$$\sin \alpha a = 0$$

i.e.,

$$\alpha a = n\pi, \quad n = 1, 2, 3, \ldots \tag{6.9}$$

This behavior represents a resonance phenomenon which does not occur in classical physics for particles but does occur in wave optics. The transmission probability is shown in Fig. 6.4b.

Let us now consider the case where $E < V_0$ and where the particle cannot transmit through region II classically. The problem is solved in the same manner as above, except that in region II the wavevector α is purely imaginary. We define a variable

$$\gamma = \sqrt{\frac{2m(V_0 - E)}{\hbar^2}} \equiv \frac{1}{2d} \tag{6.10}$$

The reflection coefficient becomes

$$R = \frac{V_0^2 \sinh^2 \gamma a}{V_0^2 \sinh^2 \gamma a + 4E\left(V_0 - E\right)} \tag{6.11}$$

The tunneling coefficient becomes

$$T = \left|\frac{F}{A}\right|^2 = \frac{4E(V_0 - E)}{V_0^2 \sinh^2 \gamma a + 4E\left(V_0 - E\right)} \tag{6.12}$$

The tunneling probability does not go to zero when E becomes less than V_0 as expected from classical physics. A typical tunneling probability is plotted in Fig. 6.4b. Note that as the barrier height or width increases the tunneling probability decreases.

The analytical expressions developed for the tunneling probability are possible only for simple square barriers. For barriers with arbitrary spatial dependence, one has to use approximation schemes or numerical techniques. Here we will develop a simple approximation which is quite accurate and yet can be applied analytically to many problems of interest.

Consider the tunneling probability through a barrier of width a when the penetration distance d defined in Eqn. 6.10 is small compared to a. This gives a value

$$T \sim \frac{16E(V_0 - E)}{V_0^2} e^{-2\gamma a} \tag{6.13}$$

The tunneling expression is dominated by the exponential term. As an approximation, we take the prefactor to be unity. Now consider the tunneling through a smoothly varying potential shown in Fig. 6.5. We can divide the potential into hatched regions as shown in the figure. The total tunneling probability through the barrier can be taken as the product of the tunneling probability T_i through each hatched region. This is valid only if the potential is smoothly varying, compared to the de Broglie wavelength of the particle. Using Eqn. 6.13 (with the prefactor taken as unity) we have

$$T = \prod_i T_i = \exp\left[-\frac{1}{\hbar}\sum_i \left[8m(V(x_i) - E)\right]^{1/2} \Delta x_i\right]$$

Now increasing the number of subdivisions while ensuring $d_i \ll \Delta x_i$ (note that when $E > V(x_i), T_i \sim 1$), we get

$$T = \exp\left[-\frac{1}{\hbar}\int_{x_1}^{x_2} [8m(V(x) - E)]^{1/2} dx\right] \tag{6.14}$$

where x_1 and x_2 are points where $E = V(x)$.

Two special cases of barriers are the triangular and trapezoidal barriers shown in Fig. 6.6. We can consider these barriers to be formed when an electric field is applied across the square barrier of Fig. 6.4. If the applied electric field is $\tilde{E}$, the tunneling probability becomes

$$\begin{aligned}
\text{Triangular}: \quad T &= \exp\left\{\frac{-4(2m)^{1/2}}{3e|\tilde{E}|\hbar}(V_0 - E)^{3/2}\right\} \\
\text{Trapezoidal}: \quad T &= \exp\left\{\frac{-4(2m)^{1/2}}{3\hbar e|\tilde{E}|}\left[(V_0 - E)^{3/2} - (V_0 - E - e|\tilde{E}|d)^{3/2}\right]\right\}
\end{aligned} \tag{6.15}$$

These expressions for tunneling are extremely useful in understanding many important aspects of semiconductor devices, including ohmic contacts and Zener tunneling.

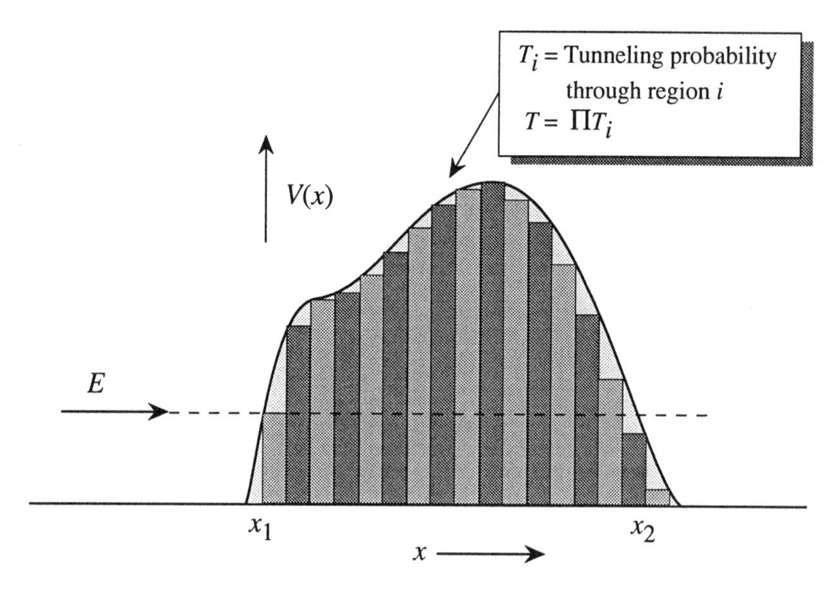

Figure 6.5: An approximation scheme used to calculate particle tunneling with energy E through a smoothly varying potential.

6.4 IMPORTANT APPLICATION EXAMPLES

We will now discuss some important application examples of particle tunneling in several important technologies.

6.4.1 Ohmic Contacts

For current flow to occur in semiconductor devices, electrons or holes must be able to flow "freely" in and out of the semiconductor. This certainly cannot happen if we try to inject or extract electrons directly through a vacuum (or air)–semiconductor interface. There is a large barrier (the work function) which restricts the flow of electrons. However, it is possible to create metal–semiconductor junctions which have a linear I-V characteristic with very low resistivity. Such junctions or contacts are called ohmic contacts.

When a metal is deposited on a semiconductor, there is a potential barrier produced for electron flow as shown in Fig. 6.7. The height of this barrier, called the Schottky barrier, is determined by the nature of the semiconductor surface and the metal. Let us say we have a built-in potential barrier, eV_{bi}. Typically, the barrier height $\phi_b = eV_{bi}$ is equal to half the bandgap of the semiconductor. The spatial extent of the barrier is determined by the carrier density in the semiconductor and is given by a simple solution of the Poisson equation. The width is called the *depletion width*, since in this region the mobile carriers are swept away by the presence of the electric field. If N_d is the doping density in the semiconductor, the depletion width

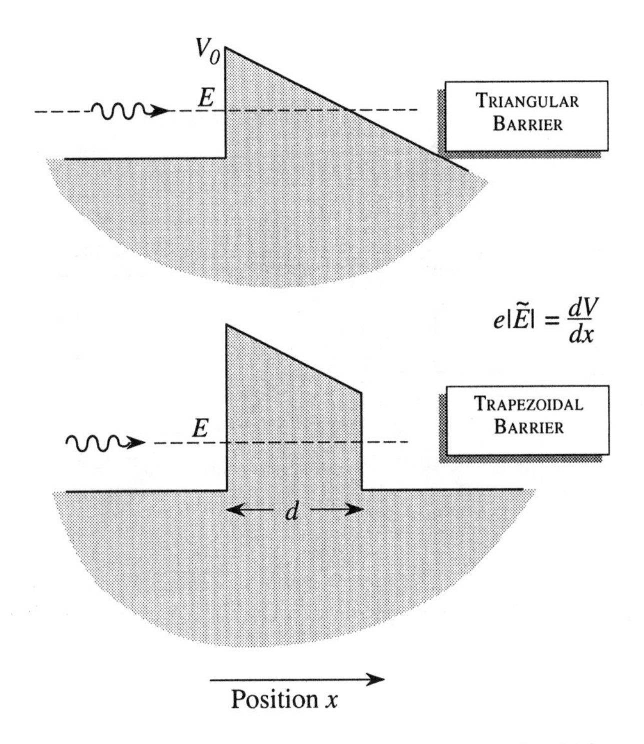

Figure 6.6: (a) Triangular and (b) trapezoidal barriers through which an electron with energy E can tunnel. Such barriers are encountered in many electronic device structures. The potential shape is described by an electric field $\tilde{E}$.

on the semiconductor side is

$$W = \left[\frac{2\epsilon V_{bi}}{eN_d}\right]^{1/2} \tag{6.16}$$

Now if the semiconductor is heavily doped near the interface region, the depletion width could be made extremely narrow. In fact, it can be *made so narrow that even though there is a potential barrier the electrons can tunnel through the barrier with ease* as shown in Fig. 6.8. The quality of an ohmic contact is usually defined through the resistance R of the contact over a certain area A. The normalized resistance is called the specific contact resistance r_c and is given by

$$r_c = R \cdot A$$

Under conditions of heavy doping where the transport is by tunneling, the specific contact resistance has the following dependence (using $V_0 - E \sim V_{bi}$ in Eqn. 6.15):

$$\ln\,(r_c) \propto \frac{1}{\ln(T)} \propto \frac{V_{bi}^{3/2}}{\mid \tilde{E} \mid}$$

where the field is

$$\mid \tilde{E} \mid = \frac{V_{bi}}{W} \propto (V_{bi})^{1/2}(N_d)^{1/2}$$

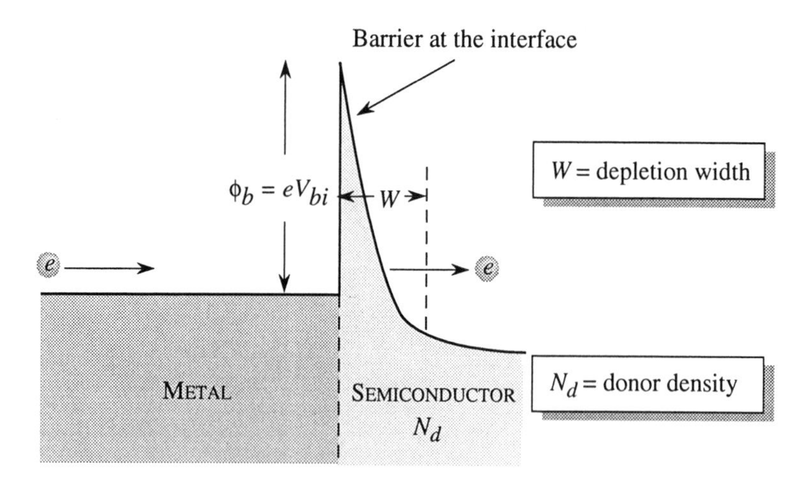

Figure 6.7: Potential profile in an ohmic contact. The contact depends upon the tunneling of electrons with very small energies ($E \sim 0$) through the narrow triangular barrier.

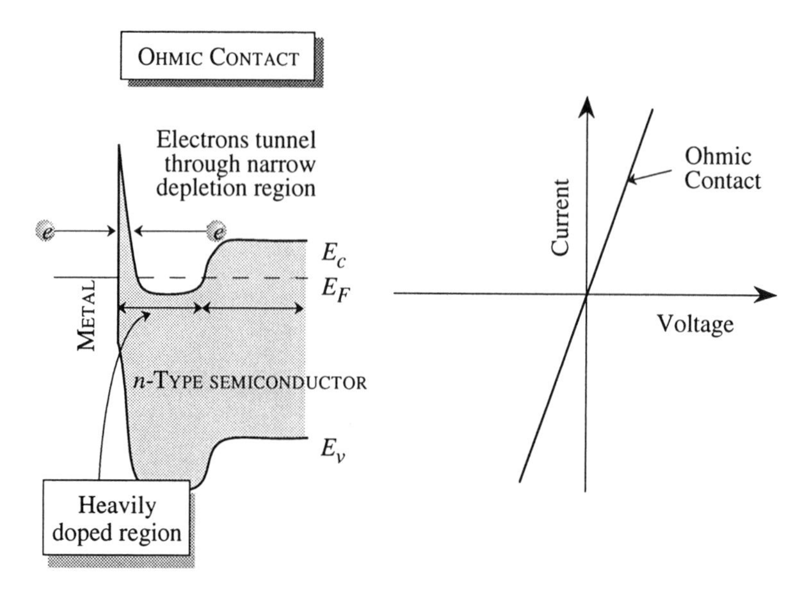

Figure 6.8: Band diagrams of metal -n^+ -n contact. The heavy doping reduces the depletion width to such an extent that the electrons can tunnel through the spiked barrier easily in either direction. This leads to a highly linear current–voltage relation.

Thus

$$\ln(r_c) \propto \frac{1}{\sqrt{N_d}} \tag{6.17}$$

The resistance can be reduced by using a low Schottky barrier height and doping the semiconductor as heavily as possible. The current–voltage relation of a good ohmic contact is highly linear, as shown in Fig. 6.8.

EXAMPLE 6.1 Assume that the potential profile of the barrier in the semiconductor side is described by an electric field $\tilde{E}$. Calculate the tunneling probability if $\tilde{E} = 10^6$ V/cm; ϕ_b = 0.5 eV and the electron mass in the semiconductor is 0.1 m_0. Note that if the tunneling probability approaches unity the electrons can go across the junction without "feeling" the barrier. This is a requirement of good ohmic contact.

The tunneling probability is given by (choosing the electron energy to be ~ 0)

$$\begin{aligned}
T &= \exp\left\{\frac{-4(2m^*)^{1/2}\phi_b^{3/2}}{3e|\tilde{E}|\hbar}\right\} \\
(2m^*)^{1/2} &= (2 \times 0.1 \times 0.091 \times 10^{-30})^{1/2} = 4.27 \times 10^{-16}(\text{kg})^{1/2} \\
\phi_b^{3/2} &= (0.5 \times 1.6 \times 10^{-19})^{3/2} = 2.26 \times 10^{-29}(\text{J})^{3/2} \\
e &= 1.6 \times 10^{-19}\text{C} \\
\tilde{E} &= 10^8 \text{ V/m}
\end{aligned}$$

This gives

$$T = \exp(-8.41) \sim 2.2 \times 10^{-4}$$

EXAMPLE 6.2 Consider a metal–Si interface on silicon doped at 10^{18} and 10^{21} cm^{-3}. The Schottky barrier height is 0.66 V. Calculate the tunneling probability for electrons with energies near the conduction band in the two doping cases. Use an effective mass of 0.34 m_0 for electrons in the semiconductor. The relative dielectric constant of Si is 11.9.

Let us first calculate the depletion width corresponding to a Schottky barrier height of 0.66 V. The depletion width is

$$\begin{aligned}
W(n = 10^{18} \text{ cm}^{-3}) &= \left[\frac{2 \times (11.9 \times 8.84 \times 10^{-14} \text{ F/cm}) \times (0.66 \text{ V})}{(1.6 \times 10^{-19} \text{ C})(10^{18} \text{ cm}^{-3})}\right]^{1/2} \\
&= 2.9 \times 10^{-6} \text{ cm} = 290 \text{ Å} \\
W(n = 10^{21} \text{ cm}^{-3}) &= \frac{W(n = 10^{18} \text{ cm}^{-3})}{\sqrt{1000}} = 9.17 \text{ Å}
\end{aligned}$$

The electrons with energy just at the conduction bandedge will have to tunnel through an approximately triangular well. We can use the tunneling probability through a triangular well, using an average electric field in the depletion region

$$\begin{aligned}
\tilde{E}(n = 10^{18} \text{ cm}^{-3}) &= \frac{0.66 \text{ V}}{2.9 \times 10^{-6} \text{ cm}} = 2.3 \times 10^5 \text{ V/cm} \\
\tilde{E}(n = 10^{21} \text{ cm}^{-3}) &= 7.2 \times 10^6 \text{ V/cm}
\end{aligned}$$

The tunneling probability is now

$$
\begin{aligned}
& T(n = 10^{18}\ \text{cm}^{-3}) \\
& = \exp\left\{ -\frac{4 \times (2 \times 0.34 \times 0.91 \times 10^{-30}\ \text{kg})^{1/2}(0.66 \times 1.6 \times 10^{-19}\ \text{J})^{3/2}}{3 \times (1.6 \times 10^{-19}\ \text{C})(2.3 \times 10^{7}\ \text{V/m})(1.05 \times 10^{-34}\ \text{J.s})} \right\} \\
& = \exp(-93) \sim 0
\end{aligned}
$$

In this case the tunneling current is essentially zero and the contact is non-ohmic.

$$
T(n = 10^{21}\ \text{cm}^{-3}) = \exp\left(-2.94\right) = 0.053
$$

At a doping of around 10^{21} cm^{-3}, the tunneling probability starts to become appreciable and the electrons can move across the junction via tunneling.

6.4.2 Field Emission Devices

The tunneling of electrons through regions where classically they are forbidden finds an important application in *field-emission*-based applications. In these devices, an electric field applied across a metal–vacuum (or semiconductor–vacuum) junction can allow electrons normally confined to the metal to escape into the vacuum. The electrons that are emitted from the metal can be exploited for a number of applications, including imaging of surfaces and in display technologies.

To understand the field emission process, we remind ourselves of the energy band picture in a typical solid-state material. As shown in Fig. 6.9a, the electron energy levels are characterized by a band which is filled up to the Fermi energy. The vacuum energy level is denoted by E_{vac} and represents the energy at which the electrons are *free* from the metal; i.e., they escape into the vacuum.

The energy difference between the vacuum level and the Fermi energy is the work function and has a value of a few electron volts in most materials. Now consider a structure, shown in Fig. 6.9b, where a potential is applied across a metal–vacuum–metal junction. In this case, an energy profile, as shown in the figure, is produced. The electrons in the metal on the left side see a triangular potential barrier and, if the applied field is strong enough, they can tunnel through the barrier into the vacuum region, causing a current flow. Such emission is called *cold emission*, in contrast to thermionic emission in which electrons with energies greater than the barrier height are emitted as a result of thermal energy.

An important application of the field emission is in "flat panel display technology." Important technologies, such as laptop computer displays, "flat" television (which is thin enough to hang on a wall), cockpit displays, etc., depend upon flat-panel display technology.

Figure 6.9c shows a display system based on field emission devices. The metal (or semiconductor) emitters are formed into sharp tips to enhance the electric field produced by the applied potential. The voltage across each tip is controlled by a driver circuitry. The phosphor screen is separated from the tips by a very small gap (submicron), so the entire system is "paper thin." Each pixel (picture element) is "illuminated" by its own set of tips, allowing a thin, low-power-consuming display system.

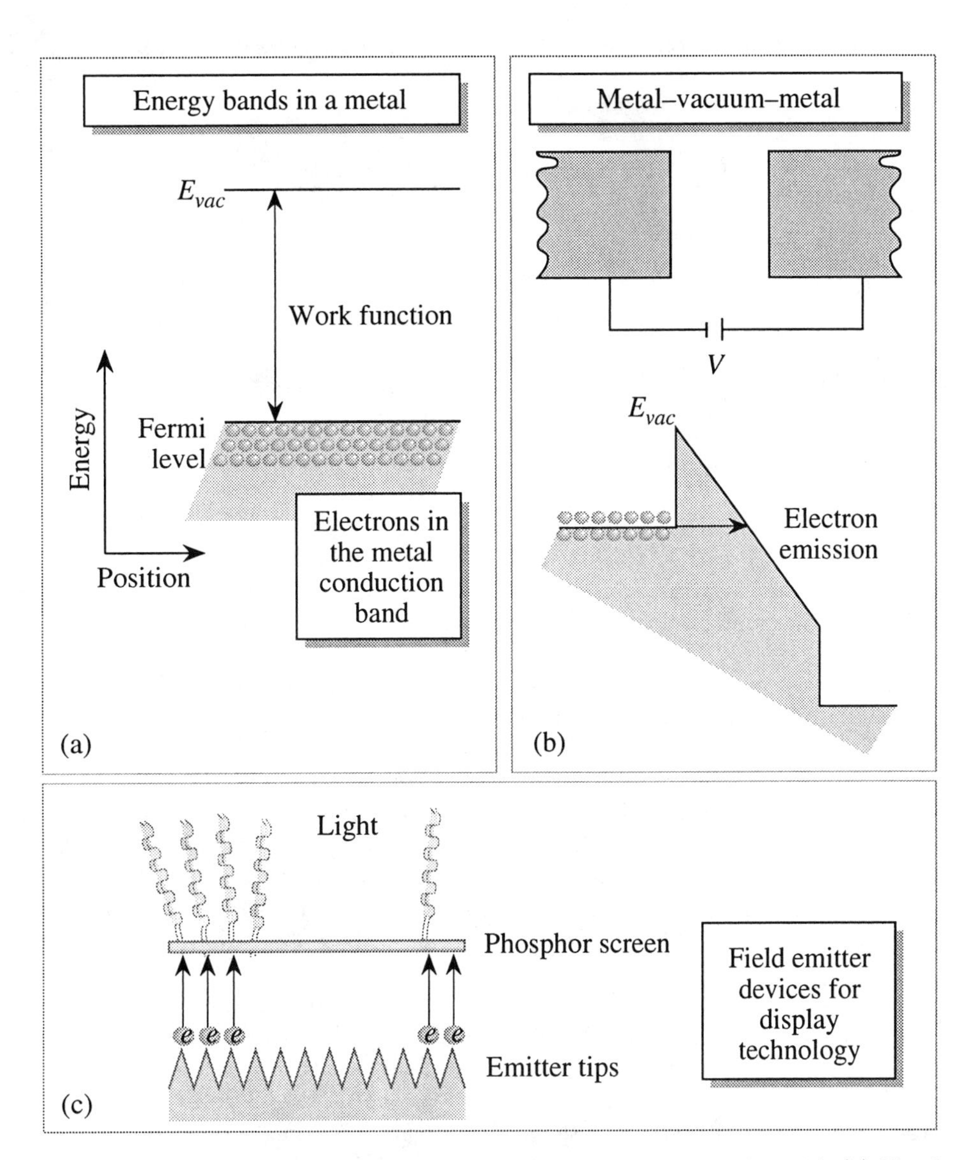

Figure 6.9: (a) Schematic of the electronic energies in a solid-state material. (b) Metal–vacuum–metal junction under applied field. (c) Schematic of a display device based on field emission tips.

EXAMPLE 6.3 A gold tip (work function 4.5 eV) is used in a field emission device. Calculate the electric field needed to allow a tunneling probability of 10^{-4}. Assume an electron effective mass equal to m_0, the free electron mass.

From the expression for the tunneling probability in a triangular barrier (Section 4.3.1 of the text), we see that for $T = 10^{-4}$, we have, for the applied field (using $E \sim 0$)

$$\begin{aligned} \tilde{E} &= \frac{4(2m)^{1/2}(V_0 - E)^{3/2}}{3e\hbar} \times \frac{1}{9.21} \\ &= \frac{4(2 \times 9.1 \times 10^{-31}\ \text{kg})^{1/2}(4.5 \times 1.6 \times 10^{-19}\ \text{J})^{3/2}}{3(1.6 \times 10^{-19}\ \text{C})(1.05 \times 10^{-34}\ \text{J.s})} \cdot \frac{1}{9.21} \\ &= 7.1 \times 10^{9}\ \text{V/m} \end{aligned}$$

6.4.3 Scanning Tunneling Microscopy

The scanning tunneling microscope (STM) exploits the tunneling of electrons from (or into) surface atoms and has become an extremely useful tool in surface characterization.

The STM depends upon injecting electrons from (to) an extremely small tip (with a radius of less than 1000 Å) into (from) atoms on the surface of a material. By proper data interpretation, the resulting current flow can be used to image surfaces with atomic resolution. The operation of the device can be understood by referring to Fig. 6.10. A tip is prepared by electrochemically etching a wire made from tungsten, gold, or some other noble metal. On an atomic scale, the edge of the tip has a protruding atom, as shown in Fig. 6.10. The tip is mounted on a tripod and *can be controlled to up to* 1 *Å precision in x-y-x space.*

When the tip surface distance is a few angstroms and a potential is applied across the tip and the material surface, current of the order of nanoamperes flows through the tip. This current involves the electrons tunneling through the potential barrier created by the work function. The current increases exponentially with the distance z, separating the tip and the surface atoms, and is therefore dominated by the atom that is at the edge of the tip and those surface atoms that are just below it.

The tunneling current (at a fixed voltage) is adjusted to equal the current from a constant current source by altering the z-spacing between the tip and the surface. As the tip is scanned on the x-y surface, one then gets a z-scan that is directly related to the positions of the atoms on the surface. Of course, due to the overlap of the electronic states in the metal tip and the surface atoms, the tunneling current is not simply from a single-surface atom. However, through careful data analysis, one can obtain the positions of single atoms on the surface.

EXAMPLE 6.4 In an STM experiment, a 50 mV potential is applied across the tip and the sample surface. Assuming a work function barrier height of 3 eV, estimate the tip-to-surface separation at which the electron tunneling probability approaches 10^{-7}. Assume the free electron mass for the calculations.

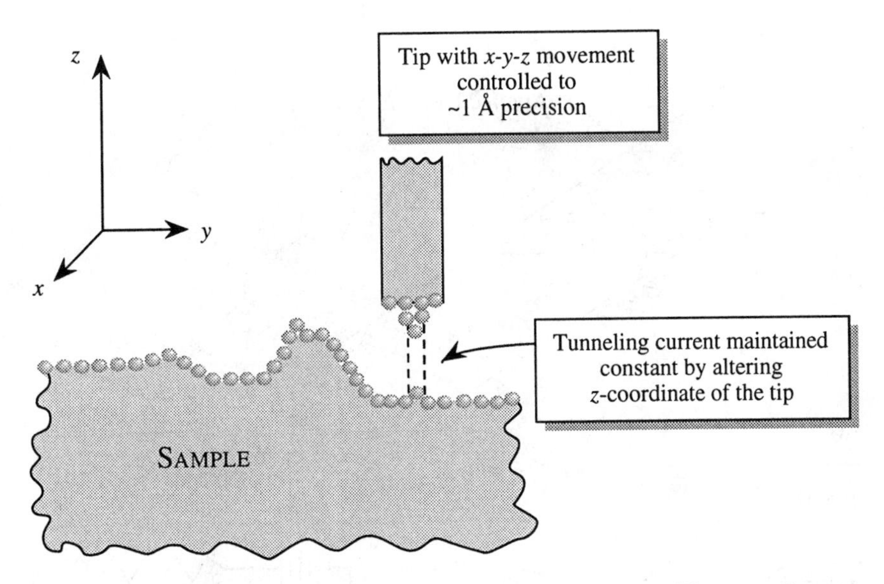

Figure 6.10: A schematic of the scanning tunneling microscope. The tunneling current from the tip is maintained at a constant level as the tip scans the material surface. This allows one to image the positions of atoms on the surface.

We will use the "cold emission" model for this problem. The electric field needed to create a tunneling probability of 10^{-6} is given by

$$\begin{aligned}\tilde{E} &= \frac{4(2m)^{1/2}(V_0)^{3/2}}{3e\hbar}\frac{1}{16.12} \\ &= 2.2 \times 10^9 \text{ V/m}\end{aligned}$$

The separation between the tip and the surface is then

$$d = \frac{50 \text{ mV}}{\tilde{E}} = 0.23 \text{ Å}$$

6.4.4 Tunneling in Semiconductor Diodes

An important semiconductor device is the p-n diode which is made by placing a p-type region next to an n-type region. In most diode applications, tunneling is not of importance, but there are some special purpose diodes that exploit tunneling across the bandgap. These include the tunnel diode (or Esaki diode) and the Zener diode. The tunnel diode has the unusual characteristic that in certain biasing regions, the current decreases with increasing voltage. This leads to negative differential resistance which can be exploited for generating microwave power. The Zener diode has the property that its resistance changes from a very large value to a very small value over a narrow biasing range.

Let us apply our tunneling theory to understand band-to-band tunneling in semiconductors. In Fig. 6.11a, we show a profile of the conduction and valence

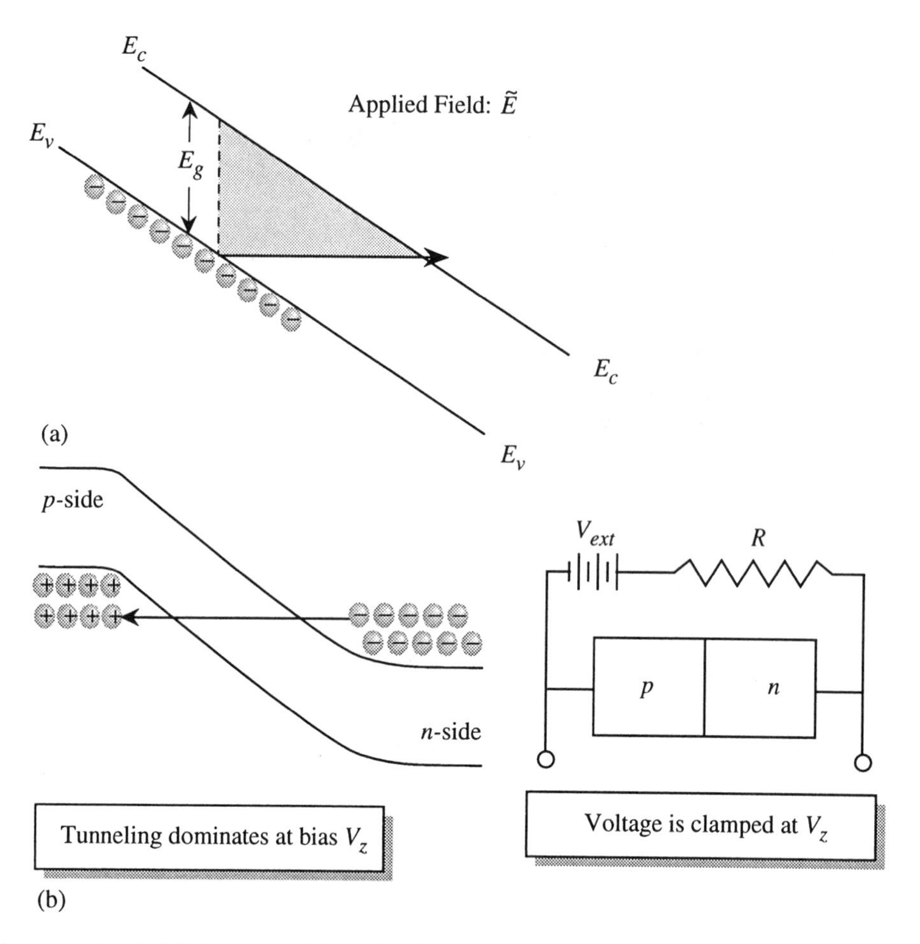

Figure 6.11: (a) Schematic of how band-to-band tunneling takes place. (b) Tunneling in a p-n diode across which a bias is applied. Also shown is the use of a reverse-biased p-n diode for a voltage-clamping circuit. The current saturates to a value determined by the external circuit resistor, while the output voltage is clamped at the diode breakdown voltage. The circuit is thus very useful as a voltage regulator.

bands in a semiconductor across which some bias is applied. Electrons from the valence band can tunnel across the forbidden bandgap and go into the conduction band. Similarly, electrons in the conduction band can tunnel into the valence band provided there are holes in the valence band. As shown in Fig. 6.11a, the electrons see a triangular barrier and the tunneling probability is

$$T = \exp\left(\frac{-4\sqrt{2m^*}E_g^{3/2}}{3e\hbar\tilde{E}}\right) \tag{6.18}$$

where $\tilde{E}$ is the field across the semiconductor.

The band-to-band tunneling is exploited to make the Zener diode. When a p-n junction is reverse biased (i.e., a negative voltage is applied to the p-side and a positive one to the n-side), a strong electric field is produced in the diode junction region, as shown in Fig. 6.11b. Once the reverse bias voltage reaches a certain value, V_z, the tunneling process becomes dominant. At this point, the diode resistance becomes negligible. The voltage across the diode is then clamped at V_z regardless of how high the external bias becomes. This voltage-clamping property of the Zener diode is very useful in many circuit applications.

The tunnel diode also exploits band-to-band tunneling, but couples it with clever energy band alignment to create negative differential resistance. The tunnel diode is designed from very heavily doped n and p sides. The high doping ensures a very large electric field in the junction so that band-to-band tunneling is possible at all biasing conditions.

The operation of the device is shown in Fig. 6.12. At a small forward bias, some of the electrons on the n-side can tunnel into the holes on the p-side. As the bias is increased, the band alignment is such that more and more of the electrons find holes to tunnel into and the current in creases as shown, Beyond the point marked B, the empty states on the p-side move higher in energy with respect to the electron energies as the n-side. Eventually, at point C, there are no empty state (holes) available for electrons to tunnel into. This causes the current to decrease to a minimum value. Eventually the current increases because the electrons and holes are able to overcome the reduced barrier due to their thermal distribution.

The negative resistance of the tunnel diode is an important feature that allows one to use it in microwave applications (eg., radar detectors).

EXAMPLE 6.5 Calculate the band-to-band tunneling probability in InAs ($E_g = 0.4$ eV; $m^* = 0.02\ m_0$) and GaAs ($E_g = 1.43$ eV; $m^* = 0.07\ m_0$) under an applied field of 5×10^5 V/cm.
The exponent in the tunneling probability has the following value for InAs:

$$\frac{-4(2 \times 0.02 \times 9.1 \times 10^{-31}\ \text{kg})^{1/2}(0.4 \times 1.6 \times 10^{-19}\ \text{J})^{3/2}}{3(1.6 \times 10^{-19}\ \text{C})(1.05 \times 10^{-34}\ \text{J.s})(5 \times 10^{7}\ \text{V/m})} = -4.9032$$

This gives

$$T = 7.423 \times 10^{-3}$$

For GaAs we get

$$T = \exp(-62) = 1.18 \times 10^{-27}$$

We see that while band tunneling can be responsible for breakdown in InAs, it is not the cause for breakdown in GaAs. For GaAs, the value of T can increase with higher fields, but there are other mechanisms that are responsible for breakdown.

6.4.5 Radioactivity

Quantum mechanical tunneling leads to a remarkable phenomenon in which the nucleus of an atom emits an electron or an alpha particle (He nucleus). These processes as well as the process where a nucleus emits a γ ray fall under the umbrella of radioactivity. Radioactivity has been widely exploited in applications ranging from nuclear reactors, atomic bombs to medical diagnostics.

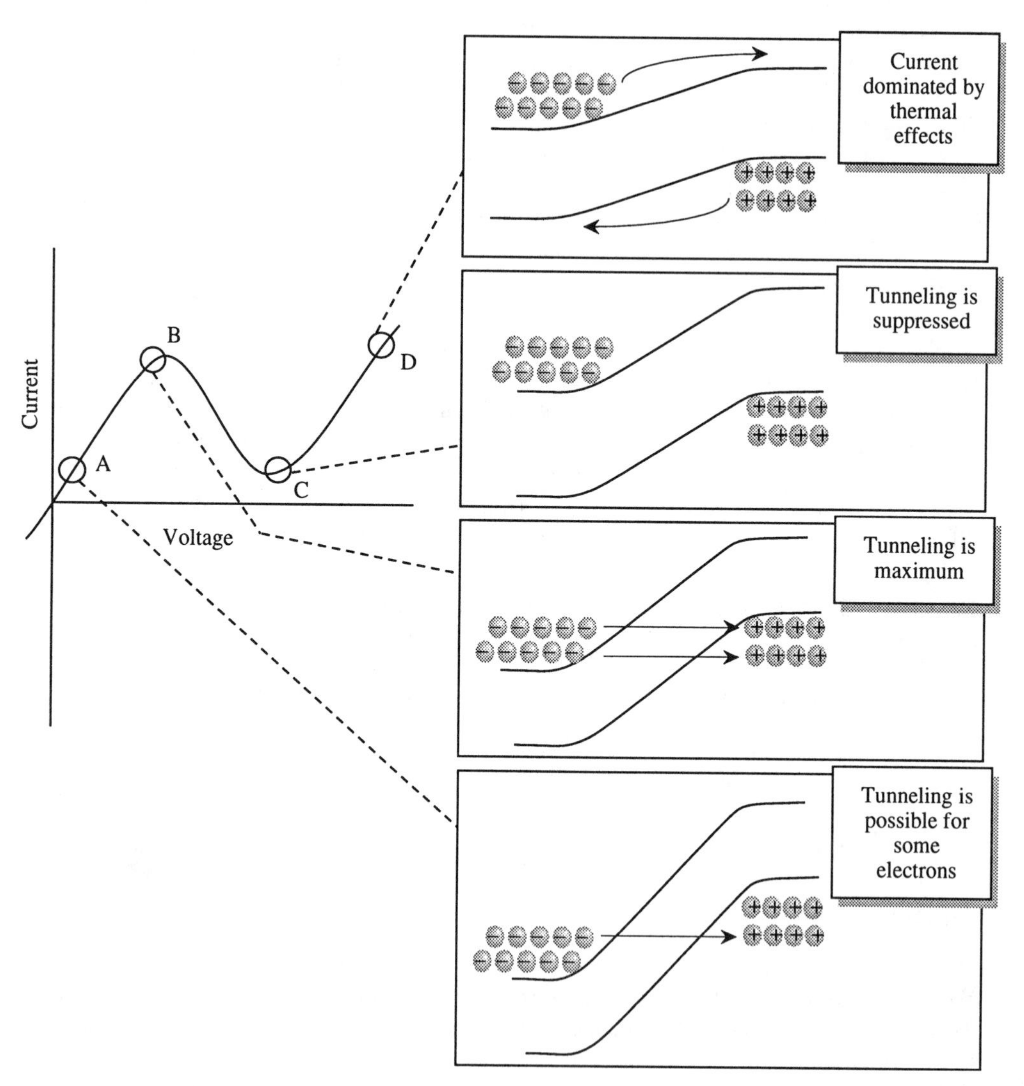

Figure 6.12: Schematic of the current voltage characteristics of a tunnel diode. Also shown are the band alignments and current flow paths at various key points.

In general, radioactivity can involve any of the following processes:
(i) α-Decay: In α-decay, a helium nucleus is emitted from the radioactive nucleus. The general reaction is written as (Z is the atomic number, A is the atomic mass)

$$(Z, A) \rightarrow (Z-2, A-4) + He$$

The original nucleus has 2 units less charge and 4 units less mass.
(ii) β-Decay: In β-decay, a negative electron is emitted from the nucleus, leaving the nucleus with one more charge and the same mass

$$(Z, A) \rightarrow (Z+1, A) + \beta^-$$

(iii) γ-Decay: this case, the original nucleus is an excited state (usually produced when another atom decays) and relaxes by emitting γ-rays.

Let us develop an understanding of radioactive decay by studying α-decay. In this process we can think of the nucleus as containing α-particles, i.e., 2 neutrons + 2 proton composites. There is a strong binding energy to keep the nucleus intact. However, in radioactive materials the nuclear structure is such that the highest energy nucleus can tunnel out of the nuclear potential.

In Fig. 6.13, we show a model potential for an α-particle as a function of its distance from the nucleus. At large distances from the nucleus, the potential energy is the electrostatic energy between a helium ion and the nucleus. At very small distances, the potential is the nuclear potential and is attractive and of the form shown in Fig. 6.13. The general potential is assumed to be of the form

$$\begin{aligned} V(r) &= -V_0, \quad r < r_0 \\ &= \frac{2Z'e^2}{4\pi\epsilon_0 r}, \quad r > r_0 \end{aligned} \tag{6.19}$$

where Z' is the charge of the final nucleus and the factor 2 is due to the charge of the helium nucleus.

Let us consider the limits of the integral for the tunneling probability. One limit is $r = r_0$ and the other can be seen to occur at (E is the energy of the α-particle) the point where $E = V(r)$. This gives for the other limit, a value r_1 where

$$r_1 = \frac{2Z'e^2}{4\pi\epsilon_0 E} \equiv \frac{A}{E} \tag{6.20}$$

Using these limits, we have, for the tunneling probability T (m is the mass of the α-particle),

$$\begin{aligned} \ln T &= -2\int_{r_0}^{r_1} \left(\frac{2m}{\hbar^2}\right)^{1/2} \left(\frac{2Z'e^2}{4\pi\epsilon_0 r} - E\right)^{1/2} dr \\ &= -2\left(\frac{2mE}{\hbar^2}\right)^{1/2} \int_{r_0}^{r_1} \left(\frac{r_1}{r} - 1\right)^{1/2} dr \end{aligned} \tag{6.21}$$

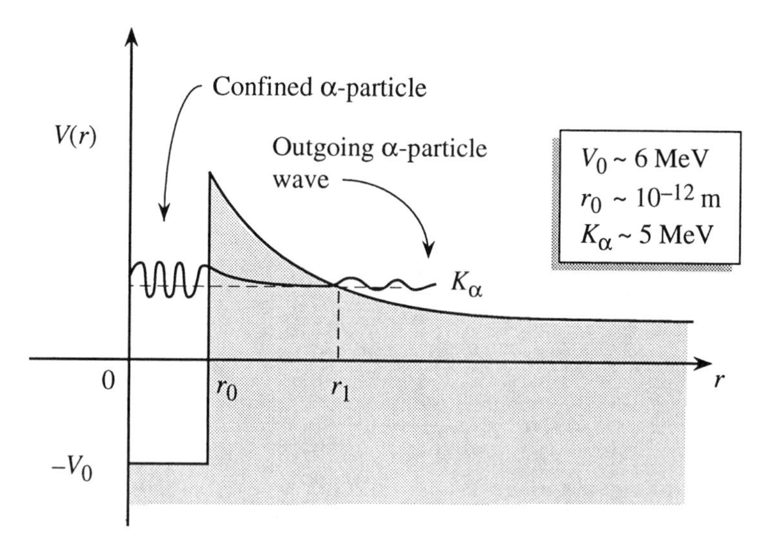

Figure 6.13: Model for the potential energy of an α-particle near a nucleus.

A solution of the integral gives the following result for the tunneling out probability of α-particles:

$$\boxed{\ln T = \frac{4e}{\hbar}\left(\frac{m}{\pi\epsilon_0}\right)^{1/2} Z'^{1/2} r_0^{1/2} - \frac{e^2}{\hbar\epsilon_0}\left(\frac{m}{2}\right)^{1/2} Z' E^{-1/2}} \tag{6.22}$$

If energy is expressed in MeV, r_0 in 10^{-15} m, we get

$$\boxed{\ln T = 2.97 Z'^{1/2} r_0^{1/2} - 3.95 Z' E^{-1/2}} \tag{6.23}$$

As α-particles leave the nucleus the nucleus transforms from one element to another. This process is known as radioactive decay. It is possible to describe the radioactive decay of an element by the equation

$$N(t) = N(0) \exp -\lambda t \tag{6.24}$$

An approximate expression for the rate λ can be obtained from the intuitive relation

$$\lambda = \nu T \tag{6.25}$$

where ν is the frequency with which the α-particle "strikes" the potential barrier. Inside a nucleus we can picture that there are N α-particles that can be produced by the proper association of protons and neutrons. The attempt frequency ν can then be written as

$$\nu = \frac{Nv}{2r_0} \tag{6.26}$$

where v is the velocity of the α-particles:

$$v = \sqrt{\frac{2E}{m}} \tag{6.27}$$

Isotope	K_α(MeV)	$t_{1/2}$	$\lambda(s^{-1})$
^{232}Th	4.01	1.4×10^{10} y	1.6×10^{-18}
^{238}U	4.19	4.5×10^{9} y	4.9×10^{-18}
^{230}Th	4.69	8.0×10^{4} y	2.8×10^{-13}
^{230}U	5.89	20.8 d	3.9×10^{-7}
^{210}Rn	6.16	2.4 h	8.0×10^{-5}
^{220}Rn	6.29	56 s	1.2×10^{-2}
^{222}Ac	7.01	5 s	0.14
^{215}Po	7.53	1.8 ms	3.9×10^{2}
^{218}Th	9.85	0.11 μs	6.3×10^{6}

Table 6.1: Half-lives and α-particle energies of some elements.

The parameter λ is related to the half-life, $t_{1/2}$ (i.e., the time it takes for half of the original atoms to decay), by the relation

$$t_{1/2} = \frac{0.693}{\lambda} \tag{6.28}$$

In Table 6.1 we show half-lives of some elements that decay by emitting α-particles. We can see that the half-life ranges from billions of years to microseconds. Radioactivity is used in a number of important technologies. We will briefly discuss a few applications.

Tracer Applications: An important application of radioactive materials is for tracer applications. This application is based on two properties: i) similarity in chemical properties of a radioactive atom (an isotope) and other atoms of an element. This allows one to follow movement of an element through chemical, physical or biological steps; ii) detection of the chemical species by examining the unique half-life of the radioactive tracer. Uses of tracer radioactive elements have been made in

- Tracking of cockroaches through city sewer systems. In this application, the insect is injected with a small radioactive material and the emission is tracked.
- Tracking of leaks in piping systems.
- In the metal industry, wear-and-tear of tools is tracked by radioactive tagging of tools and examining the radioactivity content of shavings, etc.
- A widespread use of tracers is in the field of medical diagnostics where a trace radioactive element is injected in the bloodstream and its movement is following through the body.

Treatment of Tumors: Advances in drug technology have allowed pharmaceutical companies to make drugs where radioactive elements (e.g., iodine) can be incorporated into specially designed molecules. These molecules, when injected into the bloodstream, can distinguish between healthy and tumor cells. They attach themselves to the tumor cells where the radioactive element delivers local radiation, destroying the unhealthy cells. Since the radiation is preferentially delivered, there are few side effects.

Carbon-Dating: A very useful application of radioactivity is for dating objects such as fossils, bodies buried in glaciers, and even the Turin shroud. It is known that ^{12}C and ^{13}C are stable while ^{14}C is radioactive with a half-life of 5700 years. The disintegration of ^{14}C produces ^{14}N. Cosmic rays in the atmosphere produce^{14}C which is then dispersed through the biosphere. The fraction of ^{14}C in the total carbon content in the biosphere (living organism) is fairly stable. However, when an organism dies, it no longer gets ^{14}C from the atmosphere, and the ^{14}C content starts to decrease. By measuring the residual ^{14}C in an object, it is possible to find the time of "death" in reference to contemporary time. The years before present, t, are given by

$$t = \frac{1}{\lambda} \ell n \frac{A_o}{A_s} \tag{6.29}$$

where A_o is the contemporary activity, A_s is the radioactivity in the object. Carbon dating is one of the most accurate ways of determining the age of life forms which have been dead for a long time. The accuracy of the method is such that objects can be dated to within decades.

6.5 TUNNELING THROUGH MULTIPLE BARRIERS: RESONANT TUNNELING

In the previous subsection we examined tunneling of a particle through a single barrier. As seen in Fig. 6.4b, when $E < V_0$, the tunneling probability monotonically increases as either E increases or the barrier thickness shrinks. There is no interesting resonant feature in the tunneling probability. If we examine the tunneling of particles through a potential region made up of not one single barrier, but two (or more) barriers as shown in Fig. 6.14, we find that the transmission or tunneling coefficient has resonant features.

A typical resonant tunneling structure has a form shown in Fig. 6.14. It consists of regions A and B where the particle is "free," enclosing a series of potential barriers and wells. An important resonant tunneling structure is the double-barrier resonant tunneling structure, which has the form shown in Fig. 6.14b. This structure is used for many important device applications. As shown in Fig. 6.14b, the quantum well enclosed by the two barriers has "quasi-bound" levels. If the barriers were infinitely thick, these levels would simply be the bound states in a quantum well, as discussed in Chapter 4. However, because of the finite barrier thickness, the quasi-bound states have wavefunctions that "leak" out of the well region, but the wavefunctions are primarily confined to the well region. When the energy of a particle (say, approaching from region A) approaches one of these quasi-bound energies, the transmission coefficient approaches unity.

The general problem of the resonant tunneling structure requires numerical computation, but if we assume that the structure is made up of a series of regions over which the potential energy is constant, it is possible to get analytical results. The approach is quite similar to the one we used in the last subsection—one writes

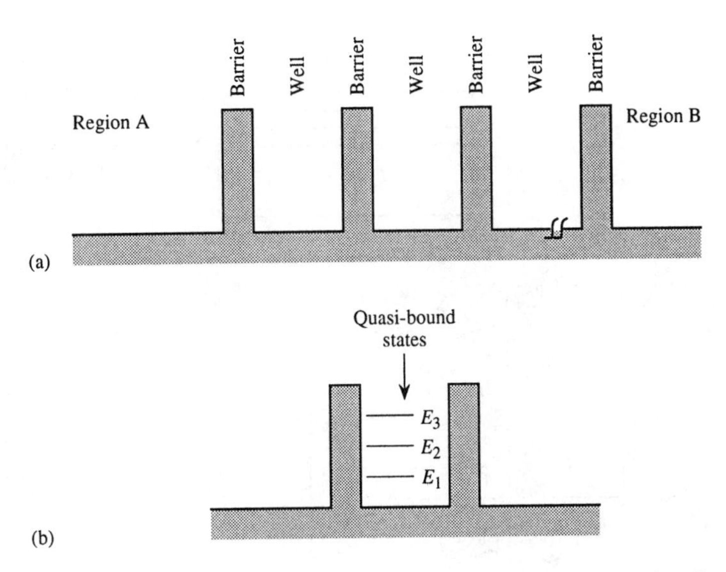

Figure 6.14: (a) A general potential profile which will show resonant tunneling behavior. (b) A double-barrier resonant tunneling structure. In the well formed between the two barriers, one has quasi-bound states with energies $E_1, E_2, E_3, \ldots$ As the particle energy approaches these energies, the transmission coefficient approaches unity.

out the general solutions for the wavefunction in each region, matches boundary conditions, eliminates the coefficients involving the wavefunction of the intermediate regions, and then obtains the transmission and reflection coefficients.

In optics, we know that if light passes through multiple media, interference effects allow transmission peaks at certain frequencies. A similar effect occurs when particles impinge on more than one barrier. An interesting barrier is the double barrier structure shown in Fig. 6.15. which can be fabricated from two semiconductors. As shown in Fig. 6.15, the conduction band profile of such a structure produces a quantum well of size W enclosed by two thin barriers of thickness a and barrier height V_0.

We may think of the quantum well formed by the double-barrier structure as having quasi-bound or resonant states at energy E_1, E_2, E_3, etc., as shown in Fig. 6.15. If the barriers were very thick the electrons in the levels E_1, E_2, E_3 would not be able to "leak out" of the well. However, since the barriers have a finite thickness, the electron wavefunctions leak out through the barrier. Now consider an electron with energy E impinging on the double-barrier structure (resonant tunneling structure) as shown from the left. If the energy of the electron aligns with the well energies E_1, E_2, E_3, etc., the electron is able to tunnel through the structure with unit probability. Otherwise, the tunneling probability is very small. This leads to sharp structures or resonances in the transmission probability versus electron energy curves, as shown in Fig. 6.15.

The tunneling probability for the double-barrier structure is given by an approach similar to the one used in Section 6.3 for a single-barrier. However, in

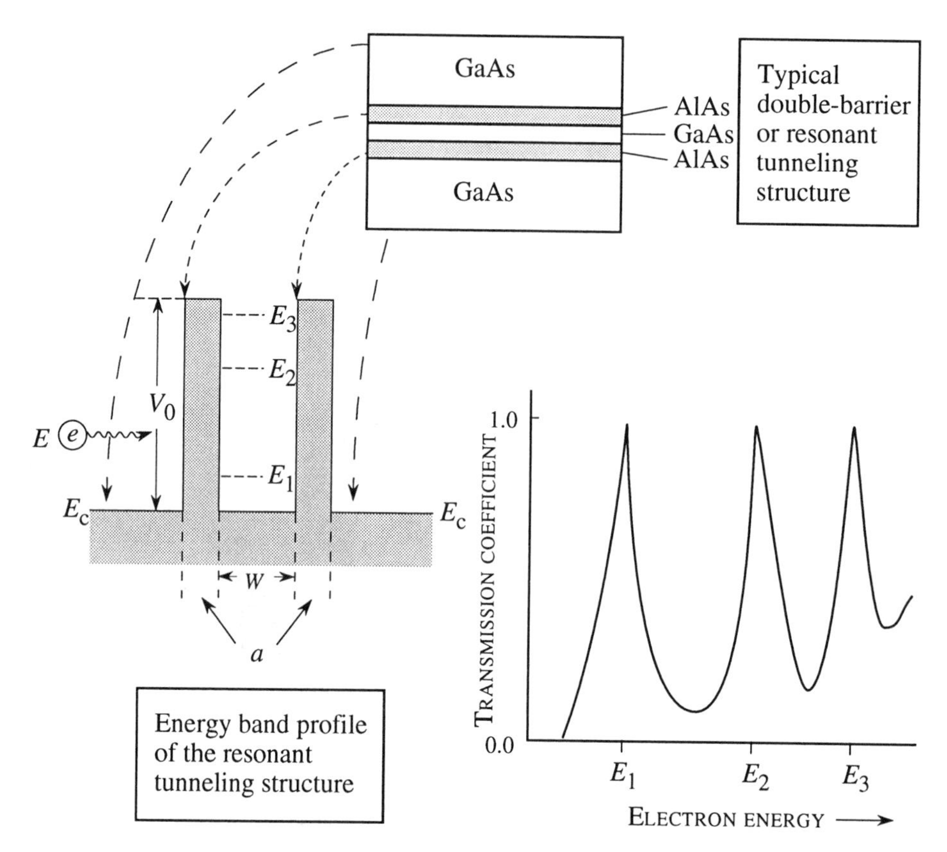

Figure 6.15: Schematic of a typical double barrier resonant tunneling structure. The energies E_1, E_2, E_3, etc. represent the energies in the well of width W. The transmission coefficient goes to unity when the electron energy coincides with these resonant energies.

this case, the boundary conditions have to be matched at both the barriers. The detailed problem is straightforward, but involves tedious algebra. We will simply give the solution here. We first reexamine the single-barrier results.

The tunneling probability through a single barrier of height V_0 and width a is given by

$$T_{1B} = \frac{4E(V_0 - E)}{V_0^2 \sinh^2 \gamma a + 4E(V_0 - E)} \tag{6.30}$$

with

$$\gamma = \frac{1}{\hbar}\sqrt{2m(V_0 - E)}$$

which is the result obtained in Section 6.3.1.

Also, as in Section 6.3.1, we have, for the reflection amplitude from the first barrier,

$$r_{1B} = \frac{\left(k_1^2 + \gamma^2\right)\left(1 - e^{-2\gamma a}\right)}{(k_1 + i\gamma)^2 - (k_1 - i\gamma)^2 e^{-2\gamma a}}$$

$$= \frac{(k_1^2+\gamma^2)\sinh\gamma a}{(k_1^2-\gamma^2)\sinh\gamma a + i2k_1\gamma\cosh\gamma a}$$

with a reflection probability

$$R_{1B} = |r_{1B}|^2 = \frac{V_0^2\ \sinh^2\ \gamma a}{V_0^2\ \sinh^2\ \gamma a + 4E(V_0-E)} \tag{6.31}$$

where

$$k_1 = \sqrt{\frac{2mE}{\hbar^2}}$$

We may define (as in optics) the phase change θ produced in the reflected wave due to the reflection from a single barrier through the relation

$$r_{1B}^2 = R_{1B}e^{-2i\theta} \tag{6.32}$$

The angle θ is given by

$$\tan\theta = \frac{2k_1\gamma\cosh\gamma a}{(k_1^2-\gamma^2)\sinh\gamma a} \tag{6.33}$$

These results have been repeated here, since the transmission probability through the double-barrier structure is expressed in terms of R_{1B}, T_{1B} and θ, given above.

For the double-barrier structure shown in Fig. 6.15, the transmission probability has sharp resonances as a function of the particle energy. In the case of a symmetric barrier, we get the following relation:

$$T_{2B} = \left[1 + \frac{4R_{1B}}{T_{1B}^2}\sin^2(k_1W-\theta)\right]^{-1} \tag{6.34}$$

where R_{1B} and T_{1B} are the single-barrier reflection and tunneling probabilities, and θ is the angle defined in Eqns. 6.32 and 6.33. From the form of Eqn. 6.34, we can see that the transmission probability becomes unity at certain well-defined particle energies. Away from these *resonant* energies, the transmission coefficient can become very small. Thus the transmission probability has very sharp features.

6.5.1 Application Example: Resonant Tunneling Diode

Semiconductor diodes incorporating the double-barrier resonant structure are known as *resonant tunneling diodes*. Resonant tunneling devices have potential for very-high-speed applications and are therefore being studied in great detail. The operation of a resonant tunneling structure is understood conceptually by examining Fig. 6.16a. For small applied voltages (case A in Fig. 6.16a) the tunneling probability of electrons with energies near the Fermi level energy is very small and, as a result, very little current flows through the structure. At point B, when the Fermi energy lines up with the quasi-bound state, a maximum amount of current flows through the structure, since the tunneling probability reaches unity. Further increasing the bias results in the energy band profile (point C), where the current through the structure decreases with increasing bias. Applying a larger bias results in a strong

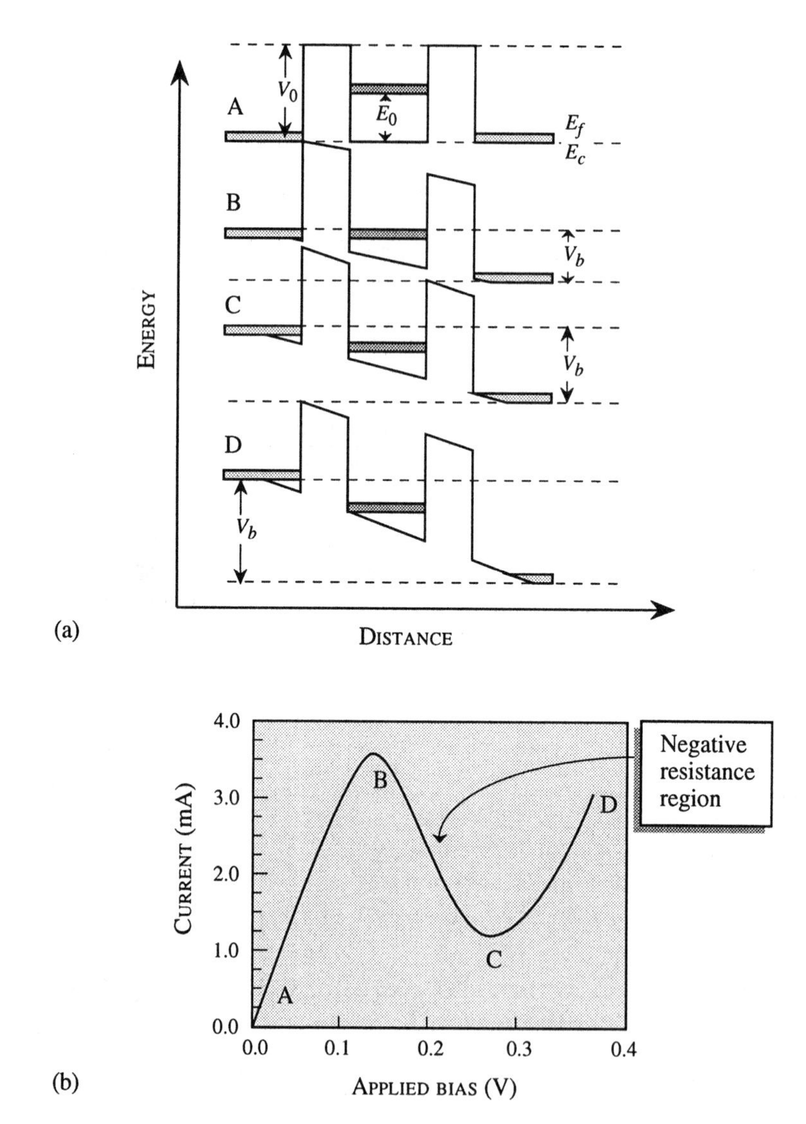

Figure 6.16: (a) Schematic explanation of the operation of resonant tunneling devices showing the energy band diagram for different bias voltages. (b) Typical current–voltage characteristic for the resonant tunneling diode.

thermionic emission current and thus, the current increases substantially, as shown at point D. A typical current–voltage relation for the diode is shown in Fig. 6.16b. Of particular interest is the negative-resistance region, which can be exploited for low-power-high-speed digital devices, as well as for microwave devices.

6.6 TUNNELING IN SUPERCONDUCTING JUNCTIONS

In Chapter 4, Section 4.6.2, we have briefly described how materials can act as superconductors. In these materials at low temperatures electrons pair up to form Cooper pairs. These pairs are bosons and not fermions. The lowest energy state where electron pairs are bosons is separated from normal electron states by a small energy gap where no states are allowed. This gap is only about a millielectron volt and plays a very important role in how electrons tunnel into superconductors.

Let us consider a thin insulator sandwiched between two metals, as shown in Fig. 6.17a. Normal metals are characterized by two important properties: i) the Fermi level is in the middle of the band and electrons occupy states essentially up to the Fermi level. There is no energy gap in this conduction band; ii) In normal metals, electrons act as fermions and as they travel they suffer scattering.

When a bias is applied to the metal–insulator–metal junction electrons can tunnel from one side to the other and current flows. At small applied bias the current–voltage characteristics are essentially linear as shown in Fig. 6.17a.

Now let us consider a junction in which an insulator is sandwiched between two superconductors as shown in Fig. 6.17b. In this case we remind ourselves of two important properties of superconductors; i) Carriers move in pairs of charge $2e$ in the superconducting state; ii) the pairs suffer *no scattering* during transport. Thus long-range phase coherence is maintained in the wavefunction of electrons. This allows us to describe the carrier transport and current flow by simply solving the Schrödinger equation.

For the superconducting junction if the insulator is thin electron pairs can tunnel through the barrier. However, unlike the case in normal metals these electron pairs can move around a closed wire to complete the current, since there is no scattering path. As a result, current flows *even in the absence of any applied bias.* When a bias is applied the current is found to be constant up to a critical voltage V_c. Beyond this voltage electrons tunnel across the superconducting gap and normal Ohmic conduction occurs. The current–voltage relations for a superconductor are shown in Fig. 6.17b. The highly non-linear current–voltage relations are very useful for devices such as mixers where two different microwave signals can be "mixed" to produce sum and difference frequencies.

The detailed analysis of the current in a superconductor tunnel junction (called Josephson junction) is somewhat involved, but is given below for completeness. Let us consider first the case where no potential is applied across the junction.

We consider the simple case in which the two superconductors are identical. Let ψ_1 represent the probability amplitude for the electron pairs on one side of the junction, and ψ_2 the amplitude on the other side. The time-dependent Schrödinger equation $i\hbar\partial\psi/\partial t = H\psi$ can be used to describe the coupling and time-dependence of the electron pairs on one side across the junction. The equations for ψ_1 and ψ_2

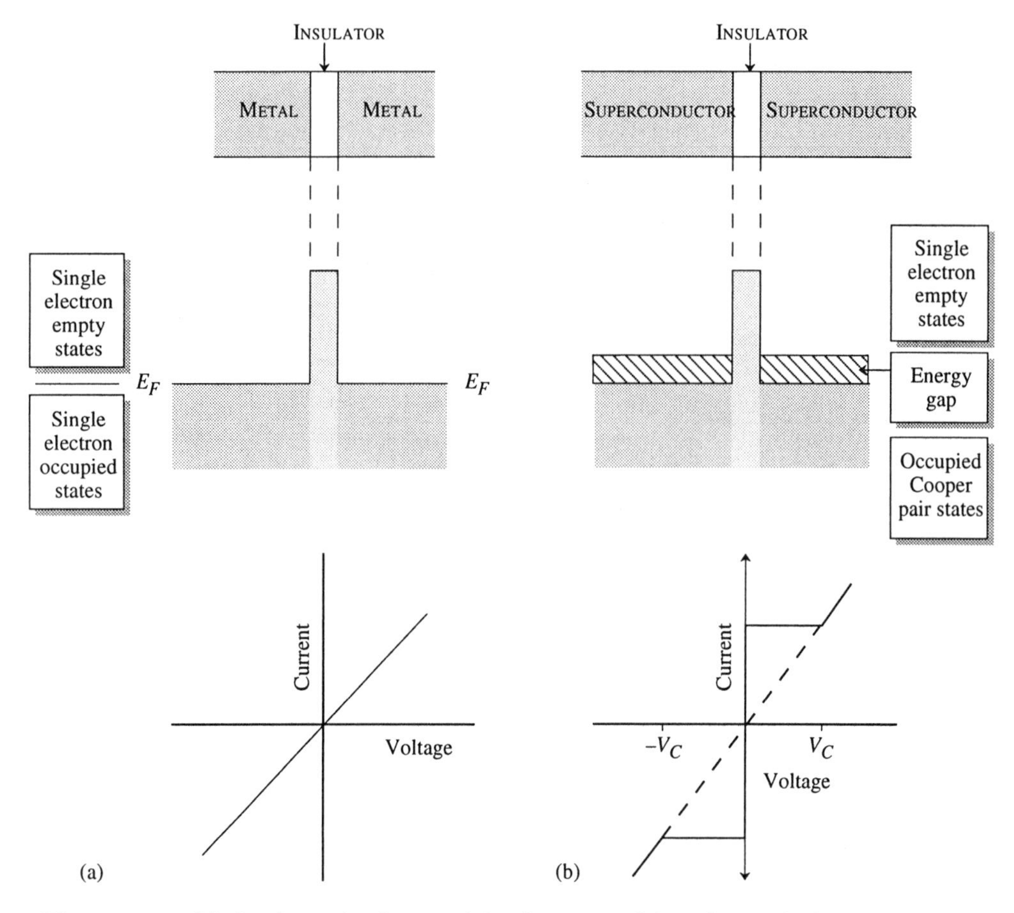

Figure 6.17: (a) A schematic of a metal–insulator–metal junction with the band profile seen by electrons and the current–voltage relation. (b) A schematic of a superconductor–insulator–superconductor junction.

are (choosing the energy of the pairs as zero)

$$
\begin{aligned}
i\hbar\frac{\partial\psi_1}{\partial t} &= H_{12}\psi_2 \\
i\hbar\frac{\partial\psi_2}{\partial t} &= H_{12}\psi_1
\end{aligned}
\tag{6.35}
$$

where the matrix element $H_{12} = \langle\psi_1|H|\psi_2\rangle = \langle\psi_2|H|\psi_1\rangle$ represents the coupling of the electron pairs across the insulator. This matrix element is proportional to the tunneling probability across the barrier.

If n_1 and n_2 are the pair densities on the two sides of the junction, we may write

$$
\psi_1 = \sqrt{n_1}e^{i\theta_1}; \quad \psi_2 = \sqrt{n_2}e^{i\theta_2} \tag{6.36}
$$

This gives, from Eqn. 6.35,

$$\frac{\partial \psi_1}{\partial t} = \frac{1}{2\sqrt{n_1}} e^{i\theta_1} \frac{\partial n_1}{\partial t} + i\psi_1 \frac{\partial \theta_1}{\partial t} = -\frac{i}{\hbar} H_{12} \psi_2 \tag{6.37}$$

$$\frac{\partial \psi_2}{\partial t} = \frac{1}{2\sqrt{n_2}} e^{i\theta_2} \frac{\partial n_2}{\partial t} + i\psi_2 \frac{\partial \theta_2}{\partial t} = -\frac{i}{\hbar} H_{12} \psi_1 \tag{6.38}$$

We multiply Eqn. 6.37 by $\sqrt{n_1} e^{-i\theta_1}$ to obtain (we define $\delta = \theta_2 - \theta_1$)

$$\frac{1}{2} \frac{\partial n_1}{\partial t} + i n_1 \frac{\partial \theta_1}{\partial t} = -\frac{i}{\hbar} H_{12} \sqrt{n_1 n_2} e^{i\delta}$$

Similarly, we have

$$\frac{1}{2} \frac{\partial n_2}{\partial t} + i n_2 \frac{\partial \theta_2}{\partial t} = -\frac{i}{\hbar} H_{12} \sqrt{n_1 n_2} e^{-i\delta}$$

Equating the real and imaginary terms within each of these equations, we get

$$\frac{\partial n_1}{\partial t} = \frac{2H_{12}}{\hbar} \sqrt{n_1 n_2} \sin \delta \; ; \; \frac{\partial n_2}{\partial t} = -\frac{2H_{12}}{\hbar} \sqrt{n_1 n_2} \sin \delta$$

$$\frac{\partial \theta_1}{\partial t} = -\frac{H_{12}}{\hbar} \left(\frac{n_2}{n_1} \right)^{1/2} \cos \delta \; ; \; \frac{\partial \theta_2}{\partial t} = -\frac{H_{12}}{\hbar} \left(\frac{n_1}{n_2} \right)^{1/2} \cos \delta$$

We now assume that $n_1 \sim n_2$, which gives

$$\frac{\partial \theta_1}{\partial t} = \frac{\partial \theta_2}{\partial t} \Rightarrow \frac{\partial \delta}{\partial t} = 0$$

and

$$\frac{\partial n_2}{\partial t} = -\frac{\partial n_1}{\partial t}$$

The current flow across the junction is

$$J \propto \frac{\partial n_2}{\partial t} = -\frac{\partial n_1}{\partial t}$$

or

$$\boxed{J = J_0 \sin \delta = J_0 \sin(\theta_2 - \theta_1)} \tag{6.39}$$

The prefactor J_0 is proportional to the coupling matrix element or the tunneling probability of the pairs across the junction. An interesting feature of this result is that *there is a current flow across the junction in the absence of an applied voltage.* The current density lies between J_0 and $-J_0$, depending upon the phase difference $\theta_2 - \theta_1$.

Let us now consider what happens when a small voltage V is applied across the junction. As a result of this voltage, there is a net energy difference of 2 eV as a Cooper pair crosses the junction. As a result, one side of the junction has an energy of −eV and the other has an energy eV. These terms lead to the following equations:

$$i\hbar \frac{\partial \psi_1}{\partial t} = H_{12} \psi_2 - eV \psi_1$$

$$i\hbar \frac{\partial \psi_2}{\partial t} = H_{12} \psi_1 + eV \psi_2 \tag{6.40}$$

Proceeding as we did for the $V = 0$ case we get

$$\frac{\partial n_1}{\partial t} = \frac{2H_{12}}{\hbar}\sqrt{n_1 n_2}\sin\delta; \quad \frac{\partial n_2}{\partial t} = -\frac{2H_{12}}{\hbar}\sqrt{n_1 n_2}\sin\delta \tag{6.41}$$

$$\frac{\partial \theta_1}{\partial t} = \frac{eV}{\hbar} - \frac{H_{12}}{\hbar}\left(\frac{n_2}{n_1}\right)^{1/2}\cos\delta; \quad \frac{\partial \theta_2}{\partial t} = -\frac{eV}{\hbar} - \frac{H_{12}}{\hbar}\left(\frac{n_1}{n_2}\right)^{1/2}\cos\delta \tag{6.42}$$

Once again, using $n_1 \sim n_2$, we get

$$\frac{\partial(\theta_2 - \theta_1)}{\partial t} = \frac{\partial \delta}{\partial t} = -\frac{2eV}{\hbar}$$

Upon integrating this equation, we get

$$\delta(t) = \delta(0) - \frac{2eVt}{\hbar}$$

The current across the junction is now

$$\boxed{J = J_0 \sin\left(\delta(0) - \frac{2eVt}{\hbar}\right)} \tag{6.43}$$

The current oscillates with a frequency

$$\boxed{\omega = \frac{2eV}{\hbar}} \tag{6.44}$$

This observation is used for precise measurement of $\frac{e}{\hbar}$. If the applied voltage is increased beyond a critical voltage, as shown in Fig. 6.17b, the current reverts to an ohmic behavior, since the pairs are able to excite into the single-particle electronic spectra where the transport is not by Cooper pairs.

The tunneling properties of Josephson junctions can be exploited, along with the use of magnetic fields, to produce low-power digital logic devices and extremely sensitive magnetic field measurement instruments. These applications are discussed in Appendix C.

6.7 CHAPTER SUMMARY

In this chapter we have discussed the very important problem of particle tunneling through potential barriers. This remarkable phenomenon is manifested in very important problems as we have seen in this chapter. We summarize the key findings in Table 6.2.

6.8 PROBLEMS

Problem 6.1 Calculate and plot the reflection and tunneling probability of an electron striking potential barrier of height 1 eV and width 2 Å. The energy of the incident electrons range from 0.1 eV to 1.5 eV.

Problem 6.2 An electron with energy 1.1 eV strikes a potential of height 1.0 eV.

TOPICS STUDIED	KEY OBSERVATIONS
Tunneling of particles	In quantum mechanics, when particles are treated as waves, they can traverse through regions where their energy is less than the potential energy. Their trajectories are completely different from what one expects from classical physics.
Tunneling as a stationary state problem	The time-independent problem can be used to calculate the tunneling behavior of particles. Physically, the time-independent problem corresponds to a constant flux of particles in space. A number of analytical and numerical approaches have been developed to solve such problems.
Applications of tunneling	Tunneling of particles has many important applications in technology. These include: • Scanning tunneling microscopes for studying precise positions of atoms on a surface. • Producing low-resistance ohmic contacts for semiconductor devices. • Tunnel diodes for microwave applications. • Field emission devices. • Radioacitve materials for medical and anthropological uses.
Tunneling through multiple barriers	When particles tunnel through multiple barriers, the tunneling probability has strong resonances as a function of particle energy. This concept is exploited in a number of devices.

Table 6.2: Summary table.

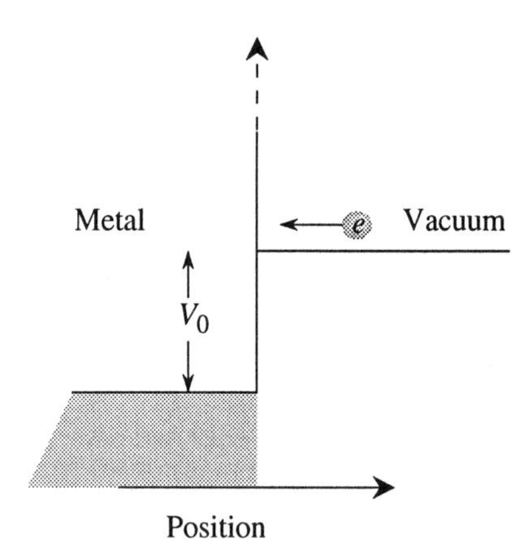

Figure 6.18: Potential profile of a metal–vacuum interface region.

Calculate the probability that the electron will be reflected as the width of the potential varies from 1 Å to 20 Å.

Problem 6.3 Consider an electron impinging upon a metal surface. Assume that the potential profile seen by the electron has the form shown in Fig. 6.17. Calculate the reflection and transmission probabilities of free electrons impinging on the metal, assuming that the energy of the free-space electron is E and the metal provides a step potential V_0. Also calculate the reflection probability for conduction electrons that strike the metal–vacuum interface with energies greater than V_0.

Problem 6.4 Electrons in a metal see a potential barrier which prevents them from escaping to the outside "vacuum"-free electron states. This barrier is the *work function*. Assume that electrons see a potential profile, shown in Fig. 6.18. Consider a beam of electrons impinging upon a metal with energy (measured from the vacuum level) 0.2 eV. If V_0 is 4.0 eV, calculate the reflection probability of the electrons.

Problem 6.5 Referring to the discussion in Problem 6.3, consider a case, as shown in Fig. 6.19, where an electric field is applied between the metal and another elec-

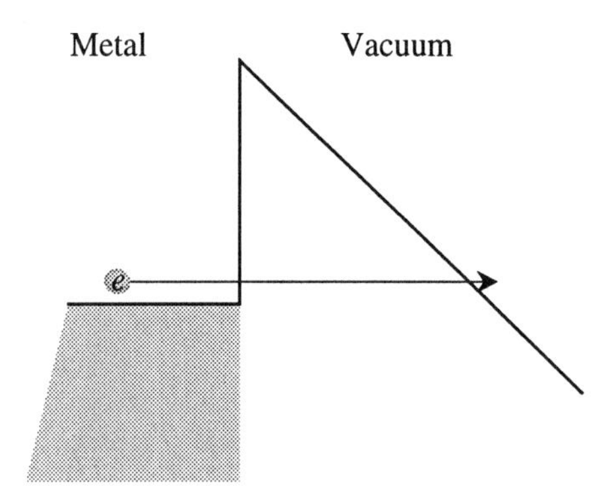

Figure 6.19: Potential profile of a metal–vacuum interface with a field across the junction.

trode separated by a "vacuum gap." Electrons are emitted from the metal (this is called "cold" emission). Calculate the cold emission currents as a function of applied field, assuming a work function V_0, as shown in Fig. 6.18.

Problem 6.6 A gold tip (work function 4.5 eV) is used in a field emission device. Calculate the electric field needed to allow a tunneling probability of 10^{-4}. Assume an electron effective mass equal to m_0, the free electron mass.

Problem 6.7 In an STM experiment, a 100 mV potential is applied across the tip and the sample surface. Assuming a work function barrier height of 3 eV, estimate the tip-to-surface separation at which the electron tunneling probability approaches 10^{-6}. Assume the free electron mass for the calculations.

Problem 6.8 In GaAs microelectronic technology, an ohmic contact is sometimes prepared by depositing a gold-germanium alloy on the surface. When the surface is annealed, the germanium dopes the semiconductor n-type. Calculate the doping density needed to create a potential barrier so that the tunneling probability across the barrier is 10^{-3}. Assume a barrier height of 0.8 eV. Relative dielectric constant is 13.18.

Problem 6.9 The metal-semiconductor barrier energy (Schottky barrier) in most semiconductors is approximately half the bandgap. Discuss why it is difficult to make ohmic contacts in large-bandgap semiconductors. Calculate the tunneling probability across a contact made to diamond when the surface region of the semiconductor is doped at 10^{21} cm^{-3}. Compare this with a contact with the same doping on InAs. ($E_g(C) = 5.5$ eV; $m^*(C) = 0.2\ m_0$; E_g (InAs) $= 0.4$ eV; m^*(InAs) $= 0.02\ m_0$). The relative dielectric constants of C and InAs are 5.57 and 15.15, respectively.

Problem 6.10 Calculate the band-to-band tunneling probability as a function of the electric field and the material bandgap.

Problem 6.11 Calculate the band-to-band tunneling probability in InAs ($E_g =$ 0.4 eV; $m^* = 0.02\ m_0$) and GaAs ($E_g = 1.43$ eV; $m^* = 0.07\ m_0$) under an applied field of 5×10^5 V/cm.

Problem 6.12 A sample of rock solidified 4.55×10^7 years ago. The rock now has 10^{24} ^{238}U atoms. How many atoms did the original rock have?

Problem 6.13 A piece of wood from a recently dead tree shows 12 ^{14}C decays per minute. A sample from an old tree shows 3.5 decays per minute. Calculate the age of the old tree.

Problem 6.14 Consider a resonant tunneling structure with the following parameters:

$$\begin{aligned} \text{Barrier height}, V_0 &= 0.3 \text{ eV} \\ \text{Well size}, W &= 60 \text{ Å} \\ \text{Barrier width}, a &= 25 \text{ Å} \\ \text{Effective mass}, m^* &= 0.07\ m_0 \end{aligned}$$

Calculate and plot the tunneling probability of electrons as a function of energy for $0 < E < V_0$.

Problem 6.15 Consider a one-dimensional quantum well formed by a 100 Å region of GaAs surrounded by AlGaAs barrier on two sides. On one side of the well, the

AlGaAs barrier is very thick, but on the other side, it is only 20 Å thick. An electron is placed in the ground state of the GaAs well. Calculate the time it will take for this electron to tunnel out through the thin barrier region. The potential profile the electron sees is given by the following:

$$\begin{aligned} V(z) &= 0.3 \text{ eV} \quad z < 0 \\ &= 0.0 \quad 0 < z < 100 \text{ Å} \\ &= 0.3 \text{ eV} \quad 100 \text{ Å} < z < 120 \text{ Å} \\ &= 0.0 \quad z > 120 \text{ Å} \end{aligned}$$

CHAPTER 7

APPROXIMATION METHODS

Chapter at a Glance

MAGIC OF TECHNOLOGY

• A chemical company is developing an organic compound to be used as a dye for a clothing company. This color is not available in any dye. After *tweaking* some organic molecules the scientists are able to get the precise color desired.
• A high-precision time-keeping company has been asked by a space communication and tracking company to come up with a clock that keeps time to a precision of 1 second in a *million years*! The scientists examine the "natural frequencies" of a number of atoms and molecules. Finally they supply a clock, which beats the *specs* by three orders of magnitude!
• As pollution from automobiles and other gasoline-driven motors chokes our cities, a chemical company is working on new inexpensive ways to remove carbon monoxide, oxides of nitogen, and certain hydrocarbons known to cause health problems. Scientists at the company come up with a new class of metals which can be sprayed onto honeycomb structures. As the exhaust gases go through the honeycomb *catalytic convertor*, the culprit gases are converted into harmless compounds.

7.1 INTRODUCTION

Very few problems in quantum mechanics can be solved exactly. It is therefore extremely important to develop approaches that can be used to address difficult quantum problems that do not have analytical solutions. In recent years, computer software has allowed scientists to solve extremely complicated quantum problems—problems which, up to a few years ago, were considered unsolvable. For most problems of interest, the Schrödinger equation can be solved by using a fast enough computer with a large enough memory. Enormously powerful computer libraries now exist which have subroutines designed to solve a variety of matrix equations as well as different classes of differential equations. Scientists wishing to solve quantum mechanics problems relevant to today's practical applications must acquaint themselves with the available software.

In this chapter we will not discuss the advances in computer software to solve quantum problems. We will focus on the many important approximation methods that are used to solve complex problems in terms of simpler problems that can often be solved analytically or with simple numerical techniques. This chapter will address problems which are time-independent. Time-dependent problems will be discussed in the next chapter.

This chapter will primarily deal with *perturbation methods* to address quantum problems which have discrete and bound state solutions. These methods separate the Hamiltonian of the difficult problem to be solved into a simple part and a correction part, or a perturbation part. The simple part is chosen so that it is solvable. The exact solutions of the simple Hamiltonian are then used as a starting point or a testing point for the approximate solutions. Various levels of approximations can then be made, depending upon the accuracy needed and the level of the perturbation.

Another useful method which finds important uses is the *variational method*. In this approach, one *guesses* the general form of a solution and, using simple theorems that place constraint on the energy expectation values, obtains the approximate solution.

7.2 STATIONARY PERTURBATION THEORY

Perturbation theory is one of the most widely used approaches for solving difficult quantum mechanical problems. A very large number of problems have a general character which allows us to use this method. The overall approach is outlined in Fig. 7.1. The "difficult" problem is identified by a Hamiltonian, which is separated into the Hamiltonian H_0 and a correction, or perturbation, part $H^{'}$. The choice of H_0 is extremely important in the successful implementation of the approach. H_0 should satisfy the following properties:

- Solutions to H_0 should be known.
- H_0 should be such that $H^{'}$ is "small" compared to it. In other words, $H^{'}$ should be a small correction so that the solution of the full problem is very close to the solution of the simpler problem.

The perturbation approach is similar in concept to the binomial expansion of a function $f(x)$ around a point x_o. We know that this expansion allows us to write the function at a point x in terms of the function at point x_o as

$$f(x) = f(x_0) + (x - x_0) \left.\frac{\partial f}{\partial x}\right|_{x_0} + \frac{(x-x_0)^2}{2} \left.\frac{\partial^2 f}{\partial x^2}\right|_{x_0} + \cdots$$

where the first term is the zeroeth-order approximation, the second term gives the first-order correction, the third term gives the second-order correction, etc.

We will now provide the results needed to apply this approach. As noted, we write the Hamiltonian as

$$H = H_0 + H^{'} \tag{7.1}$$

where the solutions of H_0 are known and are given by

$$H_0 \psi_k = E_k \psi_k \tag{7.2}$$

We consider the case where the allowed energies of the "easy" problem are E_1, ψ_1, E_2, ψ_2, $E_3, \psi_3 \cdots$ as shown in Fig. 7.1. Let us assume that we have a level m with energy E_m and wavefunction ψ_m and there is only one such allowed energy for the simple problem. When the perturbation is applied, the energy E_m changes as a result of the perturbation. The new energy can be shown to have the value

$$E = E_m + \int \psi_m^* H' \psi_m d^3r + \sum_{n=1, n\neq m}^{N} \frac{\left|\int \psi_n^* H' \psi_m d^3r\right|^2}{(E_m - E_n)} + \cdots \tag{7.3}$$

The first-order energy correction to the original energy E_m is

$$\Delta E^{(1)} = \int \psi_m^* H' \psi_m d^3r \tag{7.4}$$

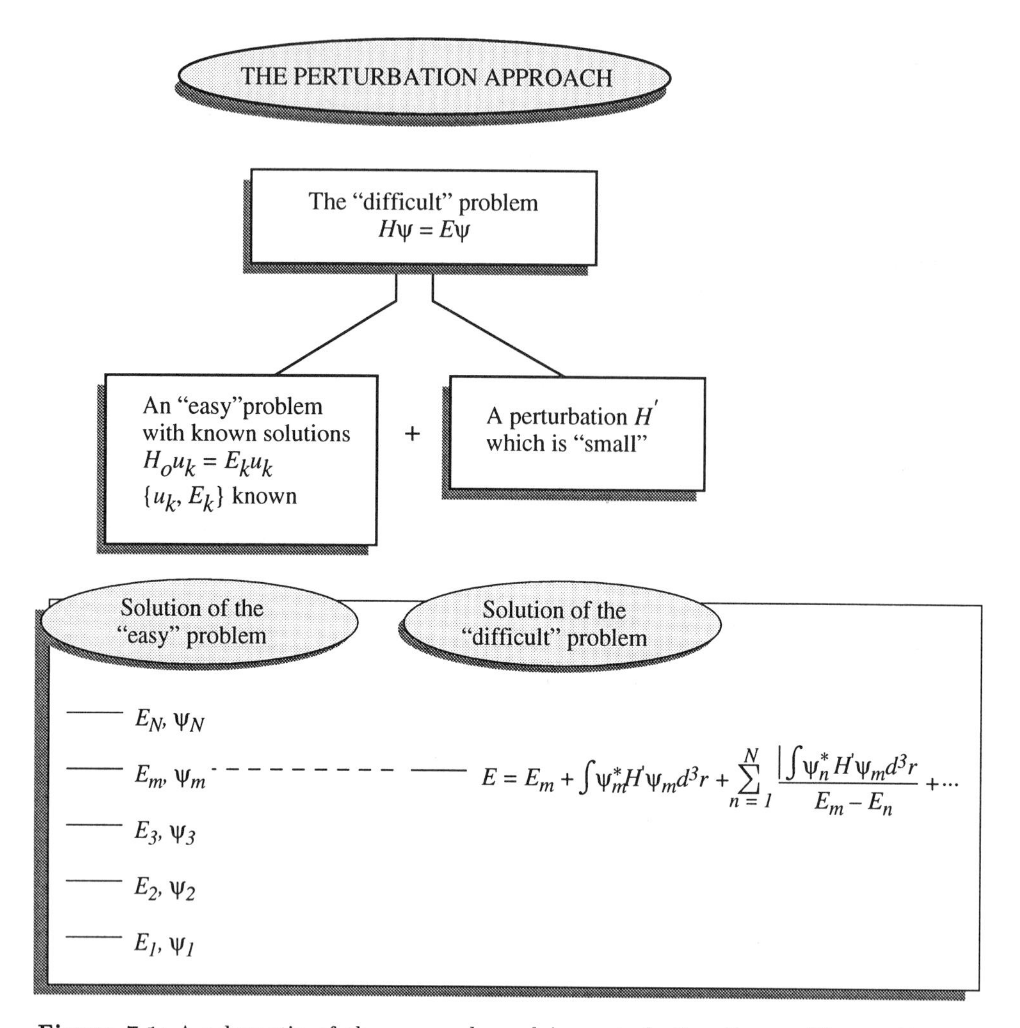

Figure 7.1: A schematic of the approach used in perturbation theory. The results are shown for a level which has only one energy solution (i.e., it is non-degenerate) for the easy problem.

while the second-order correction is

$$\Delta E^{(2)} = \sum_{n=1, n \neq m}^{N} \frac{\left| \int \psi_n^* H' \psi_m d^3r \right|^2}{E_m - E_n} \tag{7.5}$$

Higher-order corrections can also be obtained but are usually not significant.

We note that to calculate the effect of the perturbation on a problem we need to evaluate the expectation value or matrix element given by the integrals of Eqns. 7.4 and 7.5. From Eqn. 7.5 we also see that, in the second-order correction summation, we only need to include those levels for which $E_m - E_n$ is small. Levels that are well separated in energy from E_m have little influence on the correction.

7.2.1 Application Example: Stark Effect in Quantum Wells

In Chapter 5 we have discussed the properties of semiconductor quantum wells. We have seen that, as shown in Fig. 7.2a, the effective bandgap of a semiconductor quantum well is

$$E_g(eff) = E_g + E_1^e + E_1^{hh} \tag{7.6}$$

where E_1^e and E_1^{hh} are the ground state electron and heavy hole energy levels, and E_g is the bandgap of the well region. If we assume that the barrier height is infinite we have the result

$$E_g(eff) = E_g + \frac{\pi^2\hbar^2}{2W^2}\left(\frac{1}{m_e^*} + \frac{1}{m_{hh}^*}\right) \tag{7.7}$$

Now consider a situation where a uniform electric field is applied across the quantum well. The field produces a potential energy which is the perturbation H'. We have

$$H' = e\tilde{E}z \tag{7.8}$$

where $\tilde{E}$ is the field. This causes the quantum well to "bend," as shown in Fig. 7.2b. The effect of this perturbation on the ground state energy levels can be calculated using the expressions derived above. Note that since the ground state function ψ_1 is even, the first-order correction is zero

$$\Delta E^{(1)} = e\tilde{E}\int \psi_1(z) z \psi_1(z) dz = 0 \tag{7.9}$$

The second-order correction gives a result

$$\Delta E^{(2)} = -\frac{1}{24\pi^2}\left(\frac{15}{\pi^2} - 1\right)\frac{m^* e^2 \tilde{E}^2 W^4}{\hbar^2} \tag{7.10}$$

Thus the effective bandgap of the quantum well becomes

$$E_g(eff) = E_g + \frac{\pi^2\hbar^2}{2W^2}\left(\frac{1}{m_e^*} + \frac{1}{m_{hh}^*}\right) - \frac{1}{24\pi^2}\left(\frac{15}{\pi^2} - 1\right)\frac{e^2 \tilde{E}^2 W^4}{\hbar^2}(m_e^* + m_{hh}^*) \tag{7.11}$$

The effect described above is called quantum confined Stark effect and is used for a number of important optoelectronic applications. One important application is to use an electrical signal to modulate an optical signal intensity. It turns out that there is a peak (due to exciton formation) in the absorption coefficient at the point where the photon energy is near the effective gap of a quantum well, as shown in Fig. 7.2c. Due to the Stark effect, when a field is applied, this peak moves to lower energies. If an optical signal is tuned to an energy a little below the zero field peak energy, when a field is applied, the light is absorbed. Thus the electrical signal controls how much light is absorbed. A device based on this effect is called a modulator and is used in optical communications.

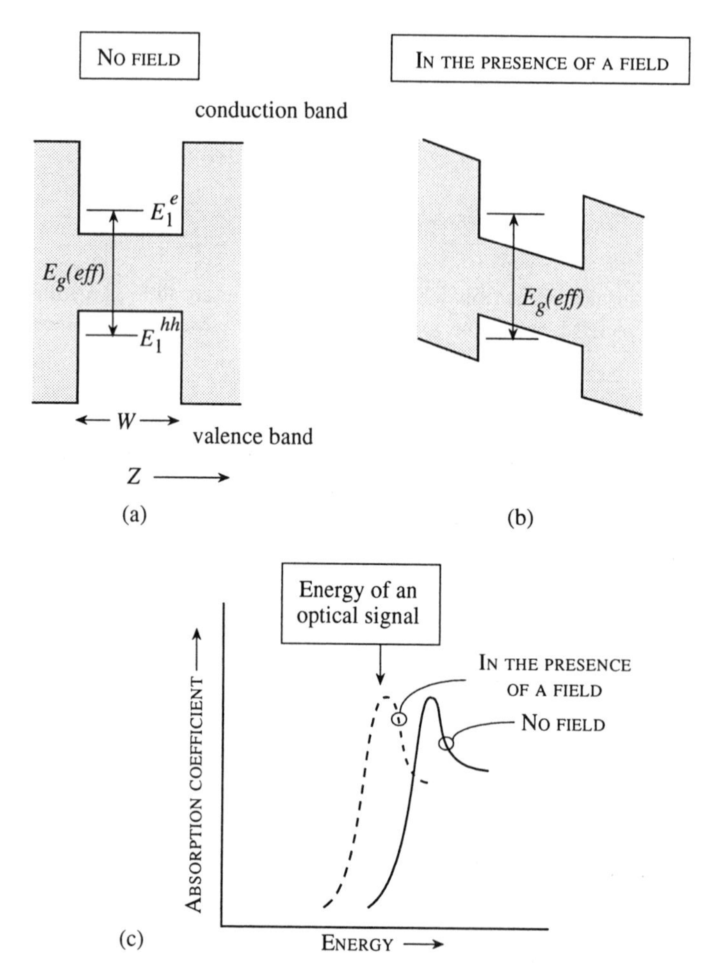

Figure 7.2: (a) A semiconductor square quantum well in the absence of any electric field. The lowest levels in the conduction band E_1^e and valence band E_1^{hh} are shown. (b) The same quantum well in the presence of a uniform electric field. The net separation between the ground state levels is changed. (c) A schematic showing how an optical signal can be modulated by an electric field in a quantum well.

7.2.2 Application Example: Magnetic Field Effects in Materials

We will now discuss what happens to electronic systems when a magnetic field is applied. The magnetic field produces a perturbation and we can use the perturbation theory to study properties of electrons (and materials) in a magnetic field.

Before examining the magnetic properties of materials we will review some important magnetic parameters. If $E(B)$ is the energy of a system in the presence of a magnetic field B, the magnetic moment is defined as

$$\mu = -\frac{\partial E}{\partial B} \tag{7.12}$$

If M is the magnetization of a system (i.e., magnetic moment per unit volume), the magnetic susceptibility per unit volume is (in SI units)

$$\chi = \frac{M}{H} = \frac{\mu_0 M}{B} \tag{7.13}$$

In cgs units

$$\chi = \frac{M}{B} \tag{7.14}$$

Material systems where χ is positive are called paramagnetic, while those where χ is negative are called diamagnetic. In ferromagnetic materials, the magnetization is positive, even in the absence of any external magnetic field.

When an electron is moving in a closed orbit with an angular momentum $\boldsymbol{L}$, it has an interaction energy with a magnetic field given by

$$\Delta E(L) = \frac{-e}{2m_0}\boldsymbol{L}\cdot\boldsymbol{B} \tag{7.15}$$

We know that an electron has an intrinsic spin in addition to angular momentum produced by its spatial motion. The interaction energy of the spin with magentic field is found to be

$$\Delta E(S) = -\frac{e}{m_0}\boldsymbol{S}\cdot\boldsymbol{B} \tag{7.16}$$

Note that the spin of an electron has a value or $-\hbar/2$. We define a quantity called the Bohr magneton given by

$$\mu_B = \frac{e\hbar}{2m_0} \tag{7.17}$$

In terms of the Bohr magneton the interaction energy of an electron with spatial angular momentum L and spin S is

$$\Delta E = \frac{-\mu_B}{\hbar}(\boldsymbol{L} + 2\boldsymbol{S})\cdot\boldsymbol{B} \tag{7.18}$$

Note that $\boldsymbol{L}$ takes values of $0, \hbar, 2\hbar, 3\hbar$, etc. Let us now consider the effect of the magnetic-field-induced perturbations on some important phenomena.

Zeeman Effects in Atoms

The Zeeman effect is a very important characterization of experiments carried out on atoms. It provides us information on the nature of electronic states in an atom. In Chapter 4 we have seen that the electronic states in an atom are represented b y three quantum numbers: i) n, the principle quantum number, $n = 1, 2, 3\cdot$; ii) ℓ, the orbital quantum number giving the orbital angular momentum for the state. In units of $\hbar$, ℓ takes values of 0, 1, 2...; iii) m, the magnetic quantum number giving the projection of the orbital angular momentum on a reference axis. m takes values $-\ell, -\ell+1, \ldots 0, \ldots, \ell-1, \ell$.

The spin of an electron is $\hbar/2$, but the spin of a state of an atom that has many electrons may be 0 or some larger value, depending upon how the electron spins add up.

In the Zeeman effect, an atom is subjected to a magnetic field, say along the z-axis. The effect of this field is that the energy of the original level is altered. The effect of the perturbation is to produce an energy change given in first-order perturbation theory by

$$\langle \Delta E \rangle = \frac{\mu_B}{\hbar} \langle \boldsymbol{L} + 2\boldsymbol{S} \rangle \cdot \boldsymbol{B} \tag{7.19}$$

Let us consider a simple case where $\boldsymbol{S}$ is zero for the atoms, i.e., the spin-up and spin-down electrons cancel each other's spin. The Zeeman effect is called normal Zeeman effect when $\boldsymbol{S} = 0$. The effect of the magnetic field is then

$$\langle \Delta E \rangle = \frac{\mu_B}{\hbar} \langle \boldsymbol{L} \rangle \cdot \boldsymbol{B} = \mu_B m B \tag{7.20}$$

We see that if we start with an s-state where $\ell = 0$, there is no effect of the magnetic field. But if we start with a p-state where $\ell = 1$, then the level is split into three levels corresponding to $m = 1, 0$, and -1. In Fig. 7.3a we show how typical p and s states are influenced by the magnetic field. The energy separation between the p and s states can be observed through absorption or emission of signals. In the absence of a magnetic field, we expect to see a single absorption line, but this splits into three when a magnetic field is applied.

An interesting effect of the influence of magnetic field on electronic levels in atoms is manifested in the Stern-Gerlach experiment shown in Fig. 7.3b. A heated oven sends out a beam of atoms which pass through a slit into a region and are deflected to different points on the screen, as shown. The Stern-Gerlach experiment clearly shows that the orbital angular momentum of electrons in atoms is quantized. In Example 7.2 we examine the details of the Stern-Gerlach experiment.

Diamagnetic Effect

In certain atomic systems, the magnetic susceptibility is negative, leading to diamagnetism. Diamagnetism can be understood on the basis of first-order perturbation theory by examining the effect of magnetic field on an electron in an atom. According to the Larmor theorem, an electron in a magnetic field undergoes a precession with a frequency

$$\omega = \frac{eB}{2m_0} \tag{7.21}$$

This precession may be considered to lead to current around the nucleus given by

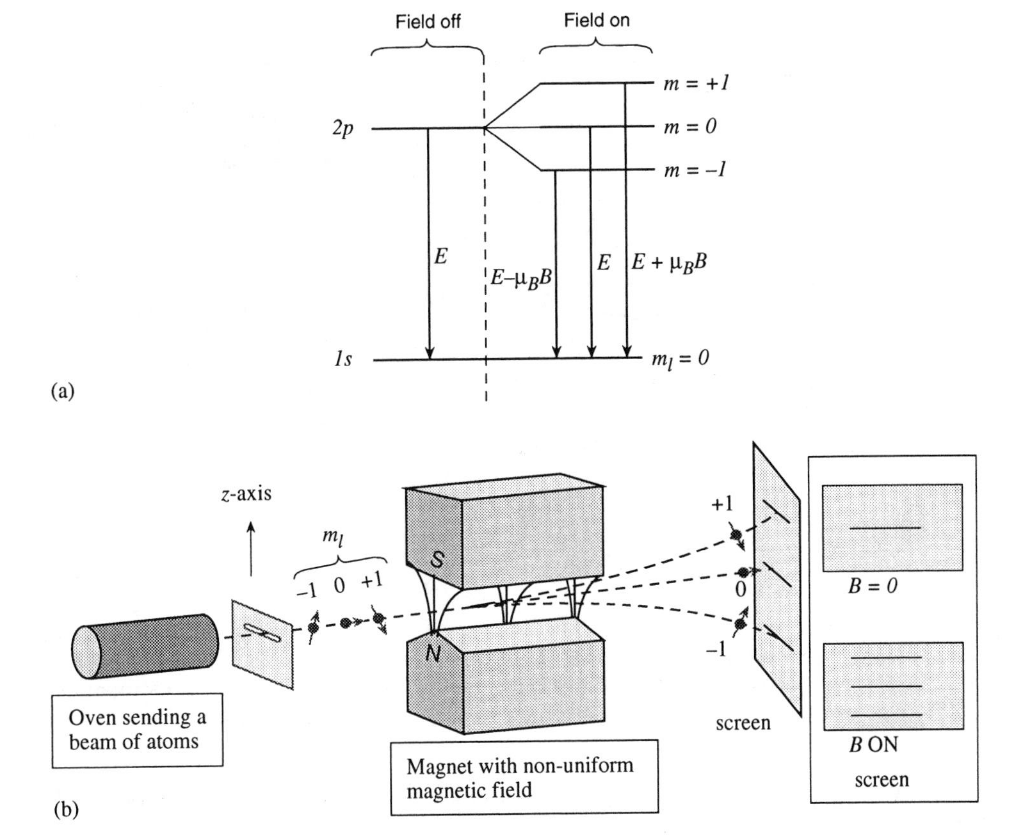

Figure 7.3: (a) A schematic of how s and p states split in the presence of a magnetic field. (b) Schematic diagram of Stern-Gerlach experiment. A beam of atoms from an oven passes through a slit and then enters a region where there is a nonuniform magnetic field. Atoms with different magnetic moments experience different forces.

(Z is the number of electrons)

$$I = (\text{charge})\ (\text{revolutions per unit time})\ = -Ze \cdot \frac{1}{2\pi} \cdot \frac{eB}{2m_0} \tag{7.22}$$

If $\langle \rho^2 \rangle = \langle x^2 \rangle + \langle y^2 \rangle$ is the mean area of the orbit around the nucleus, the magnetic moment produced by this current is

$$\mu = \text{current} \times \text{area of the loop} \ = \ \frac{-Ze^2 B}{4m} \langle \rho^2 \rangle \tag{7.23}$$

Note that in cgs units

$$\mu = \frac{-Ze^2 B}{4mc^2} \langle \rho^2 \rangle \tag{7.24}$$

If $\langle r^2 \rangle = \langle x^2 \rangle + \langle y^2 \rangle + \langle z^2 \rangle$ is the mean square distance of the electrons from the nucleus, we have

$$\langle r^2 \rangle = \frac{3}{2} \langle \rho^2 \rangle \tag{7.25}$$

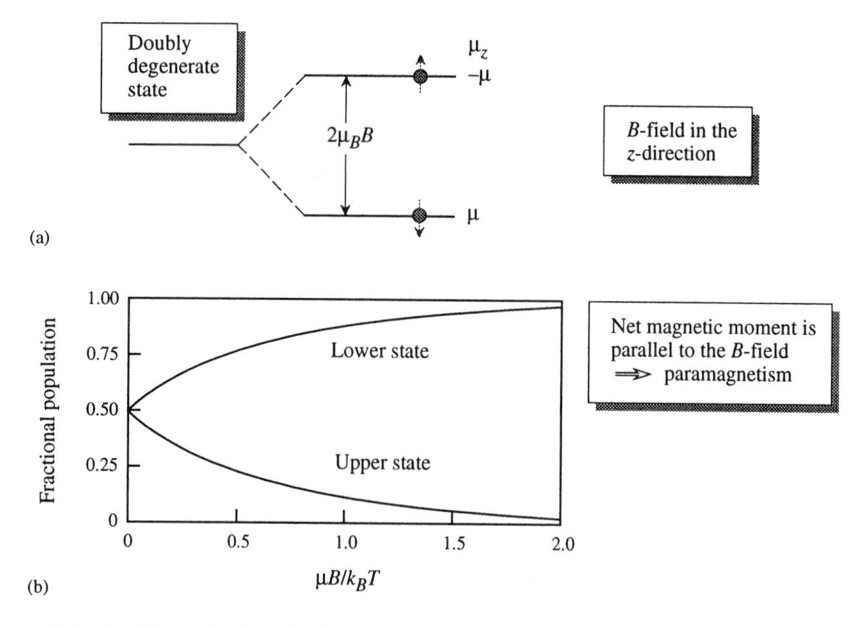

Figure 7.4: (a) The splitting of electronic levels with $L = 0$. The magnetic field is along the z-direction. Note that for electrons, the magnetic moment is opposite in sign to the spin. In the lower energy state, the magnetic moment is parallel to the field. (b) Occupation of the two-level system as a function of the ratio of the field to the thermal energy.

The diamagnetic susceptibility per volume is (N is the number of atoms per unit volume)

$$\chi = \frac{N\mu}{B} = \frac{-NZe^2}{6m_0} \langle r^2 \rangle \tag{7.26}$$

or, in cgs units

$$\chi = \frac{-NZe^2}{6m_0 c^2} \langle r^2 \rangle \tag{7.27}$$

Paramagnetic Effect

In our discussion of the Zeeman effect, we have seen that when a magnetic field is applied, the atomic levels split according to the value of $< \boldsymbol{L} + 2\boldsymbol{S} > \cdot \boldsymbol{B}$. Let us understand the paramagnetic effect, assuming that $\ell = 0$ so that an energy level will split into two states corresponding to spin-up or spin-down as shown in Fig. 7.4. The unperturbed level that is doubly degenerate (corresponding to the two spin states) is split. The lower energy state has its spin antiparallel to the field, while the magnetic moment is parallel to the field as shown in Fig. 7.4b.

The relative equilibrium populations of the two levels is, at equilibrium,

$$\begin{aligned} \frac{N_1}{N} &= \frac{\exp(\mu_B B/k_B T)}{\exp(\mu_B B/k_B T) + \exp(-\mu_B B/k_B T)} \\ \frac{N_2}{N} &= \frac{\exp(-\mu_B B/k_B T)}{\exp(\mu_B B/k_B T) + \exp(-\mu_B B/k_B T)} \end{aligned} \tag{7.28}$$

where N_1 and N_2 are the populations of the lower energy and upper energy states and N is the total population. The populations are plotted in Fig. 7.4b as a function of the field. We see that, depending upon the temperature, there is a greater number of electrons with magnetic moment along the field. The net magnetization is given by (writing $x = \mu_B B/k_B T$)

$$M = (N_1 - N_2)\mu_B = N\mu_B \frac{e^x - e^{-x}}{e^x + e^{-x}} = N\mu_B \tanh x \tag{7.29}$$

If

$$\frac{\mu_B B}{k_B T} \ll 1$$

we have

$$\boxed{M \cong \frac{N\mu_B^2 B}{k_B T}} \tag{7.30}$$

This is the Curie law for paramagnetism applied to the case where $\ell = 0$.

Ferromagnetism

One of the most interesting and important magnetic properties of materials is ferromagnetism—a property where a material has a net magnetization in the absence of any external magnetic field. Only a handful of materials display ferromagnetism, some of the important materials being Fe, Co, and Ni.

Ferromagnetism occurs in materials in which atoms have net non-zero spin for the electrons and the spins on different atoms have an interaction with each other. If the interaction between spins is zero or small, the spins are arranged randomly and the net spin of the material is zero, as shown in the top panel of Fig. 7.5. The magnetic moment of the material is thus zero and becomes non-zero only when an external magnetic field aligns the spins.

In ferromagnetic materials, the neighboring spins have a strong interaction with each other, which causes the spins in the system to align along a certain direction, as shown in Fig. 7.5. It is possible to use an external field to align the spins along a chosen direction. Then when the external field is removed, the spins stay aligned. Use of perturbation theory shows that the magnetic susceptibility in ferromagnets is given by

$$\chi = \frac{C}{T - T_c} \tag{7.31}$$

where T_c is a critical temperature. Above T_c, the material loses its ferromagnetism. In Fig. 7.5 we show the values of T_c for some materials.

EXAMPLE 7.1 Consider a hydrogen atom in which an electron jumps from the $2p$ state to the $1s$ state emitting light. Calculate the change in the wavelength of light emitted when the atom is subject to a magnetic field of $2T$.

In the absence of the field the energy of the photon is

$$E_{2p \to 1s} = -13.6\left(\frac{1}{2^2} - \frac{1}{1^2}\right) \text{ eV} = 10.2 \text{ eV}$$

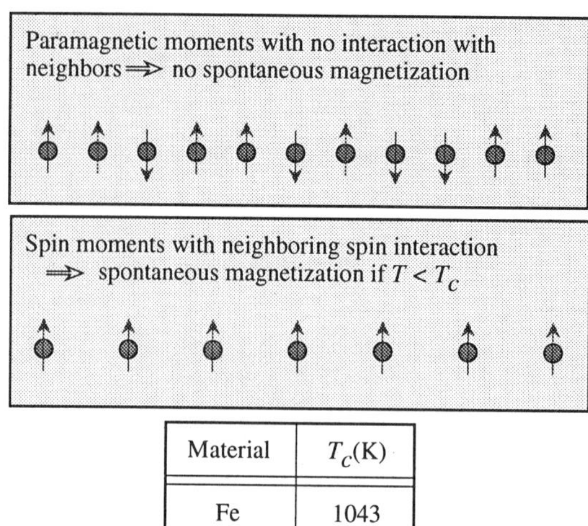

Material	T_C(K)
Fe	1043
Co	1388
Ni	627
Gd	293
CrO_2	387

Figure 7.5: Schematic of a collection of paramagnetic magnetic moments with no interactions between the moments (top figure) and a spin–spin interaction which tends to align the moments. Values of T_c for some materials are given.

The wavelength of the photon is

$$\lambda = \frac{hc}{E} = \frac{\left(6.6 \times 10^{-34} \text{ J.s}\right)\left(3 \times 10^{10} \text{ m/s}\right)}{\left(10.2 \times 1.6 \times 10^{-19} \text{ J}\right)} = 1.22 \times 10^{-5} \text{ m}$$

The change in the energy of the $2p$ level is

$$\begin{aligned} \Delta E(2p) &= \mu_B B = \left(9.27 \times 10^{-24} \text{ JT}^{-1}\right)(2T) \\ &= 1.85 \times 10^{-23} \text{ J} = 1.16 \times 10^{-4} \text{ eV} \end{aligned}$$

The change in wavelength is

$$\Delta\lambda = \frac{\lambda^2}{hc}\Delta E = 0.0139 \text{ Å}$$

EXAMPLE 7.2 Consider a Stern-Gerlach experiment in which silver atoms travel through the magnet at a speed of 700 m/s. The magnet extends over a distance of 3 cm and provides a field gradient of 1.5 T/mm. The field is along the z-axis. Calculate the separation of the two beams of silver atoms when they emerge from the magnet. The mass of silver atoms is 1.8×10^{-25} kg and the magnetic moment is 1 Bohr magneton.

The potential energy of the silver atom is

$$U = -\mu \cdot B = -\mu_z B_z$$

The force on the atoms is given by

$$F_z = \frac{-\partial U}{\partial z} = \mu_z \frac{\partial B}{\partial z}$$

The time the atoms spend in the magnet cavity is (x is the cavity length)

$$t = \frac{x}{v}$$

The separation of each component of the silver atoms is then ($\Delta z - \frac{1}{2}\frac{F}{m}t^2$)

$$\begin{aligned} d = 2\Delta z &= \mu_z \left(\frac{dBz}{dz}\right) \frac{x^2}{mv^2} \\ &= \frac{\left(9.27 \times 10^{-24}\ J/T\right)\left(1.5 \times 10^3\ T/m\right)\left(3 \times 10^{-2}\ \mathrm{m}\right)^2}{\left(1.8 \times 10^{-25}\ \mathrm{kg}\right)\left(100\ \mathrm{m/s}\right)^2} \\ &= 1.4 \times 10^{-4}\ \mathrm{m} = 0.14\ \mathrm{mm} \end{aligned}$$

EXAMPLE 7.3 The molar susceptibility of He is (in cgs units) 1.9×10^{-6} cm^3/mole. Calculate the expectation value $\langle r^2 \rangle$.

Using Eqn. 7.27, we have ($Z = 2$)

$$\begin{aligned} \langle r^2 \rangle &= \frac{\left(1.9 \times 10^{-6}\ \mathrm{cm^3/mole}\right) 6 \left(9.1 \times 10^{-28}\ \mathrm{gm}\right)\left(3 \times 10^{10}\ \mathrm{cm/s}\right)^2}{\left(6.022 \times 10^{23}\right) 2 \left(4.8 \times 10^{-10}\ \mathrm{esu}\right)^2} \\ &= 3.36 \times 10^{-17}\ \mathrm{cm}^2 \end{aligned}$$

If we were to take the square root of this value, it gives us a typical electron–nucleus distance of 0.58 Å.

7.2.3 Application Example: Magnetic Resonance Effects

We have seen in the previous subsection that the spin degeneracy of elementary particles (nuclei and electrons) is removed when a magnetic field is applied. It is possible to have magnetic effects where a variety of transitions are induced between these spin-split states. One of the most important applications of the technique is the well-known nuclear magnetic resonance (NMR), which has become a powerful medical diagnostic tool. A range of other resonance phenomena, including electron paramagnetic resonance (EPR), ferromagnetic resonance (FMR), antiferromagnetic resonance (AFMR), etc., can be understood by the principles described here.

Nuclear Magnetic Resonance

Nuclear magnetic resonance has become a very useful tool for characterizing new materials and for imaging the presence of known materials. In medical diagnostics, the proton in water molecules (in the H atom) is imaged to reveal soft tissue and thus provide an image with very high resolution and with no radiation side effects (as occur in X-ray imaging). NMR has become a powerful technique to detect tumors in people.

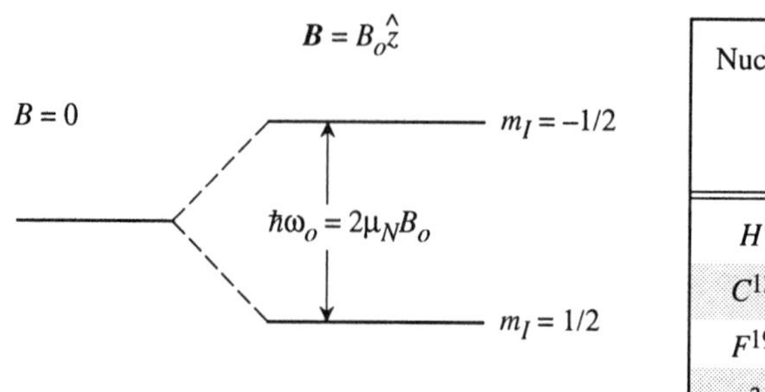

Nucleus	Nuclear magnetic moment $\frac{e\hbar}{2M_P}$
H'	2.793
C^{13}	0.702
F^{19}	2.627
P^{31}	1.131

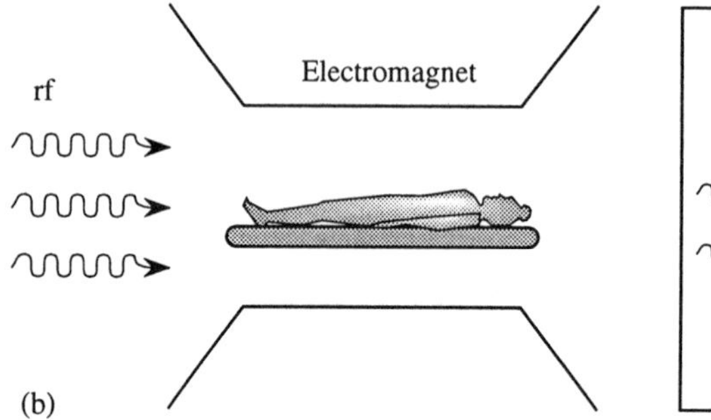

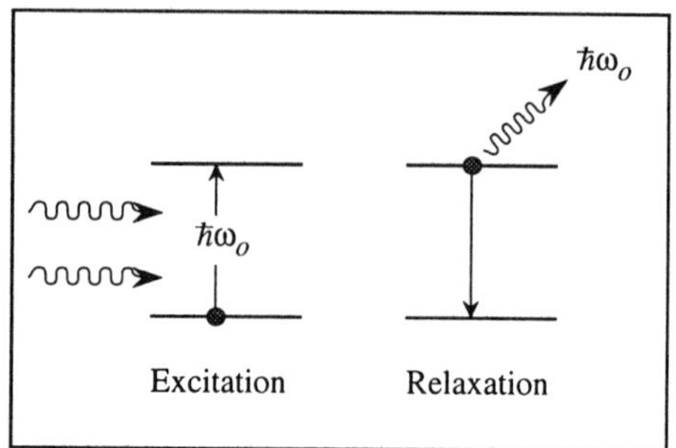

Figure 7.6: (a) The spin splitting of a nuclear state by a magnetic field. Shown in the inset are the values of the nuclear magnetic moments of some nuclei. (b) Arrangement used for magnetic resonance experiments. The excitation and relaxation processes are shown for the two levels.

The basis of NMR can be understood by starting with the Zeeman effect. In Fig. 7.6, we show the spin-split levels for a nucleus with total angular momentum 1/2. A field $B = B_o\hat{z}$ produces a splitting, given by

$$E_{12} = 2\mu_N B_o \tag{7.32}$$

where μ_N is the magnetic moment of the nuclei. In Fig. 7.6a we show the values of the nuclear magnetic moments μ_N for some important nuclei in units of the nuclear magneton, $e\hbar/2M_P$, where M_P is the proton mass. The nuclear magneton has a value of 5.051×10^{-27} JT^{-1}.

Resonance experiments involve putting the sample in a magnetic field and exciting it with rf signal. When the rf energy $\hbar\omega_o$ becomes equal to E_{12}, there is an excitation of the nuclear system, as shown in Fig. 7.6b. For protons, the frequency of the rf signal at resonance is given by

$$\begin{aligned} \nu(MHz) &= 4.258 B_o \text{ (Kilogauss)} \\ &= 42.58 B_o \text{ (Tesla)} \end{aligned}$$

For a given nuclei, there is a well-defined value of E_{21} and this can be used as a signature to identify the presence of the nuclei.

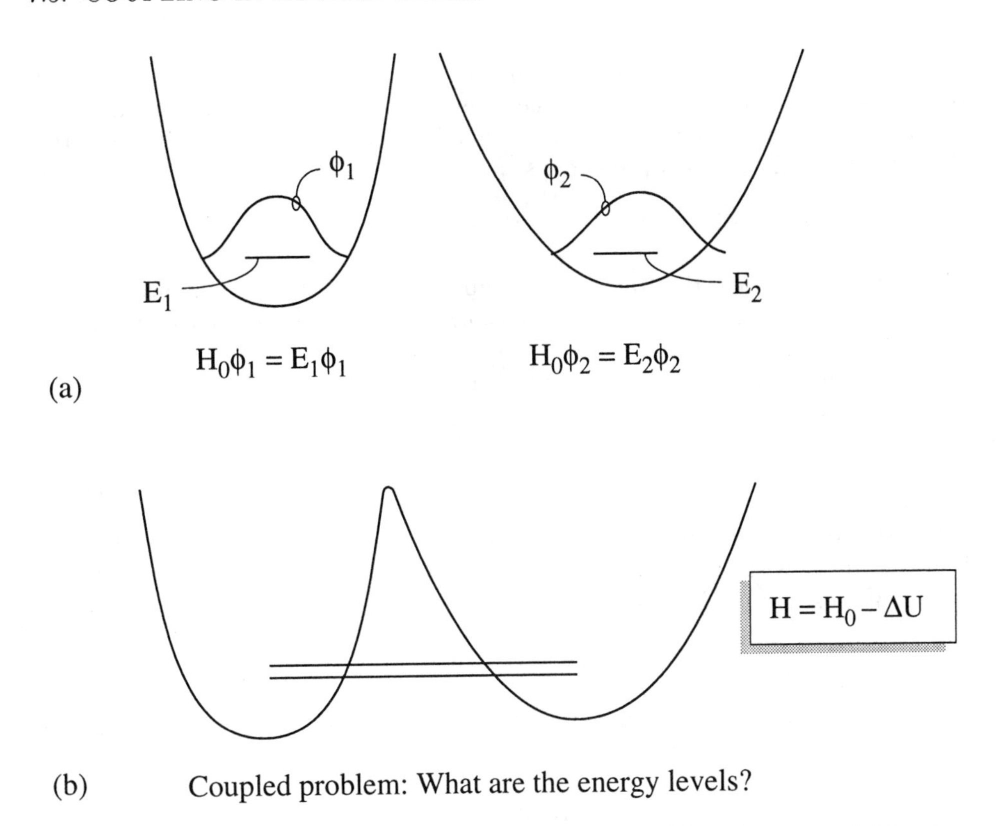

Figure 7.7: (a) A schematic of two uncoupled wells separated by a large potential barrier. (b) A schematic of two coupled potential wells.

7.3 COUPLING IN DOUBLE WELLS

Consider the following questions: Why do atoms attract each other and form molecules? Why do atoms stick to the surface of a substrate during crystal growth? How do catalysts work? How does the atomic clock work? The answers to such questions can be given by examining the physics of the very interesting problem of two *coupled quantum wells*. This problem will be addressed using the perturbation theory approach. Consider first two quantum wells (they could be identical or dissimilar) separated by a large distance so that there is no *coupling* between them. Thus electrons occupying the bound states of each quantum well have no overlap with electrons in the other well, as shown in Fig. 7.7a. Now let us assume that the wells are brought closer to each other so that electrons in each well can sense the presence of the neighboring well, as shown in Fig. 7.7b. What happens to the allowed electronic levels in the coupled quantum wells problem?

When the quantum wells are spaced far apart and are uncoupled, let us say they each have states with energies E_1 and E_2, respectively, and wavefunctions ϕ_1 and ϕ_2, as shown in Fig. 7.7a. There may be other states in the quantum wells, but we will assume that their energies lie far from E_1 and E_2. We will see later that states far removed from the ones under consideration have a smaller effect on the solutions. We are interested in the solutions of the problem where the wells are brought closer so that there is some coupling between the wells. We write the Hamiltonian of the coupled problem as

$$H = H_0 - \Delta U \tag{7.33}$$

where ΔU is the correction due to the coupling of the well. The potential energy is reduced in the region between the wells, compared to the uncoupled system, so that ΔU is positive.

In the absence of any coupling, we have

$$\begin{aligned} H_0\phi_1 &= E_1\phi_1 \\ H_0\phi_2 &= E_2\phi_2 \end{aligned} \tag{7.34}$$

When the wells are coupled, the functions ϕ_1 and ϕ_2 are no longer the solutions. However, the new solutions can be expressed in terms of the uncoupled solutions to a good approximation. Let us write the solution for the coupled problem as

$$\psi = a_1\phi_1 + a_2\phi_2 \tag{7.35}$$

where a_1 and a_2 are unknown parameters which we will solve. If the system is uncoupled we have the two solutions

$$\psi_1 = \phi_1; \quad a_1 = 1, \ a_2 = 0$$

$$\psi_2 = \phi_2; \quad a_1 = 0, \ a_2 = 1$$

Coming to the coupled problem, the equation to be solved has the form

$$H(a_1\phi_1 + a_2\phi_2) = E(a_1\phi_1 + a_2\phi_2) \tag{7.36}$$

where H represents the full Hamiltonian of the coupled problem. We now multiply this equation from the left by ϕ_1^* and integrate over space to get

$$a_1 \int \phi_1^*(H_0 - \Delta U)\phi_1 d^3r + a_2 \int \phi_1^*(H_0 - \Delta U)\phi_2 d^3r = Ea_1 \int \phi_1^*\phi_1 d^3r + Ea_2 \int \phi_1^*\phi_2 d^3r$$

Using the following equations:

$$\begin{aligned} \int \phi_1^*(H_0 - \Delta U)\phi_1 d^3r &= E_1 \int \phi_1^*\phi_1 d^3r - \int \phi_1^*\Delta U \phi_1 d^3r \cong E_1 \\ \int \phi_1^*\phi_1 d^3r &= 1 \\ \int \phi_1^*\phi_2 d^3r &= 0 \end{aligned}$$

we get

$$a_1(E_1 - E) + a_2 H_{12} = 0 \tag{7.37}$$

where we have defined

$$H_{12} = -\int \phi_1^* \Delta U \phi_2 d^3r$$

The quantity H_{12} is called the matrix element of the Hamiltonian between the two states ϕ_1 and ϕ_2. In case the quantum wells are separated by a large distance the matrix element is zero. It increases as the coupling increases.

If we repeat the process described above but multiply Eqn. 7.36 by ϕ_2^* instead of ϕ_1^*, we get

$$H_{21}a_1 + a_2(E_2 - E) = 0 \tag{7.38}$$

where

$$H_{21} = -\int \phi_2^* \Delta U \phi_1 d^3r$$

Assuming that the energy operator is real (energy is conserved) we have,

$$H_{21} = H_{12}$$

The two coupled equations (Eqns. 7.37 and 7.38) can be written as a matrix vector product

$$\begin{vmatrix} E - E_1 & H_{12} \\ H_{21} & E - E_2 \end{vmatrix} \begin{vmatrix} a_1 \\ a_2 \end{vmatrix} = 0$$

To get non-trivial solutions of this equation, the determinant of the matrix must vanish. This gives us a quadratic equation with the following solutions:

$$E = \frac{E_1 + E_2}{2} \pm \sqrt{\frac{(E_1 - E_2)^2}{4} + H_{12}^2} \tag{7.39}$$

The coefficients a_1 and a_2 can now be solved for and are

$$\frac{a_1}{a_2} = \frac{H_{12}}{E - E_1} \tag{7.40}$$

or

$$\frac{a_2}{a_1} = \frac{E - E_2}{H_{12}} \tag{7.41}$$

The simple equation we have derived has very useful implications for understanding many interesting and important physical systems.

7.3.1 Coupling of Identical Quantum Wells

Let us examine in some detail the case where the two coupled quantum wells are identical and have the same initial energies and states, as shown in Fig. 7.8. It is important to keep in mind that when we talk of "quantum wells" we are simply referring to any problem with bound states. Thus the quantum wells may be atoms and molecules as well as potential wells created by use of semiconductors. Let us write

$$\begin{aligned} E_1 &= E_2 = E_0 \\ H_{12} &= H_{21} = -A \end{aligned} \tag{7.42}$$

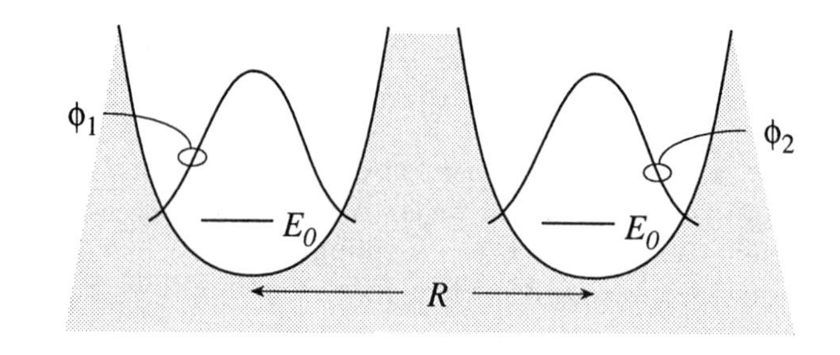

(a)

$\Psi_A = \frac{1}{\sqrt{2}}(\phi_1 - \phi_2)$

$E_0 + A$

$E_0 - A$

R

$\Psi_S = \frac{1}{\sqrt{2}}(\phi_1 + \phi_2)$

(b)

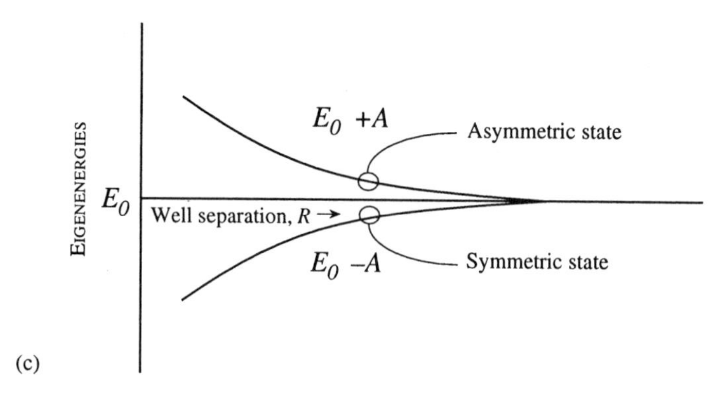

(c)

Figure 7.8: (a) A schematic of two identical uncoupled wells; (b) Coupled identical wells with energy levels and eigenfunctions; and (c) dependence of the symmetric and asymmetric eigenenergies on well separation or coupling.

The quantity A represents the coupling coefficient between the wells.

From the derivation given above, the energy eigenvalues of the coupled system is now

$$\begin{aligned} E_S &= E_0 - A \\ E_A &= E_0 + A \end{aligned} \tag{7.43}$$

For the state with energy E_S the coefficients of the state are, from Eqn. 7.40,

$$a_1 = a_2 \tag{7.44}$$

while for the state with energy E_A we have

$$a_1 = -a_2 \tag{7.45}$$

If we normalize the state using

$$a_1^2 + a_2^2 = 1$$

we get the following solutions:

Symmetric state:

$$\Psi_S = \frac{1}{\sqrt{2}}[\phi_1 + \phi_2]; \quad E_S = E_0 - A \tag{7.46}$$

Asymmetric state:

$$\Psi_A = \frac{1}{\sqrt{2}}[\phi_1 - \phi_2]; \quad E_A = E_0 + A \tag{7.47}$$

The states are shown schematically in Fig. 7.8b. We see that as a result of the coupling between the wells, the degenerate states E_0 are split in to two states— one with energy below the uncoupled state and one with a higher energy. Note that as the wells are brought closer to each other the coupling strength will increase and the symmetric state energy will continuously decrease, as shown in Fig. 7.8c. *The system behaves as if the coupling creates an attactive interaction in the symmetric state.*

7.3.2 Application Example: Hydrogen Molecule Ion

As an example of the coupled quantum well problem, let us consider two hydrogen nuclei with a single electron. As shown in Fig. 7.9, when the two nuclei are far apart the electron can be in either one of the nuclei. Let these states, which are the ground states of the hydrogen atom, be denoted as before by ϕ_1 and ϕ_2, respectively. The energy of these states is just

$$E_0 = E_H = -13.6 \text{ eV}$$

As the nuclei are brought closer together, the electron on one atom feels the potential due to the attractive potential of the other nucleus. This gives a matrix element, which is a function of the inter-nucleus separation, R

$$H_{12} = -A(R)$$

As noted in the previous subsection, the original state will now split into a lower energy symmetric state and a higher energy asymmetric state.

In Fig. 7.10a we show a plot of the change in the electronic energy as a function of the inter-nucleus separation. In the symmetric state, where the electronic function has a high probablility of occupying the space between the two nuclei, the system feels an attractive interaction, since the energy is reduced as the nuclei come closer. This is the reason the ion H_2^+ is stable. For the asymmetric state, where the electron is pushed away from the center of the two nuclei, there is an effective

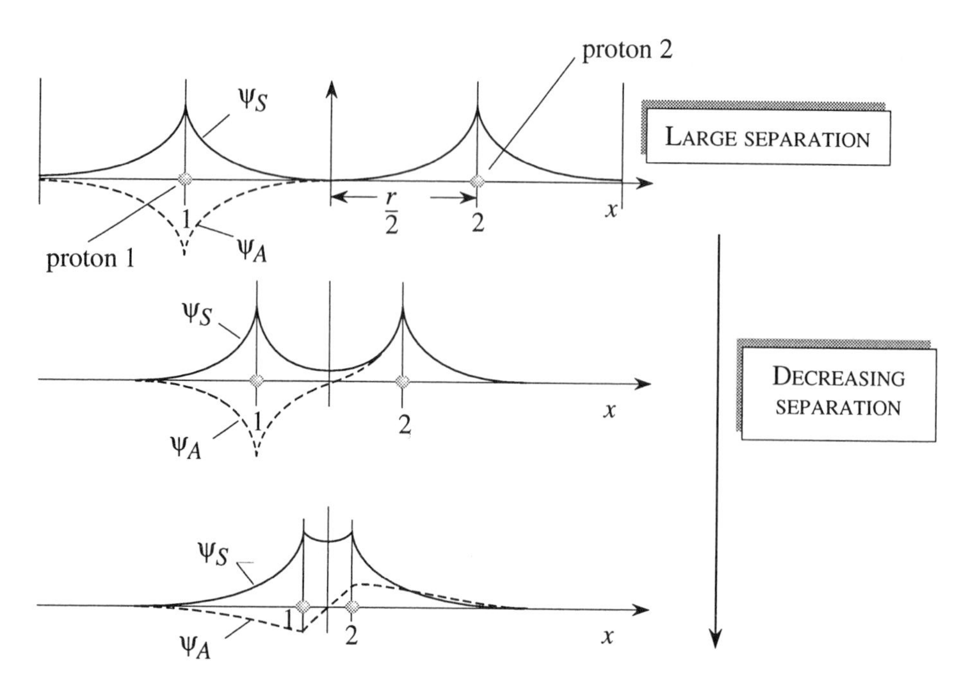

Figure 7.9: A schematic of the symmetric and antisymmetric wavefunction constructed from the ground sate u_{100}. In the symmetric state, the electron is closer to the nuclei.

repulsive interaction, since the energy is larger than the energy when the nuclei are separated by a large distance.

To obtain the total energy of the H_2^+ ion as a function of inter-nucleus separation we need to add the repulsive interaction between the positively charged nuclei. This amounts to an energy

$$U_{rep} = \frac{1}{4\pi\epsilon_0}\frac{e^2}{R}$$

This energy is plotted in Fig. 7.10a. The total energy in the ion due to the presence of the two nuclei is now (in the ground state)

$$E(H_2^+) = E_0 - A(R) + \frac{1}{4\pi\epsilon_0}\frac{e^2}{R} \tag{7.48}$$

This total energy is plotted in Fig. 7.10b. We see that the energy minimizes at an inter-nucleus separation of R_{min}. Numerical calculations show that

$$R_{min} = 1.0 \text{ Å} \tag{7.49}$$

The binding energy of the ion is found to be (E_H is the magnitude of the ground state energy)

$$E_b = -0.2E_H = -2.7 \text{ eV} \tag{7.50}$$

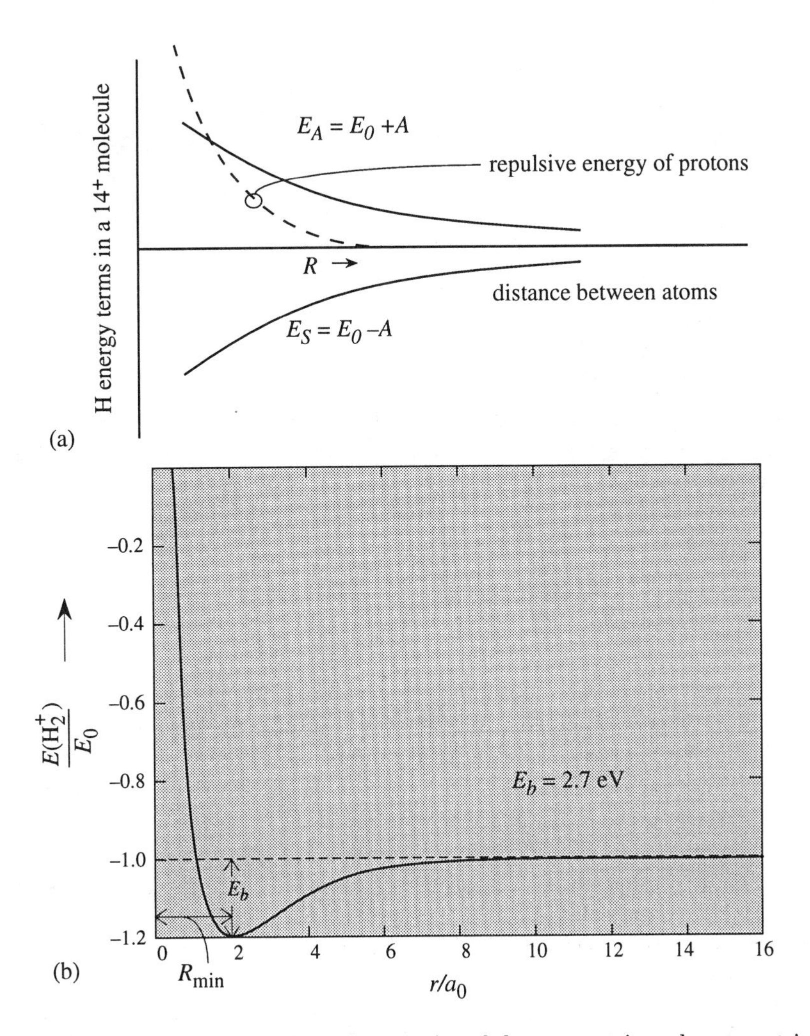

Figure 7.10: (a) A schematic of how the energies of the symmetric and asymmetric states vary with separation of the H-atom nuclei. Also shown is the repulsive energy arising from proton–proton repulsion; and (b) change in the energy of the H_2^+ ion as a function of inter-proton separation.

The example in this subsection shows how chemical bonds are formed by *sharing* of electrons between two nuclei. We have seen how the attractive interaction due to the coupling of the two bound states and the repulsive potential due to the nuclear charge play a role in setting the equilibrium bond distance.

7.3.3 Coupling Between Dissimilar Quantum Wells

Let us examine how the attractive and repulsive interactions due to coupling of wells are influenced when the two starting potetial wells are dissimilar. This would occur if, for example, we had a H and a Na nucleus coming together instead of two

H nuclei, as considered in the previous subsection. Let the starting energies of the two potentials be E_1 and E_2, and let E_2 be larger than E_1. From the derivation given (see Eqn. 7.39) we have for the symmetric and asymmetric state

$$\begin{aligned} E_S &= E_1 - \frac{A^2}{E_2 - E_1} \\ E_A &= E_2 - \frac{A^2}{E_2 - E_1} \end{aligned} \tag{7.51}$$

If the value of $E_2 - E_1$ is much larger than the coupling coefficient A, we see that the effect of the coupling is very small. This is the reason bonding between dissimilar atoms is weak.

7.3.4 H_2 Molecule

Let us now move on to the problem of how atoms attract each other to form a chemical bond. This is obviously an important question in chemistry and material science. The problem of H_2^+ ion discussed previously sheds some light on the problem. However, in that problem we only had to consider one electron and two nuclei. What happens when there are two electrons?

As before, let us consider two H atoms initially far apart, as shown in Fig. 7.11a. In this uncoupled state each atom has an electron cloud around its nucleus, just as we expect in an isolated atom. The two states ϕ_1 and ϕ_2 are created through an exchange of the two electrons, as shown in Fig. 7.11a. For clarity we will call the electrons 1 and 2.

As the atoms come closer to each other, there is an interaction between the atoms as each electron senses the attractive potential of the neighboring nucleus and the repulsive potential of the neighboring electron. The overall coupling is again represented by $A(R)$. Once again we have a symmetric state and an asymmetric state made from the original ϕ_1 and ϕ_2. In the terminology of chemical bonds, the symmetric state is called the bonding state and the asymmetric state is called the antibonding state. In the symmetric state the overall interaction is attractive and the energy of the system decreases in comparison with the energy it has when the atoms are well separated. In the asymmetric state there is a repulsive interaction. At very close spacing the repulsive potential of the charged particles dominates. The overall energy is shown in Fig. 7.11b. We see that in the symmetric state there is an equilibrium spacing between the atoms where the energy is minimum.

Detailed calculations show that the spacing between the atoms in equilibrium is 0.74 Å. This is the proton–proton spacing in the H_2 molecule. The binding energy of the molecule, i.e., the energy difference between the lowest energy state and the energy of two isolated H atoms is 4.52 eV, as shown in Fig. 7.11. Also shown are the binding energies (or dissociation energies) of several molecules.

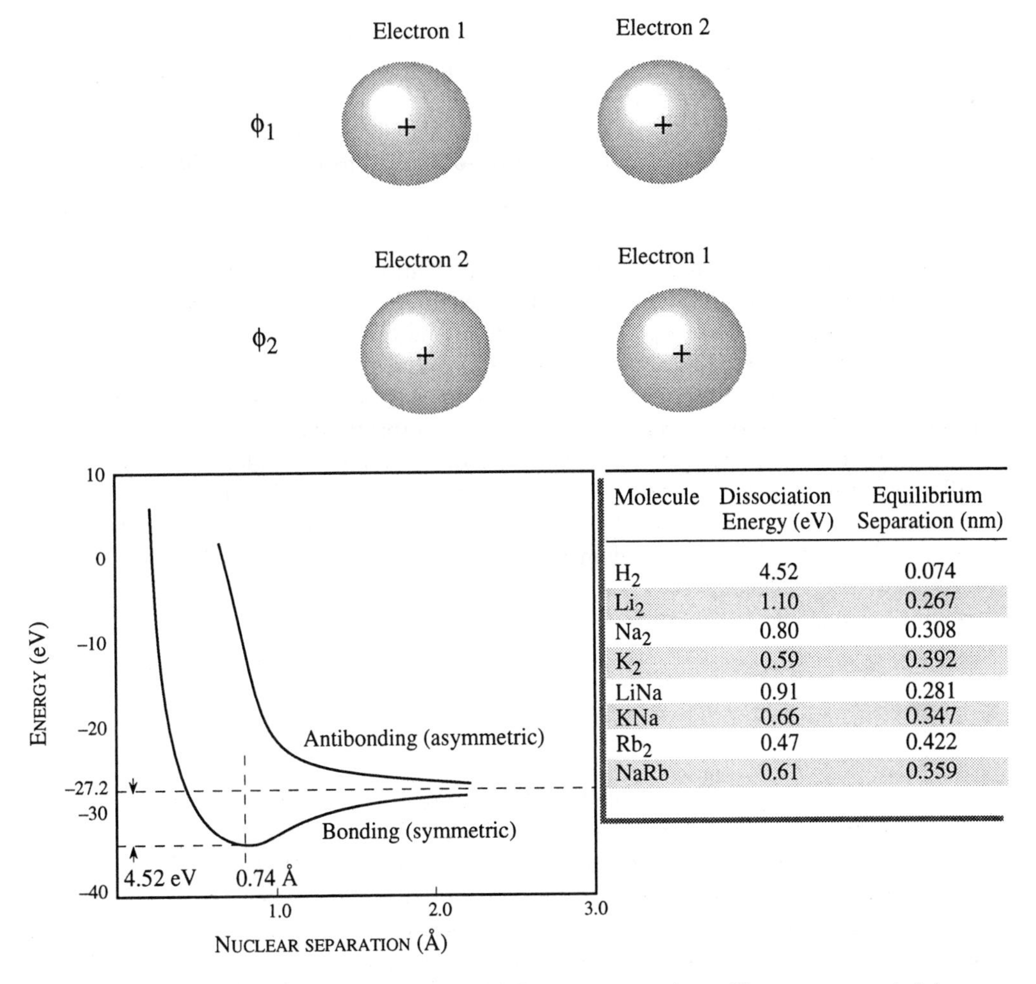

Molecule	Dissociation Energy (eV)	Equilibrium Separation (nm)
H_2	4.52	0.074
Li_2	1.10	0.267
Na_2	0.80	0.308
K_2	0.59	0.392
LiNa	0.91	0.281
KNa	0.66	0.347
Rb_2	0.47	0.422
NaRb	0.61	0.359

Figure 7.11: (a) A schematic of the two different states of two H atoms separated by a large distance; and (b) change in the symmetric (bonding) and asymmetric (antibonding) states as a function of nuclear spacing. Also shown are dissociative energies of several molecules.

7.3.5 Application Examples: Ammonia Molecules and Organic Dyes

The problem of the two degenerate states represented by the potential wells coupled through a small interaction is manifested in many important problems. These include stabilization of benzene molecules, oscillations of an ammonia molecule as used in masers, etc. Many of these phenomena are extremely useful for building important devices such as masers used to generate microwaves, atomic clocks, and special dyes.

We have seen that if we start with two states that are degenerate and allow some coupling between them, the original energy splits into two energy levels,

E_1 and E_2. In most systems, this energy difference is fixed by nature, but it is also possible to design systems where this energy can be tailored by altering the coupling between the original degenerate states.

Once the two levels E_1 and E_2 are present, it is possible to have electrons go from one state to another and emit or absorb photons of energy given by

$$\hbar\omega = E_2 - E_1 \tag{7.52}$$

If the separation between E_1 and E_2 is $\sim 10^{-4}$ meV, the photons frequency is in the range of microwave frequencies used, for example, in radar detectors or wireless communications. If the energy difference is of the order of 1 eV, the photons are in the visible regime. The ability of the coupled systems to emit or absorb light in these important regimes makes such systems very useful in technology. We will briefly discuss the NH_3 maser, organic dyes, and atomic clocks.

In Fig. 7.12 we show the two states of the NH_3 molecule. The two states differ by the way the N atom is oriented. The N atom can flip from one side to another and go from one state to another. This phenomenon can be described by the coupled potential well problem discussed earlier and shown in Fig. 7.12b. As discussed, the molecular state oscillates between the two configurations, and the oscillation frequency is found to be

$$\begin{aligned} \omega &= 1.508 \times 10^{11} \text{ radians/s} \\ \nu &= 24 \text{ GHz} \end{aligned}$$

This corresponds to a splitting $E_2 - E_1$ of 10^{-4} eV. The 24 GHz frequency is very useful for many microwave applications and is exploited in fabricating the maser operating at this frequency. It is also important to note that the NH_3 molecule carries an electric dipole moment, as shown in Fig. 7.12a. When the state of the molecule changes, the dipole moment is reversed. *This effect can be used to alter $E_2 - E_1$ by placing the NH_3 molecule in an electric field.*

The resonance frequency associated with the coupled well problem is exploited in organic molecules to produce "dyes" that can absorb light at a well-defined color or optical frequency. Due to the symmetry present in the structure of organic molecules, there are two equivalent states of the molecule. The flipping of electrons between these two states (or potential wells) results in values of E_2 (energy for the asymmetric state) and E_1 (energy for the symmetric state), which are such that $E_2 - E_1$ is about 1.0 to 2.0 eV, i.e., in the visible photon regime. Such dyes are used as fabric whiteners.

7.3.6 Application Example: Atomic Clock

The atomic clock uses the internal resonance frequency of atoms (or molecules) to measure time. The resonant structure generates pulses at regular intervals and this is combined with a pulse counter (an electronic device) to produce the clock. As discussed in this section, the resonant frequency is given by

$$\hbar\omega = E_2 - E_1$$

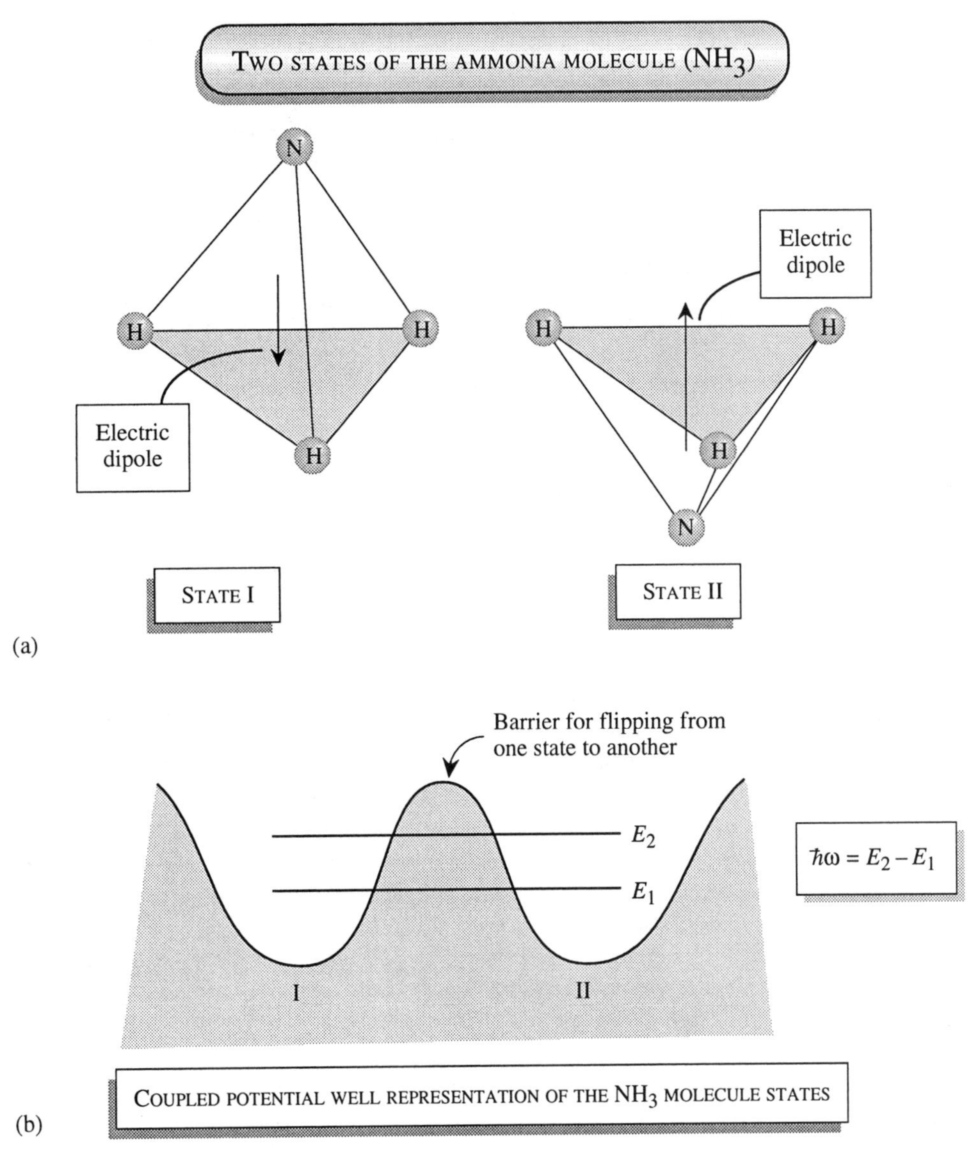

Figure 7.12: (a) The two states of an ammonia molecule. The orientation of the N atom can "flip" from one state to another. (b) A schematic of the potential energy profile of the ammonia molecule. The potential is similar to the problem of two coupled wells. Transitions between the two split levels can produce microwave radiation.

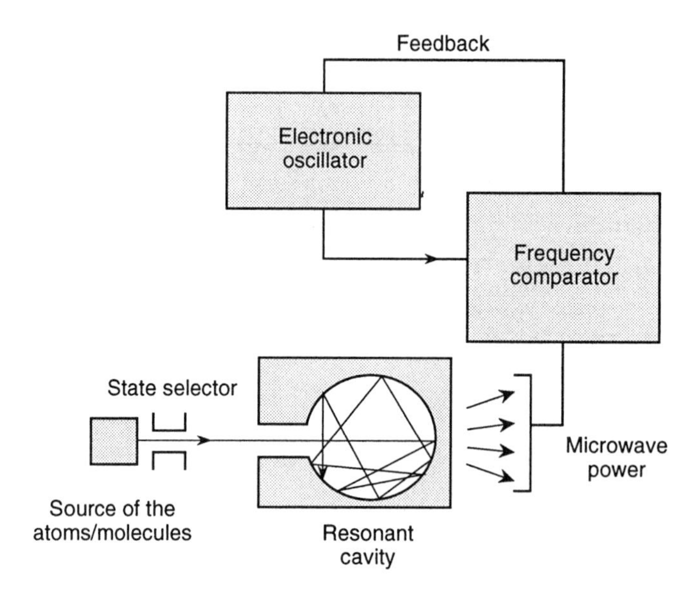

Figure 7.13: A typical setup for an atomic clock.

where E_2 and E_1 are the energy levels of the two states of the system.

The atomic clock is extremely stable and easy to reproduce anywhere. The atomic frequency standard (i.e., the atomic clock) can be of two types: active or passive. An active standard uses as a reference the electromagnetic radiation emitted by molecules as they decay from a higher state to the lower state. A passive clock attempts to match the frequency of the atomic (molecular) system.

The first demonstration of the atomic clock was based on the NH_3 maser using the 24 GHz transition of the molecule. A number of other clocks have been demonstrated, including those based on cesium atoms (using a 9193 MHz energy difference in the cesium-133 atom), hydrogen maser (using a 1420 MHz resonance in the splitting of H atoms), and rubidium atoms (using a 6835 MHz resonance).

A typical schematic of the atomic clock is shown in Fig. 7.13. A state selector is used to prepare the system in its excited state. The excited atoms are then placed in a resonant cavity tuned to the resonant frequency of the atoms. The decaying atoms emit microwave power that is then compared with an electronic oscillator.

Atomic clocks are widely used in applications where precision is of paramount importance. Errors as small as one second in 10^6 years are possible to detect in these devices. Applications include the Global Positioning System (which uses cesium and rubidium clocks), communications, radio astronomy, space exploration, etc.

7.4 VARIATIONAL METHOD

An extremely useful approximation method in quantum mechanics is the *variational method* discussed in this section. There are a number of important problems in which the Hamiltonian is such that it simply cannot be separated into a solvable part and a small perturbation. For such problems, there is no good starting point to apply the perturbation theory. The variational method can be used to solve such problems and many others where other approximation methods are too difficult to employ.

The variational method is based upon simple theorems that place limits on the energy expectation values of a general Hamiltonian. Let us derive these limits for the ground state expectation energy.

Consider a general Hamiltonian H and a complete and orthonormal set of functions $\{u_E\}$. If ψ is an energy eigenfunction, we can write

$$\psi = \sum_E A_E u_E \tag{7.53}$$

where

$$Hu_E = Eu_E \tag{7.54}$$

The expectation value of the Hamiltonian in the state ψ is

$$\langle H \rangle = \int \psi^* H \psi \, d\tau = \sum_E E \left| A_E \right|^2 \tag{7.55}$$

where $d\tau$ represents the coordinates which describe ψ. We now replace the energy in the summation on the right-hand side by the smallest energy E_0. This gives us an inequality

$$\langle H \rangle \quad \geq \quad \sum_E E_0 \left| A_E \right|^2 = E_0 \sum_E \left| A_E \right|^2 \tag{7.56}$$

If ψ is normalized, we have

$$\langle H \rangle \quad \geq \quad E_0 \tag{7.57}$$

In case the function is not normalized, we have

$$\boxed{E_0 \leq \frac{\int \psi^* H \psi \, d\tau}{\int \psi^* \psi \, d\tau}} \tag{7.58}$$

Based on this inequality, we obtain the ground state energy of a problem by the following recipe: We start with a "guess" for the wavefunction. The wavefunction depends upon several parameters. We then evaluate the integral given in the equation above. The integral will depend upon the variational parameters. These parameters are varied until the expectation value of the energy is minimum. The variational method can also be applied to obtain the eigenfunctions for the higher excited states if the trial function is orthogonal to the eigenfunctions of the lower states.

EXAMPLE 7.4 The choice of the *form of the eigenfunction* used is quite important in getting accurate values for the eigenenergies. For example, let us consider the ground state of the hydrogen atom problem. Consider the following trial functions:

$$\begin{aligned}
\psi_1 &= A_1 e^{-(b/a)r} \\
\psi_2 &= A_2 \frac{1}{b^2 + (r/a)^2} \\
\psi_3 &= A_3 \frac{r}{a} e^{-(b/a)r}
\end{aligned}$$

where

$$\begin{aligned} a &= \text{Bohr radius} \\ &= \frac{4\pi\epsilon_0\hbar^2}{me^2} \end{aligned}$$

From normalization conditions we have

$$\begin{aligned} A_1 &= \left(\frac{b^3}{\pi a^3}\right)^{1/2} \\ A_2 &= \frac{1}{\pi}\left(\frac{b}{a^3}\right)^{1/2} \\ A_3 &= \left(\frac{b^5}{3\pi a^3}\right)^{1/2} \end{aligned}$$

Using the notation

$$\begin{aligned} \Delta &\equiv \nabla^2 \\ &= \frac{1}{r^2}\frac{\partial}{\partial r}\left(r^2\frac{\partial}{\partial r}\right) + \frac{1}{r^2 \sin\theta}\frac{\partial}{\partial\theta}\left(\sin\theta\frac{\partial}{\partial\theta}\right) + \frac{1}{r^2\sin^2\theta}\frac{\partial^2}{\partial\phi^2} \end{aligned}$$

in spherical coordinates, we have for ψ_1:

$$\begin{aligned} E_0 &= \min \int \psi_1^* \left(-\frac{\hbar^2}{2m}\Delta - \frac{e^2}{4\pi\epsilon_0 r}\right)\psi_1 \, d^3r \\ &= \min\left\{\frac{2\hbar^2 b^3}{ma^3}\int_0^\infty e^{-(b/a)r}\,\Delta e^{-(b/a)r} r^2\, dr \right. \\ &\qquad \left. - \frac{4e^2b^2}{4\pi\epsilon_0 a^3}\int_0^\infty e^{-2br/a}\, r\, dr\right\} \end{aligned}$$

Now,

$$\begin{aligned} \int_0^\infty e^{-(b/a)r}\,\Delta e^{-(b/a)r} r^2\, dr &= -\frac{a}{4b} \\ \int_0^\infty e^{-(2b/a)r} r\, dr &= \frac{a^2}{4b^2} \end{aligned}$$

Thus

$$\begin{aligned} E_0 &= \min\left(\frac{\hbar^2 b^2}{2ma^2} - \frac{e^2 b}{(4\pi\epsilon_0)a}\right) \\ &= -\frac{e^2}{(4\pi\epsilon_0)2a} \end{aligned}$$

which is the correct result. For the different choices of the trial functions, we get

$$\begin{aligned} \psi_1 &: b = 1; \quad E_0 = E_H = -\frac{me^4}{2(4\pi\epsilon_0)^2\hbar^2} \\ \psi_2 &: b = \frac{\pi}{4}; \quad E_0 = -0.81E_H \\ \psi_3 &: b = \frac{3}{2}; \quad E_0 = -0.75E_H \end{aligned}$$

Thus an incorrect form of the starting trial function can make a significant error in the energy solutions. The error in the wavefunctions is, however, much greater than the error in the energy.

7.4.1 Application Example: Exciton in Quantum Wells

In recent years, a number of exciting devices have been demonstrated using exciton effects in semiconductor quantum wells. The exciton is, as discussed in Section 5.4, a particle describing an electron-hole interacting via the Coulombic interaction. In a three-dimensional world, the ground state binding energy of the exciton is given by

$$E_b^{ex} = \frac{e^4 m_r^*}{2(4\pi\epsilon)^2\hbar^2}$$

where m_r^* is the e-h reduced mass and ϵ is the dielectric constant of the material. This binding energy is of the order of a few meV in most materials. The exciton binding is increased if the exciton is confined in a two-dimensional world. Let us start with a simple variational calculation to obtain the exciton binding energy in a strictly two-dimensional world.

Note that the Laplacian in a 2D system (ρ, θ) is

$$\nabla^2 \longrightarrow \frac{1}{\rho}\frac{\partial}{\partial\rho}\left(\rho\frac{\partial}{\partial\rho}\right) + \frac{1}{\rho^2}\frac{\partial^2}{\partial\theta^2} \tag{7.59}$$

Let us assume a wavefunction for the exciton

$$\psi = Ae^{-\alpha\rho} \tag{7.60}$$

Using the normalization

$$\int \psi^2 \,\rho d\rho \; d\theta = 1$$

we get

$$A = \sqrt{\frac{2}{\pi}}\alpha$$

The expectation energy of the exciton system is

$$\langle E\rangle = A^2 \int e^{-\alpha\rho}\left[-\frac{\hbar^2}{2m_r^*}\nabla^2 - \frac{e^2}{4\pi\epsilon\rho}\right] e^{-\alpha\rho}\rho \; d\rho \; d\theta$$

Solving for the integral we get

$$\langle E\rangle = 2\left[\frac{\hbar^2\alpha^2}{4m_r^*} - \frac{e^2\alpha}{4\pi\epsilon}\right]$$

Using

$$\frac{\partial\langle E\rangle}{\partial\alpha} = 0 \quad \text{gives } \alpha = \frac{2m_r^* e^2}{4\pi\epsilon\hbar^2}$$

Thus the exciton energy is

$$E_{ex}^{2D} = \frac{2m_r^* e^4}{(4\pi\epsilon)^2\hbar^2} = 4E_{ex}^{3D} \tag{7.61}$$

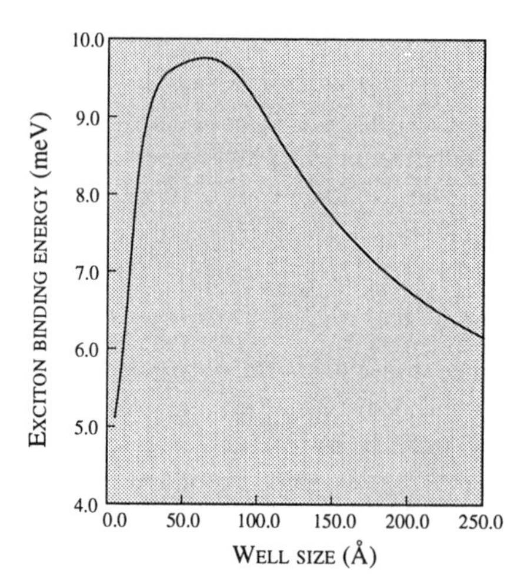

Figure 7.14: Variation of the heavy-hole exciton binding energy as a function of well size in GaAs/$Al_{0.3}Ga_{0.7}As$ wells.

The Bohr radius is given by the inverse of α,

$$a_{ex}^{2D} = \frac{1}{\alpha} = \frac{4\pi\epsilon\hbar^2}{2m_r^* e^2} = \frac{a_{ex}^{3D}}{2} \tag{7.62}$$

In real quantum well structures, the electrons and holes do not see a strictly two-dimensional world, but a quasi-two-dimensional world. This is because the envelope functions in a finite quantum well always have a finite spatial extent in the confining direction. As a result of this, the binding energy of the exciton is not four times the binding energy in three dimensions, but somewhat smaller. Variational calculations have been widely used to solve for the exciton problem in quantum well (and other low-dimensional systems). This approach gives quite reliable results, with the effects of choosing different forms of variational functions being no more than ~10% of the exciton binding energies. In Fig. 7.14, we show the effect of well size on the ground state excitonic states in GaAs/$Al_{0.3}Ga_{0.7}As$ quantum well structures. As can be seen, the exciton binding energies can increase by up to a factor of ~2.5 in quantum well structures made from this material system.

7.5 CONFIGURATION ENERGY DIAGRAM

In the previous section, we have seen several plots of how the energy of a system changes as the configuration (e.g., spacing between two atoms) changes. Such descriptions are known as *configuration-energy plots* and are extremely useful in describing how systems evolve from one state to another. These diagrams are particularly useful if a system has several possible "channels" through which it can evolve. Let us bring this concept into focus by examining the problem of a hydrogen molecule impinging on the surface of a substrate and breaking up to form a H-chemical bond with the surface atoms.

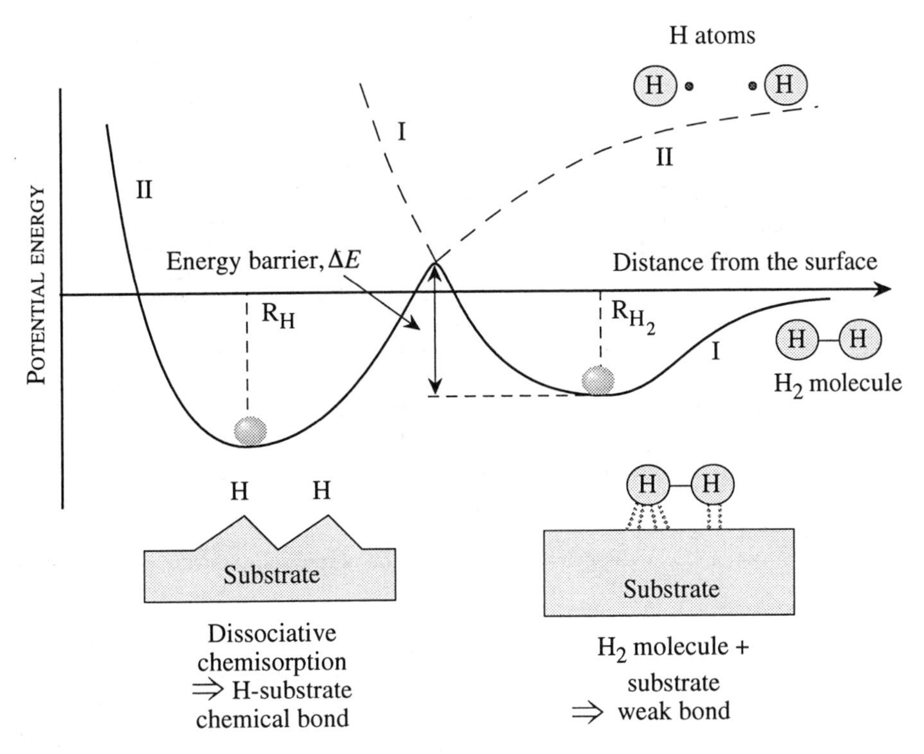

Figure 7.15: A schematic of the change in energy of a H_2 molecule and two H atoms as a function of separation from the surface of a substrate. The H_2 molecule can be in a local minimum of energy at position R_{H_2} or at a time minimum at R_H where the atoms form chemical bonds with the substrate atoms.

In Fig. 7.15 we show a schematic of how the system energy changes as a function of separation of the H molecule or H atoms form the surface of the substrate. Let us follow two paths, as shown in Fig. 7.15. In path I, a H_2 molecule is brought close to the surface. As the separation decreases, the energy of the system decreases up to a certain distance R_{H_2} and then the energy starts to increase as a result of the repulsion between the H_2 molecule and the substrate atoms. If the H_2 molecule sits at a distance R_{H_2}, i.e., in the minimum energy state of path I, the molecule forms a "bond" with the substrate. The nature of this bond is not very strong.

Let us now take the second path labeled II in Fig. 7.15. Here we consider what happens if two H atoms (not a H_2 molecule) approach the substrate. The starting energy of the two atoms is higher than the starting energy of path I, since the H_2 molecule has a lower energy state than two uncoupled H atoms. As the atoms come toward the surface of the substrate the energy lowers due to the coupling between the surface atom and H-atom. One scenario that may occur is shown by path II. The energy minimizes at a distance R_H, as shown, and then increases again due to the repulsive potential of the nuclei. At the minimum energy position, the H atoms form chemical bonds with the surface atom and the energy of the system

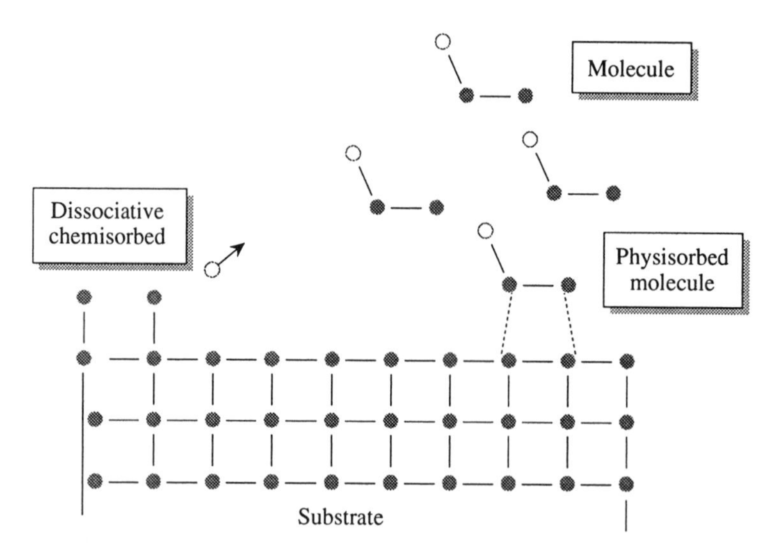

Figure 7.16: A schematic of dissociative chemisorption of molecules during crystal growth.

is lower than what it would be if the a H_2 molecule were bonded to the substrate.

When we examine Fig. 7.15, we see that when a H_2 molecule impinges on the substrate, it first finds a local minimum in energy at a distance R_{H_2}. However, the true lowest energy configuration is at R_H. To reach this true minimum, the H_2 molecule must overcome an energy barrier (shown in Fig. 7.15). Thus the *molecule must roll up this energy barrier hill* to get to the state where the two H atoms form a chemical bond to the surface. The probability that the system will make a transition from the a H_2 molecule–surface bond to ^{2}H atom–surface bonds is proportional to

$$P(H_2 \rightarrow 2H) \propto \exp\left(-\frac{\Delta E}{k_B T}\right) \tag{7.63}$$

where ΔE is the energy barrier in the configuration-energy diagram. At low temperatures this transition will not occur or will occur very slowly.

The simple qualitative picture we have given here is very important in understanding a variety of processes such as crystal growth and catalysis. We will briefly discuss physical basis of these phenomena.

7.5.1 Crystal Growth from Vapor Phase

Crystal growth is a key step in most modern information processing devices. The most controllable manner of growth involves sending molecular species containing the atom (atoms) which for the crystal. The molecular typically go into a weakly bonded state (called the *physisorbed state*) where the molecule is still intact. From this physisorbed state, the molecule usually has to overcome a barrier to undergo dissociative chemisorption, whereby one or more atoms are chemically bonded to the surface atoms. In some processes it is possible that some atom (atoms) of the incident molecule leaves the surface, as shown in Fig. 7.16.

We can see from our discussion that the dissociative chemisorption will not occur unless the temperature of the substrate is high enough to allow the molecular state to overcome the "energy hill" to form the chemical bond.

7.5.2 Catalysis

In our discussion of the configuration energy diagram, we have seen that most chemical reactions require the initial components to go through an energy barrier before the final products are realized. It is important to note that the atoms or molecules cannot tunnel through this barrier, since the tunneling probability is nearly zero. Often the reaction rate which is proportional to $\exp(-\Delta E/k_BT)$ is so small that it would take an unacceptably long time to reach the final lowest energy state. Examples may involve reactions that convert toxic materials into harmless products; environmentally damaging products into biodegradable products; fabrication of new polymers, etc.

An examination of the configuration energy diagram suggests that the solution to speeding up processes lies in being able to reduce the energy barrier that has to be overcome in going to the final state. It is also possible that in some special applications, we want to slow down a reaction rate. Examples may be degradation rate of food or other products. In this case we would like to increase the energy barrier.

The catalyst is a material that can accomplish the tasks mentioned above. A catalyst is a relatively small amount of foreign material that is able to augment or decrease a reaction rate. Symbolically, consider a reaction that is too slow:

$$A + B \rightarrow D \tag{7.64}$$

The catalyst-enhanced reaction may be written as (C is the catalyst)

$$\begin{aligned} A + C &\rightarrow AC \\ AC + B &\rightarrow D + C \end{aligned} \tag{7.65}$$

The overall reaction is thus

$$A + B \xrightarrow{C} D \tag{7.66}$$

One of the most widely applied use of catalysts is the catalytic converter used in automobiles to convert poisonous exhaust gases to harmless ones.

7.6 CHAPTER SUMMARY

The important issues discussed in this chapter are summarized in Tables 7.1–7.3.

7.7 PROBLEMS

Problem 7.1 Consider a 100 Å GaAs/AlGaAs quantum well across which an electric field is applied. Using a simple infinite potential barrier model, plot the separation of the ground states in the conduction and valence bands as the field

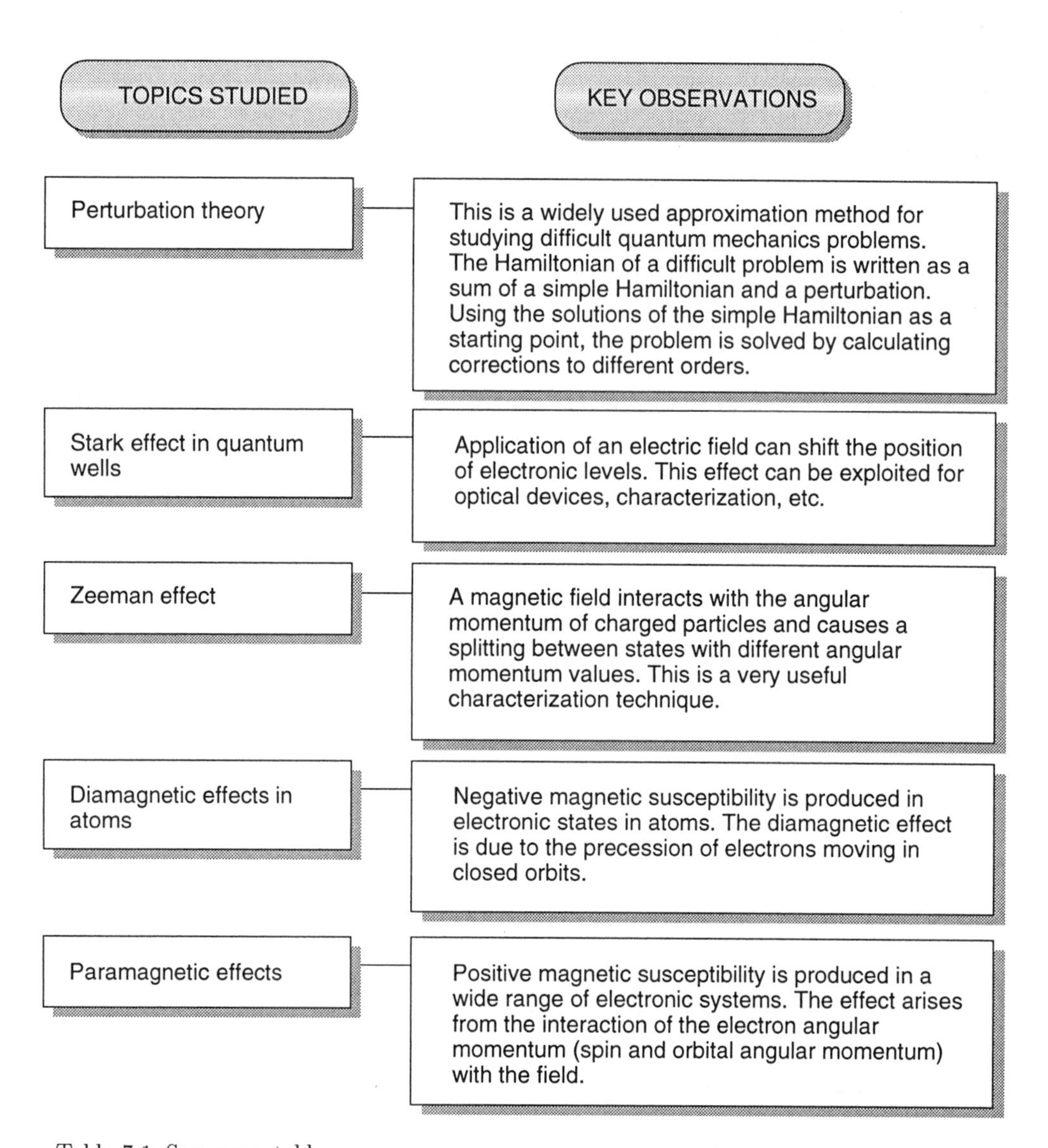

TOPICS STUDIED	KEY OBSERVATIONS
Perturbation theory	This is a widely used approximation method for studying difficult quantum mechanics problems. The Hamiltonian of a difficult problem is written as a sum of a simple Hamiltonian and a perturbation. Using the solutions of the simple Hamiltonian as a starting point, the problem is solved by calculating corrections to different orders.
Stark effect in quantum wells	Application of an electric field can shift the position of electronic levels. This effect can be exploited for optical devices, characterization, etc.
Zeeman effect	A magnetic field interacts with the angular momentum of charged particles and causes a splitting between states with different angular momentum values. This is a very useful characterization technique.
Diamagnetic effects in atoms	Negative magnetic susceptibility is produced in electronic states in atoms. The diamagnetic effect is due to the precession of electrons moving in closed orbits.
Paramagnetic effects	Positive magnetic susceptibility is produced in a wide range of electronic systems. The effect arises from the interaction of the electron angular momentum (spin and orbital angular momentum) with the field.

Table 7.1: Summary table.

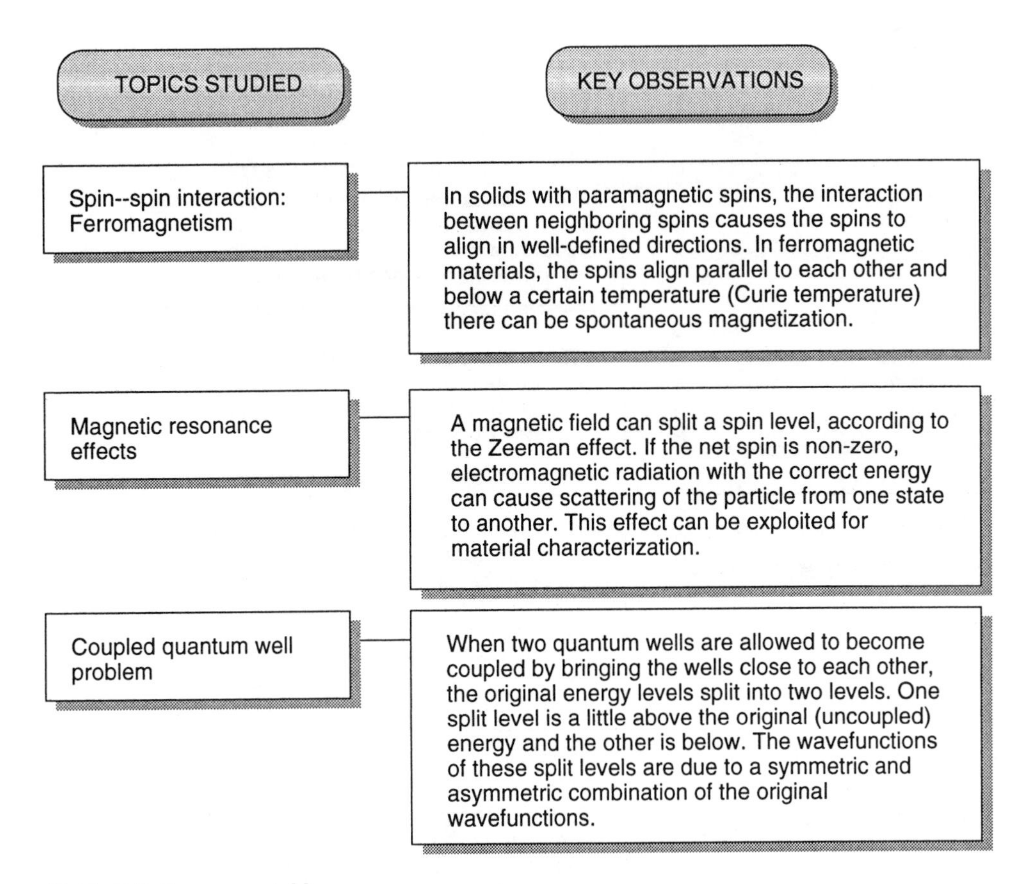

TOPICS STUDIED	KEY OBSERVATIONS
Spin--spin interaction: Ferromagnetism	In solids with paramagnetic spins, the interaction between neighboring spins causes the spins to align in well-defined directions. In ferromagnetic materials, the spins align parallel to each other and below a certain temperature (Curie temperature) there can be spontaneous magnetization.
Magnetic resonance effects	A magnetic field can split a spin level, according to the Zeeman effect. If the net spin is non-zero, electromagnetic radiation with the correct energy can cause scattering of the particle from one state to another. This effect can be exploited for material characterization.
Coupled quantum well problem	When two quantum wells are allowed to become coupled by bringing the wells close to each other, the original energy levels split into two levels. One split level is a little above the original (uncoupled) energy and the other is below. The wavefunctions of these split levels are due to a symmetric and asymmetric combination of the original wavefunctions.

Table 7.2: Summary table.

goes from 0 to 100 kV/cm. $m_e^* = 0.067\ m_0$; $m_{hh}^* = 0.45\ m_0$.

Problem 7.2 In the "square" quantum well the first-order correction term is zero when a field is applied. What would you do to ensure that the first-order term is non-zero? Discuss how a quantum well where the first or Stark effect is non-zero would be more useful as an optical modulator.

Problem 7.3 Consider a square quantum well where the well region is made from GaAs. Calculate the size of the well which leads to a device in which the effective bandgap shifts by 20 meV when an electric field of 100 kV/cm is applied.

Problem 7.4 Calculate the magnetic field needed to cause a splitting of a state with angular momentum $L = \hbar$ and spin 0 to be 1 meV.

Problem 7.5 Hydrogen atoms are placed in a magnetic field of 3.0 T. Ignoring the spin effects, calculate the wavelengths of normal Zeeman components of i) $3d$ to $3p$ transitions and ii) $3s$ to $2p$ transitions. Assume that only those transitions for which $\Delta m = 0$ or ± 1 are allowed.

Problem 7.6 Consider a Stern-Gerlach experiment in which the oven is at a temperature of 1000 K. A magnetic field with gradient 10 T/m is applied. The length of the magnetic field region is 0.5 m. Also the distance of the field free region between

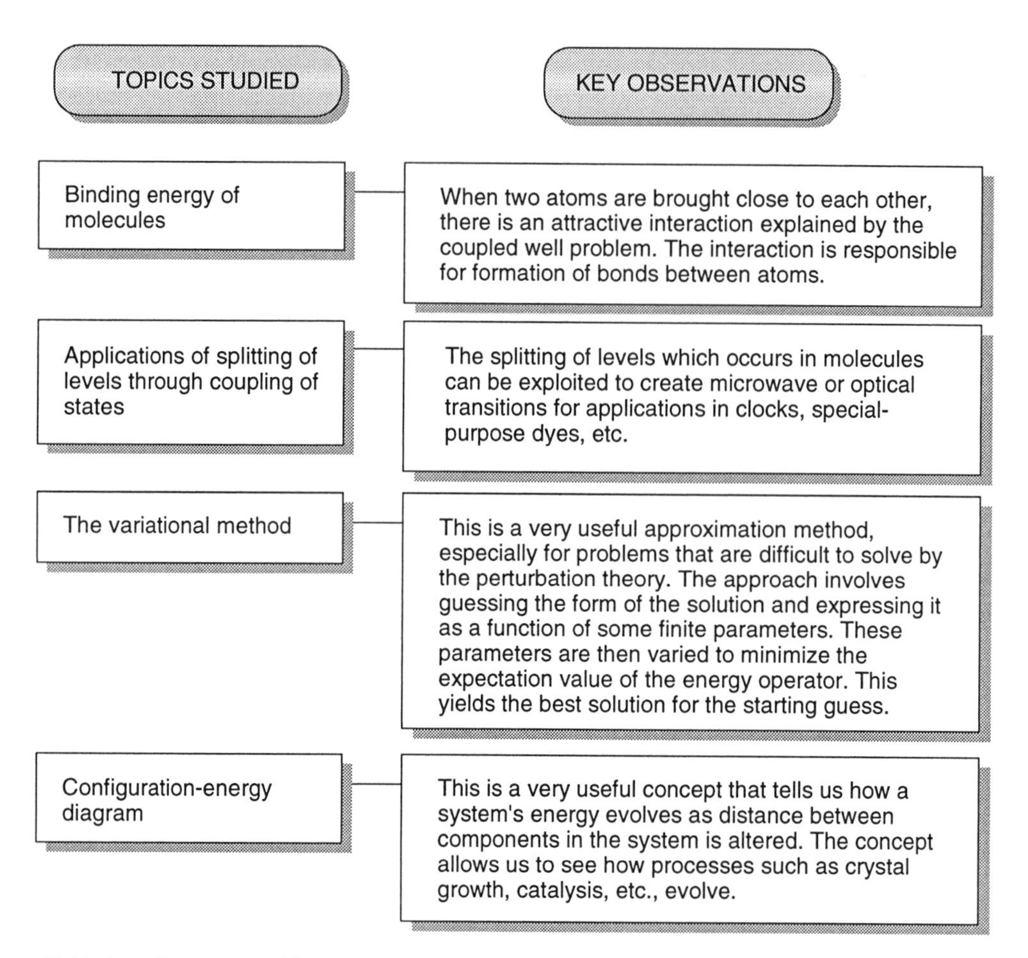

TOPICS STUDIED	KEY OBSERVATIONS
Binding energy of molecules	When two atoms are brought close to each other, there is an attractive interaction explained by the coupled well problem. The interaction is responsible for formation of bonds between atoms.
Applications of splitting of levels through coupling of states	The splitting of levels which occurs in molecules can be exploited to create microwave or optical transitions for applications in clocks, special-purpose dyes, etc.
The variational method	This is a very useful approximation method, especially for problems that are difficult to solve by the perturbation theory. The approach involves guessing the form of the solution and expressing it as a function of some finite parameters. These parameters are then varied to minimize the expectation value of the energy operator. This yields the best solution for the starting guess.
Configuration-energy diagram	This is a very useful concept that tells us how a system's energy evolves as distance between components in the system is altered. The concept allows us to see how processes such as crystal growth, catalysis, etc., evolve.

Table 7.3: Summary table.

the magnetic and the screen is 0.5 m. Estimate the separation of the split images obtained on the screen.

Problem 7.7 Consider a hydrogen atom in its ground state. Show that the molar diamagnetic susceptibility is -2.36×10^{-6} cm^3 mole^{-1}.

Problem 7.8 Calculate the Curie constant for a system of electrons with zero orbital angular momentum. The electron density is 10^{23} cm^{-3}.

Problem 7.9 Discuss why the diamagnetic result derived in this chapter is applicable to dielectric solids, but not to metals.

Problem 7.10 Consider two identical square quantum wells with a barrier height V_0, and well size W. The two wells are spaced so that the width of the barrier region between them is W_b. The coupling matrix element A defined in the text is given by

$$A = \frac{E_0}{2\pi} \exp\left(-\frac{\sqrt{2m^*(V_0 - E_0)}W_b}{\hbar} \right)$$

where E_0 is the energy position of the level in the uncoupled well measured from the bottom of the well.

Design a GaAs/$Al_{0.3}Ga_{0.7}$As quantum well (i.e., provide the well size of one of the two identical coupled wells and the separation W_b) to give a splitting of the ground state in the conduction band well of 5 meV.

Do not use the infinite barrier model for the problem of the uncoupled well.

Problem 7.11 Use the expression given in the previous problem for the coupling coefficient A to design two identical square quantum wells made from InGaAs wells and InP barriers, so that the splitting for the ground state level in the coupled well is 10 meV.

The electron mass is 0.04 m_0; the barrier height for electrons is 0.2 eV.

CHAPTER 8

SCATTERING AND COLLISIONS

CHAPTER AT A GLANCE

8.1 INTRODUCTION

Many important problems in quantum mechanics are such that the Hamiltonian describing them is time-dependent. An important example of such problems is the interaction of electrons in an atom or a solid with electromagnetic radiation. Another example is the motion of electrons in a solid in which the atoms are vibrating. In most such problems the Hamiltonian can be divided into an *easy* part that can be solved and a *small* time-dependent part that can be treated as a small perturbation, much in the spirit of the problems discussed in the previous chapter. Time-dependent perturbation theory, in particular, is of extreme importance in applied problems. So important is this formalism that the result which emerges is called the *Fermi golden rule*.

The electron–photon interactions in solids are responsible for a large number of devices, such as lasers, detectors, display screens, etc., and the time-dependent perturbation theory developed here can be successfully applied to understand these devices. In this chapter we will study this important problem.

One of the most important classes of problems in physics involves the collisions and scattering of "particles". We could have a situation where two (or more) particles scatter (or collide) off each other, interacting via some mutual interaction. We could also have a situation where a particle scatters from a potential created by some disturbance in space. Such problems are encountered when elementary particles are flung at each other with extremely high energies in giant accelerators. They are also encountered by electrons as they move under an applied electric field in a computer memory bit. Focusing on two particles scattering from each other, after the scattering has occurred (i.e., the particles are far from each other with negligible interaction), the state of the particles will, in general, change from the initial state. In this chapter we will calculate what the effects of scattering are on the final state. We will also apply this theory to study how electrons (holes) move in materials and respond to electric fields.

8.2 TIME-DEPENDENT PERTURBATION THEORY

The time-dependent perturbation theory is used extensively to address the problem of scattering of electrons by phonons or photons and many other "particles" arising from the quantization of wavefields. The general issues in such problems are described in Fig. 8.1. The general Hamiltonian of interest is of the form

$$H = H_0 + H^{'} \tag{8.1}$$

where H_0 is a simple Hamiltonian with known solutions

$$H_0 u_k = E_k u_k \tag{8.2}$$

and E_k, u_k are known. *In the absence of $H^{'}(t)$, if a particle is placed in a state u_k, it remains there forever. The effect of $H^{'}$ is to cause time-dependent transitions between the states u_k.*

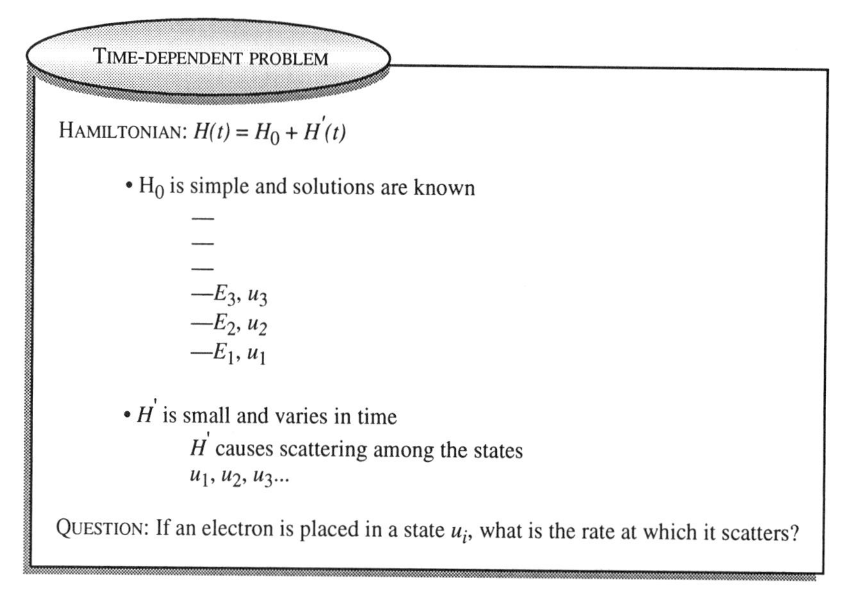

Figure 8.1: The time-dependent perturbation problem in quantum mechanics.

Appendix A gives a detailed solution of the derivation of the Fermi golden rule, which tells us how the perturbation $H'(t)$ causes scattering of particles. Readers who are interested in the derivation of this powerful result should go through Appendix A.

As noted above, if there is no perturbation, a particle placed in any state u_k will remain in that state. The perturbation is responsible to nudge the particle, causing it to scatter into a different state. Of course, in quantum mechanics, we only talk about the probability of finding the particle in any allowed state. We can ask the question: What is the rate at which a particle initially in a state u_i scatters into a final state u_f as a result of the perturbation? The Fermi golden rule answers this question.

Let us assume that the perturbation is of the form

$$H'(t) = 2H(0)\sin\omega t \tag{8.3}$$

Such a perturbation could be due to electromagnetic radiation impinging on a material, or due to the vibration of atoms in a solid. Let us remind ourselves that, as discussed in Chapter 3, the vibrations of such classical waves is represented in quantum mechanics by the presence of quanta of energy $\hbar\omega$.

According to quantum mechanics, the quanta of perturbation can carry the particle from its initial state to a final state that is at energy $\hbar\omega$ above the initial state, as shown in Fig. 8.2. In such a case, if initially there are n quanta in the perturbation, in the final state the number of quanta is $n - 1$. Conservation of energy ensures that the missing quantum has been used to increase the energy of the scattered electron by $\hbar\omega$. Such a process is called absorption and is shown schematically in the top panel of Fig. 8.2.

It is also possible to have a scattering process in which after scattering, the electron loses energy and goes to a state with energy E_g, which is $\hbar\omega$ below the initial energy E_i. In such a case, a quantum is created. Such processes are called emission processes. In the lower panel of Fig. 8.2, we show an emission process. Initially, there are n quanta and after scattering there are $n+1$ quanta.

The Fermi golden rule gives us the following expression for the rate W_{if} of scattering:

$$W_{if} = \frac{2\pi}{\hbar}\left|\int u_i^* H'(0) u_f d^3r\right|^2 \delta\left(E_i - E_f \pm \hbar\omega\right) \tag{8.4}$$

Denoting the matrix element (the integral) by the simpler Dirac notation

$$\int u_i^* H^{'}(0) u_f d^3r = \langle i|H^{'}(0)|f\rangle$$

we have

$$W_{if} = \frac{2\pi}{\hbar}\left|\langle i|H^{'}(0)|f\rangle\right|^2 \delta\left(E_i - E_f \pm \hbar\omega\right) \tag{8.5}$$

The Dirac delta function ensures that energy is conserved in the scattering process, i.e., as discussed above, the final energy is either (absorption process)

$$E_f = E_i + \hbar\omega \tag{8.6}$$

or (emission process)

$$E_f = E_i - \hbar\omega \tag{8.7}$$

We can also ask what the total rate of scattering out of a state u_i is. This is given by

$$\boxed{W_m = \frac{2\pi}{\hbar} \sum_{\text{final states}} \mid \langle k|H^{'}|i\rangle \mid^2 \delta\left(E_f - E_i \pm \hbar\omega\right)} \tag{8.8}$$

The signs + and − in the Dirac function represents two different scattering processes, i.e., either absorption of a quantum or emission. Note that the sum over all final states at the same energy E_f is just the density of states at energy E_f.

We will discuss the applications of the Fermi golden rule to several important problems. It is interesting to note that if the perturbation has no time dependence, i.e., the frequency ω goes to zero, the scattering rate derived above still holds, and the expression is known as the Born approximation.

8.3 OPTICAL PROPERTIES OF SEMICONDUCTORS

We will now discuss how electrons in a semiconductor respond to electromagnetic fields or photons. The interaction of electrons and photons is the basis of all semiconductor optoelectronic devices. We have discussed in Section 8.2 the physical basis of the scattering of electrons from photons. There are two classes of scattering: i) absorption of photons, where the electron gains energy by absorbing a photon; and ii) emission, where the electron emits a photon and loses energy. The emission process itself is characterized as spontaneous emission and stimulated emission. Spontaneous

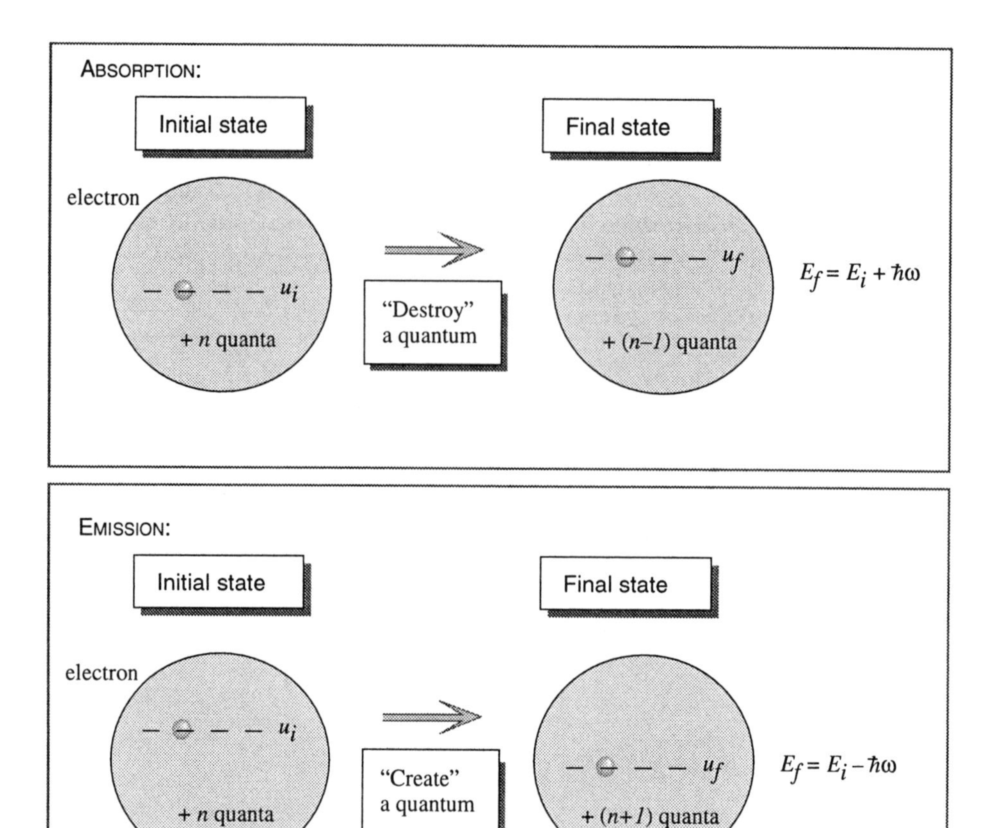

Figure 8.2: Scattering processes can be classified as *absorption processes* and *emission processes*. In absorption, a quantum of energy is absorbed, while in emission, a quantum is emitted.

emission occurs even if there are no photons present while stimulated emission occurs because of the presence of photons. In this section we will give quantitative expressions for the various processes characterizing electron–photon interactions. These expressions will not be derived in detail, but will be simply introduced to the reader. The operation of the various optoelectronic devices is governed by the expressions presented in this and the next few sections.

Light is represented by electromagnetic waves as discussed in Chapter 1. Electromagnetic waves travelling through a medium like a semiconductor are described by Maxwell's equations, which show that the waves have a form given by the electric field vector dependence

$$\mathbf{E} = \mathbf{E}_o \exp\left\{i\omega\left(\frac{n_r}{c} - t\right)\right\} \exp\left(-\frac{\alpha z}{2}\right) \tag{8.9}$$

Here z is the propagation direction, ω the frequency, n_r the refractive index (the real part), and α the absorption coefficient of the medium. If α is zero, the wave propagates without attenuation with a velocity $\frac{c}{n_r}$. However, for nonzero α, the photon flux I ($\sim E^* E$) falls as

$$I(z) = I(0)\ \exp\ \{-\alpha z\} \tag{8.10}$$

The absorption of light can occur for a variety of reasons, including absorption by impurities in the material, or intraband absorption, where electrons in the conduction band or holes in the valence band absorb the radiation. *However, the most important optoelectronic interaction in semiconductors as far as devices are concerned is the band-to-band transition* shown in Fig. 8.3. In the photon absorption process, a photon scatters an electron in the valence band, causing the electron to go into the conduction band. In the reverse process, the electron in the conduction band recombines with a hole in the valence band to generate a photon. These two processes are of obvious importance for light-detection and light-emission devices. As has been noted above, the detailed scattering theory is beyond the scope of this text. However, the important expressions which control the performance of detectors and lasers will be examined in this section. The rates for the light emission and absorption processes are determined by quantum mechanics. The scattering involves the following issues:
i) **Conservation of Energy**: In the absorption and emission process we have for the initial and final energies of the electrons E_i and E_f

$$\text{Absorption}: \quad E_f = E_i + \hbar\omega \tag{8.11}$$

$$\text{Emission}: \quad E_f = E_i - \hbar\omega \tag{8.12}$$

where $\hbar\omega$ is the photon energy. Since the minimum energy difference between the conduction and valence band states is the bandgap E_g, *the photon energy must be larger than the bandgap for absorption to occur.*
ii) **Conservation of Momentum**: In addition to energy conservation, one also needs to conserve the momentum $\hbar k$ for the electrons and the photon system. The photon k_{ph} value is given by

$$k_{ph} = \frac{2\pi}{\lambda} \tag{8.13}$$

Since 1 eV photons correspond to a wavelength of 1.24 μm, the k-values of relevance are $\sim 10^{-4}$ Å^{-1}, which is essentially zero compared to the k-values for electrons. Thus k-conservation ensures that the initial and final electrons should have the same k-value. Another way to say this is that *only "vertical" k transitions are allowed* in the bandstructure picture, as shown in Fig. 8.3.

Because of this constraint of k-conservation, in semiconductors where the valence band and conduction bandedges are at the same $k = 0$ value (the direct semiconductors), the optical transitions are quite strong. In indirect materials like Si, Ge, etc., the optical transitions are very weak. This makes a tremendous difference in the optical properties of these two kinds of materials. The transitions are very weak near the bandedges of indirect semiconductors, and such materials cannot be used for high-performance optoelectronic devices such as lasers.

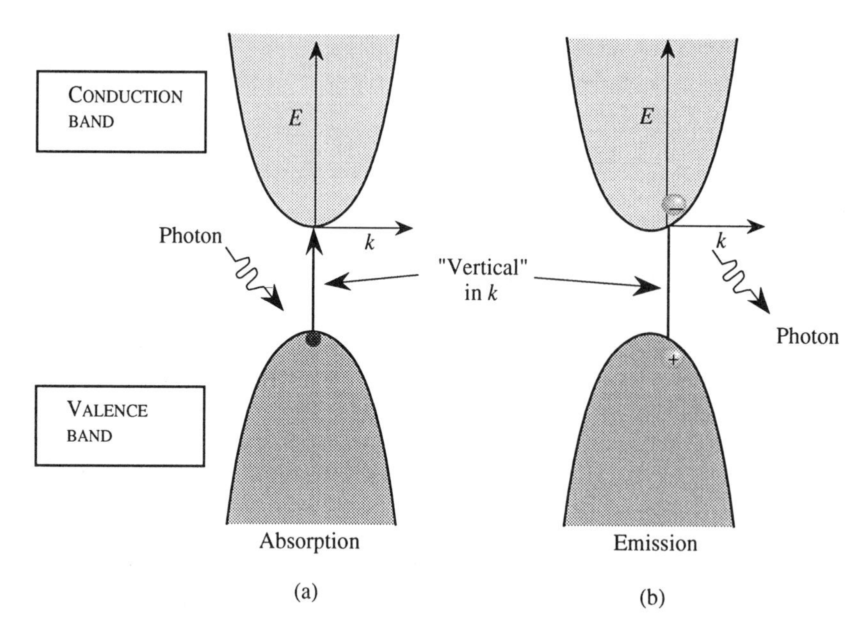

Figure 8.3: Band-to-band absorption in semiconductors. (a) An electron in the valence band "absorbs" a photon and moves into the conduction band. (b) In the reverse process, the electron recombines with a hole to emit a photon. Momentum conservation ensures that only vertical transitions are allowed.

Because the k-values for the electron and hole are the same in vertical transitions, we have, as shown in Fig. 8.4,

$$\begin{aligned} \hbar\omega &= E_g + \frac{\hbar^2 k^2}{2}\left(\frac{1}{m_e^*} + \frac{1}{m_h^*}\right) \\ &= E_g + \frac{\hbar^2 k^2}{2m_r^*} \end{aligned} \tag{8.14}$$

where m_r^* is the reduced mass of the electron-hole system. Thus the *relevant density of states function in the scattering process is that where the mass used is the reduced mass.* This density of states is called the *joint density of states.*

Keeping all these issues in mind, one can calculate the absorption coefficient for a semiconductor. For light polarized along direction **a**, the absorption coefficient turns out to be

$$\boxed{\alpha = \frac{\pi\, e^2\hbar}{m_0^2 c n_r \epsilon_o}\, \frac{1}{\hbar\omega} \mid (\mathbf{a}\cdot\mathbf{p})_{if} \mid^2 N_{cv}(\hbar\omega)} \tag{8.15}$$

With N_{cv}, the joint density of states given by

$$\boxed{N_{cv}(\hbar\omega) = \frac{\sqrt{2}(m_r^*)^{3/2}(\hbar\omega - E_g)^{1/2}}{\pi^2\hbar^3}} \tag{8.16}$$

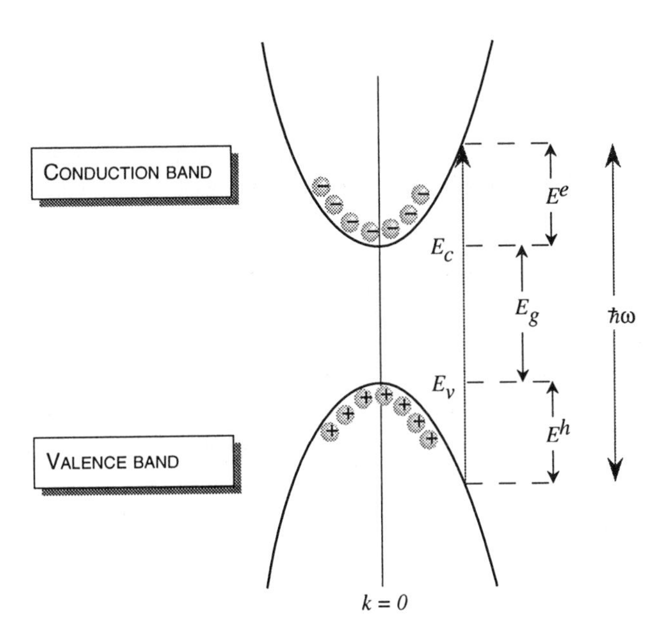

Figure 8.4: A schematic showing the relationship between photon energy and electron and hole energies in momentum conserving transitions.

The quantity $|\,(\mathbf{a}\cdot\mathbf{p})_{if}\,|^2$ is called the momentum or dipole matrix element between the conduction and the valence band. The polarization averaged matrix element turns out to be $(2/3)p_{cv}^2$, where it is found that, for most semiconductors

$$\frac{2p_{cv}^2}{m_0} \cong 20 \text{ to } 24 \text{ eV} \tag{8.17}$$

If we plug in some numbers for the absorption coefficient of unpolarized light, we get for GaAs ($E_g \cong 1.5$ eV) (see Example 8.4)

$$\boxed{\alpha(\hbar\omega) \cong 4.7 \times 10^4 \, \frac{(\hbar\omega - E_g)^{1/2}}{\hbar\omega} \ \text{cm}^{-1}} \tag{8.18}$$

where $\hbar\omega$ and E_g are in units of electron volts. The prefactor for any other direct semiconductor can be obtained by scaling the reduced mass ratio by a power of 3/2, according to the reduced density of states.

In Fig. 8.5 we show the absorption coefficient for several direct and indirect bandgap semiconductors. In the case of indirect bandgap materials, the absorption coefficient is considerably smaller at the bandedges.

From our discussion of the harmonic oscillator problem in Chapter 5 we know that the energy in an electromagnetic field in mode ω is

$$E = (n_{ph} + 1/2)\,\hbar\omega$$

where n_{ph} is the photon occupation number. Even if n_{ph} is zero, there is an energy in the mode (this is called the vacuum energy).

If we have an electron in a conduction band state and a hole in the valence band state with the same k-value, the two can recombine and emit a photon. There are two important classes of emission processes. *In spontaneous emission, as discussed in Section 8.2.1, an electron recombines with a hole, even though no photons are present in the region, and emits a photon.* The rate for this emission process is given by

$$W_{em}(\hbar\omega) = \frac{e^2 n_r \hbar\omega}{3\pi\epsilon_0 m_o^2 c^3 \hbar^2} \mid p_{cv} \mid^2 \tag{8.19}$$

If photons with frequency ω are already present in the cavity with the semiconductor, the recombination rate is enhanced by the presence of these photons. If $n_{ph}(\hbar\omega)$ is the photon occupation (i.e., the number of photons in a particular mode), the emission (called stimulated emission) rate is given by

$$W_{em}^{st}(\hbar\omega) = \frac{e^2 n_r \hbar\omega}{3\pi\epsilon_o m_o^2 c^3 \hbar^2} \mid p_{cv} \mid^2 \cdot n_{ph}(\hbar\omega) \tag{8.20}$$

Thus the rate is increased in proportion to the photon density already present in the cavity.

In spontaneous emission, the photons that are emitted have no particular phase relationship with each other and are thus incoherent. Light emission in light-emitting diodes (LEDs) is due to spontaneous emission. In stimulated emission, however, the photons that are emitted are in phase with the original photons that are present. The radiation is thus coherent. Laser diodes depend upon stimulated emission. The acronym *laser* stands for *light amplification* by *stimulated emission radiation.*

The radiative recombination rate of an electron having a momentum $\hbar k$ with a hole (in absence of photons) having the same momentum is

$$\tau_o = \frac{1}{W_{em}} \tag{8.21}$$

In the definition of τ_o it is assumed that the electron can find a hole with which to recombine. If the probability of finding the hole is small, the radiative time can be much longer, as discussed in Section 8.5. For materials like GaAs, the value of τ_o is about 1 ns (see Example 8.5). However, for indirect materials, the recombination time can be as large as 1 μs. The recombination process is not only important for optical emission devices, but the rate also plays a key role in the speed of many electronic devices, e.g., bipolar devices, diodes, etc.

The electron-hole recombination during stimulated emission can be quite a bit smaller than τ_o, depending upon photon intensity present.

EXAMPLE 8.1 A 1.6 eV photon is absorbed by a valence band electron in GaAs. If the bandgap of GaAs is 1.41 eV, calculate the energy of the electron and heavy hole produced by the photon absorption.

The electron, heavy-hole, and reduced mass of GaAs are 0.067 m_0, 0.45 m_0, and 0.058 m_0, respectively. The electron and the hole generated by photon absorption have

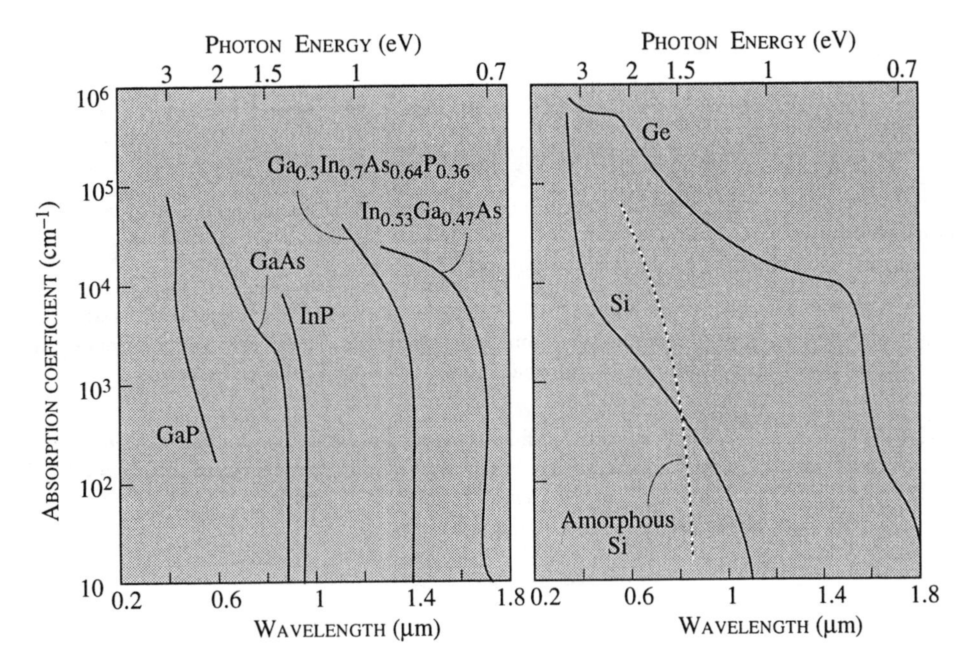

Figure 8.5: Absorption coefficient of several direct and indirect semiconductors. For the direct gap material, the absorption coefficient is very strong once the photon energy exceeds the bandgap. For indirect materials, the absorption coefficient rises much more gradually. Once the photon energy is more than the direct gap, the absorption coefficient increases rapidly.

the same momentum. The energy of the electron is

$$\begin{aligned} E^e &= E_c + \frac{m_r^*}{m_e^*}(\hbar\omega - E_g) \\ E^e - E_c &= \frac{0.058}{0.067}(1.6 - 1.41) = 0.164 \text{ eV} \end{aligned}$$

The hole energy is

$$\begin{aligned} E^h - E_v &= -\frac{m_r^*}{m_h^*}(\hbar\omega - E_g) = -\frac{0.058}{0.45}(1.6 - 1.41) \\ &= -0.025 \text{ eV} \end{aligned}$$

The electron, by virtue of its lower mass, is created with a much greater energy than the hole.

EXAMPLE 8.2 In silicon, an electron from the top of the valence band is taken to the bottom of the conduction band by photon absorption. Calculate the change in the electron momentum. Can this momentum difference be provided by a photon?

The conduction band minima for silicon are at a k-value of $\frac{2\pi}{a}$ (0.85, 0, 0). There are five other similar bandedges. The top of the valence band has a k-value of 0. The change in the momentum is thus

$$\begin{aligned} \hbar\Delta k &= \hbar\frac{2\pi}{a}(0.85) = (1.05\times10^{-34})\left(\frac{2\pi}{5.43\times10^{-10}}\right)(0.85) \\ &= 1.03\times10^{-24}\text{kg.m.s}^{-1} \end{aligned}$$

A photon which has an energy equal to the silicon bandgap can only provide a momentum of

$$\hbar k_{ph} = \hbar\cdot\frac{2\pi}{\lambda}$$

The λ for silicon bandgap is 1.06 μm and thus the photon momentum is about a factor of 1800 too small to balance the momentum needed for the momentum conservation. The lattice vibrations produced by thermal vibration are needed for the process.

EXAMPLE 8.3 The absorption coefficient near the bandedges of GaAs and Si are $\sim 10^4$ cm^{-1} and 10^3 cm^{-1}, respectively. What is the minimum thickness of a sample in each case which can absorb 90% of the incident light?

The light absorbed in a sample of length L is

$$\frac{I_{abs}}{I_{inc}} = -\exp(-\alpha L)$$

or

$$L = \frac{1}{\alpha}\,\ell n\left(1-\frac{I_{abs}}{I_{inc}}\right)$$

Using $\frac{I_{abs}}{I_{inc}}$ equal to 0.9, we get

$$\begin{aligned} L(\text{GaAs}) = -\frac{1}{10^4}\,\ell n\,(0.1) &= 2.3\times10^{-4}\text{ cm} \\ &= 2.3\ \mu\text{m} \\ L(\text{Si}) = -\frac{1}{10^3}\,\ell n\,(0.1) &= 23\ \mu\text{m} \end{aligned}$$

Thus an Si detector requires a very thick active absorption layer to function.

EXAMPLE 8.4 Calculate the absorption coefficient of GaAs as a function of photon frequency.

The joint density of states for GaAs is (using a reduced mass of $0.058m_0$)

$$\begin{aligned} N_{cv}(E) &= \frac{\sqrt{2}(m_r^*)^{3/2}(E-E_g)^{1/2}}{\pi^2\hbar^3} \\ &= \frac{1.414\times(0.058\times0.91\times10^{-30}\text{ kg})^{3/2}(E-E_g)^{1/2}}{9.87\times(1.05\times10^{-34})^3} \\ &= 1.5\times10^{54}(E-\hbar\omega)^{1/2}\text{ J}^{-1}\text{m}^{-3} \end{aligned}$$

The absorption coefficient is for unpolarized light

$$\alpha(\hbar\omega) = \frac{\pi e^2\hbar}{2n_r c\epsilon_o m_0}\left(\frac{2p_{cv}^2}{m_0}\right)\frac{N_{cv}(\hbar\omega)}{\hbar\omega}\cdot\frac{2}{3}$$

The term $\frac{2p_{cv}^2}{m_0}$ is $\sim$23.0 eV for GaAs. This gives

$$\alpha(\hbar\omega) = \frac{3.1416\times(1.6\times10^{-19}\text{ C})^2(1.05\times10^{-34}\text{ J.s})}{2\times3.4\times(3\times10^8\text{ m/s})(8.84\times10^{-12}(\text{F/m})^2)}$$

$$\cdot \frac{(23.0 \times 1.6 \times 10^{-19}\ \text{J})}{(0.91 \times 10^{-30}\ \text{kg})} \frac{(\hbar\omega - E_g)^{1/2}}{\hbar\omega} \times 1.5 \times 10^{54} \times \frac{2}{3}$$

$$\alpha(\hbar\omega) = 1.9 \times 10^{-3} \frac{(\hbar\omega - E_g)^{1/2}}{\hbar\omega}\ \text{m}^{-1}$$

Here the energy and $\hbar\omega$ are in units of Joules. It is usual to express the energy in eV, and the absorption coefficient in cm^{-1}. This is obtained by multiplying the result by

$$\left[\frac{1}{(1.6 \times 10^{-19})^{1/2}} \times \frac{1}{100}\right]$$

$$\alpha(\hbar\omega) = 4.7 \times 10^{4} \frac{(\hbar\omega - E_g)^{1/2}}{\hbar\omega}\ \text{cm}^{-1}$$

For GaAs the bandgap is 1.5 eV at low temperatures and 1.43 eV at room temperatures. From the value of α, we can see that a few microns of GaAs are adequate to absorb a significant fraction of light above the bandgap.

EXAMPLE 8.5 Calculate the electron-hole recombination time in GaAs.

The recombination rate is given by

$$W_{em} = \frac{e^2 n_r}{6\pi\epsilon_o m_0 c^3 \hbar^2} \left(\frac{2p_{cv}^2}{m_0}\right) \hbar\omega$$

with $\frac{2p_{cv}^2}{m_0}$ being 23 eV for GaAs.

$$\begin{aligned} W_{em} &= \frac{(1.6 \times 10^{-19}\ \text{C})^2 \times 3.4 \times (23 \times 1.6 \times 10^{-19}\ \text{J})\hbar\omega}{6 \times 3.1416 \times (8.84 \times 10^{-12}\ \text{F/m}) \times (0.91 \times 10^{-30}\ \text{kg})} \\ &\quad \cdot \frac{1}{(3 \times 10^{8}\ m/s)^3 \times (1.05 \times 10^{-34}\ \text{J.s})^2} \\ &= 7.1 \times 10^{27} \hbar\omega\ s^{-1} \end{aligned}$$

If we require the value of $\hbar\omega$ in eV instead of J we get

$$\begin{aligned} W_{em} &= 7.1 \times 10^{27} \times (1.6 \times 10^{-19})\hbar\omega\ \text{s}^{-1} \\ &= 1.14 \times 10^{9}\ \hbar\omega\ \text{s}^{-1} \end{aligned}$$

For GaAs, $\hbar\omega \sim 1.5$ eV so that

$$W_{em} = 1.71 \times 10^{9}\ \text{s}^{-1}$$

The corresponding recombination time is

$$\tau_o = \frac{1}{W_{em}} = 0.58\ \text{ns}$$

Remember that this is the recombination time when an electron can find a hole to recombine with. This happens when there is a high concentration of electrons and holes, i.e., at high injection of electrons and holes or when a minority carrier is injected into a heavily doped majority carrier region.

8.4 CHARGE INJECTION AND QUASI-FERMI LEVELS

In the discussion of absorption coefficient and recombination time, we have not mentioned whether *electrons and holes are actually present in the states which are involved in the optical processes.* For example, in the absorption of photons we have assumed that the *initial state in the valence band is occupied while the final state is empty.* In general, this may not be true. If electrons and holes are injected into a semiconductor, either by external contacts or by optical excitation, the system may not be in equilibrium, and the question then arises: What kind of distribution function describes the electron and hole occupation? We know that in equilibrium the electron and hole occupation is represented by the Fermi function. A new function is needed to describe the system when the electrons and holes are not in equilibrium.

8.4.1 Quasi-Fermi Levels

Under equilibrium conditions the distribution of electrons and holes is given by the Fermi function, which is defined once one knows the Fermi level. Also the product of electrons and holes, np, is approximately constant. If excess electrons and holes are injected into the semiconductor, clearly the same function will not describe the occupation of states. Under certain assumptions the electron and hole occupation can be described by the use of *quasi-Fermi levels.* These assumptions are:

i) The electrons are in thermal equilibrium in the conduction band and the holes are in equilibrium in the valence band. This means that the carriers are neither gaining nor losing energy from the crystal lattice atoms.

ii) The electron-hole recombination time is much larger than the time for the electrons and holes to reach equilibrium within the conduction and valence band, respectively.

In most problems of interest, the time to reach equilibrium in the same band is approximately a few picoseconds, while the *e-h* recombination time is anywhere from a nanosecond to a microsecond. Thus the above assumptions are usually met. In this case, the quasi-equilibrium electron and holes can be represented by an electron Fermi function f^e (with electron Fermi level) and a hole Fermi function f^h (with a *different* hole Fermi level). We now have

$$n = \int_{E_c}^{\infty} N_e(E) f^e(E) dE \tag{8.22}$$

$$p = \int_{-\infty}^{E_v} N_h(E) f^h(E) dE \tag{8.23}$$

where

$$f^e(E) = \frac{1}{\exp\left(\frac{E-E_{Fn}}{k_BT}\right)+1} \tag{8.24}$$

If $f^v(E)$ is the electron occupation in the valence band, the hole occupation is

$$f^h(E) = 1 - f^v(E) = 1 - \frac{1}{\exp\left(\frac{E-E_{Fp}}{k_BT}\right)+1}$$

$$= \frac{1}{\exp\left(\frac{E_{Fp}-E}{k_BT}\right)+1} \tag{8.25}$$

At equilibrium $E_{Fn} = E_{Fp}$. If excess electrons and holes are injected into the semiconductor, the electron quasi-Fermi level E_{Fn} moves toward the conduction band, while the hole quasi-Fermi level E_{Fp} moves toward the valence band. The ability to define quasi-Fermi levels E_{Fn} and E_{Fp} provides us a very useful approach to solve non-equilibrium problems which are, of course, of greatest interest in devices.

By defining separate Fermi levels for the electrons and holes, one can study the properties of excess carriers using the same relationship between Fermi level and carrier density as we developed for the equilibrium problem (see Section 4.8). Thus, in the approximation where the Fermi distribution is replaced by an exponential, we have

$$\boxed{\begin{aligned} n &= N_c \exp\left[\frac{(E_{Fn}-E_c)}{k_BT}\right] \\ p &= N_v \exp\left[\frac{(E_v-E_{Fp})}{k_BT}\right] \end{aligned}} \tag{8.26}$$

In the more accurate Joyce-Dixon approximation we have

$$\begin{aligned} (E_{Fn}-E_c) &= k_BT\left[\ell n\ \frac{n}{N_c} + \frac{n}{\sqrt{8}N_c}\right] \\ (E_v-E_{Fp}) &= k_BT\left[\ell n\ \frac{p}{N_v} + \frac{p}{\sqrt{8}N_v}\right] \end{aligned} \tag{8.27}$$

The dependence of the quasi-Fermi levels on the electron and hole densities in GaAs at 300 K are shown in Fig. 8.6. Note that for the same carrier injection ($n = p$), the electron quasi-Fermi level moves a greater amount than the hole quasi-Fermi level. This is because of the smaller electron effective mass.

EXAMPLE 8.6 Using Boltzmann statistics calculate the position of the electron and hole quasi-Fermi levels when an e-h density of 10^{17} cm^{-3} is injected into pure (undoped) silicon at 300 K.

At room temperature for Si we have

$$\begin{aligned} N_c &= 2.8 \times 10^{19}\ \text{cm}^{-3} \\ N_v &= 1.04 \times 10^{19}\ \text{cm}^{-3} \end{aligned}$$

If $n_c = p_v = 10^{17}$ cm^{-3}, we obtain ($k_BT = 0.026$ eV)

$$\begin{aligned} E_{Fn} &= k_BT\ \ell n\ \left[\frac{n}{N_c}\right] + E_c \\ &= E_c - 0.146\ \text{eV} \\ E_{Fp} &= E_v - k_BT\left[\ell n\ \frac{p}{N_v}\right] \\ &= E_v + 0.121\ \text{eV} \end{aligned}$$

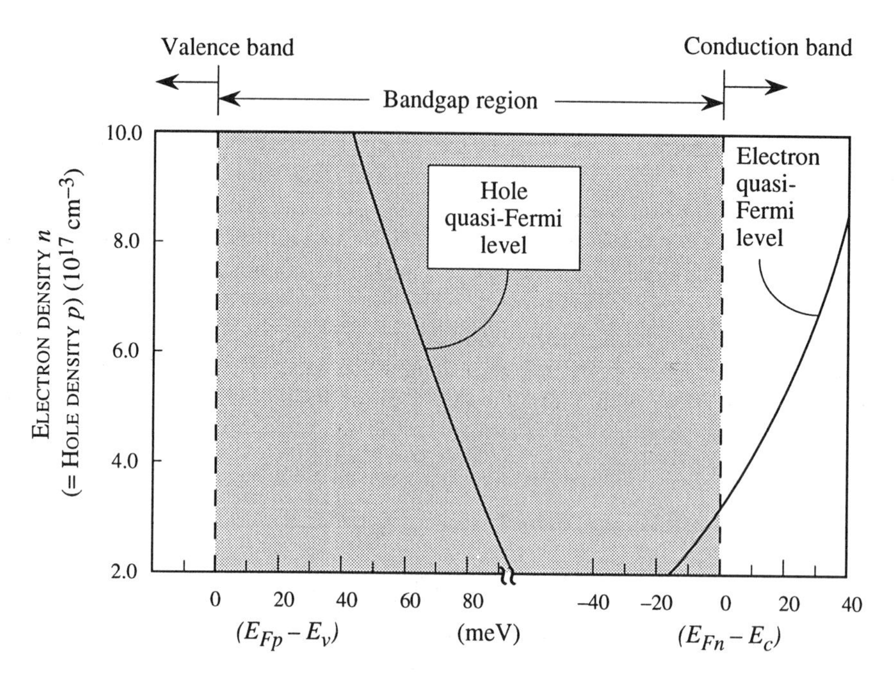

Figure 8.6: Dependence of the quasi-Fermi level positions in GaAs at 300 K on the electron and hole density.

Since in Si, the bandgap is $E_c - E_v = 1.1$eV, we have

$$\begin{aligned} E_{Fn} - E_{Fp} &= (E_c - E_v) - (0.146 + 0.121) \\ &= 1.1 - 0.267 = 0.833 \text{ eV} \end{aligned}$$

If we had injected only 10^{15} cm^{-3} electrons and holes, the difference in the quasi-Fermi levels would be

$$\begin{aligned} E_{Fn} - E_{Fp} &= (E_c - E_v) - (0.266 + 0.24) \\ &= 1.1 - 0.506 = 0.59 \text{ eV} \end{aligned}$$

Thus as the carrier injection is increased, the separation increases.

8.5 CHARGE INJECTION AND RADIATIVE RECOMBINATION

Electrons and holes can be injected into the conduction and valence band in a number of ways. We can shine light on the material and the absorption of photons creates electron-hole pairs. We can also use an external battery in a *p-n* diode and inject electrons and holes in a region. The electrons and holes will, in general, recombine with each other so that the electron in the conduction band will return to the valence band. This recombination process can be via two processes, which are denoted as radiative processes and non-radiative processes. In the radiative process the e-h pair recombines and a photon is emitted. This is the inverse of the photon absorption process.

Electron-hole pairs can also recombine without emitting light. Instead, they may emit either heat or a phonon by generating vibrations of the crystal. Such processes are non-radiative processes. The non-radiative processes usually reduce the device efficiency for optoelectronic devices and must be avoided by using pure materials.

In the previous sections we have discussed how a photon impinging upon a semiconductor can be absorbed by moving an electron from the valence to conduction band. The photon beam thus decays as it moves through the semiconductor. The process of absorption requires the valence band states to have electrons present and conduction band states to have no electrons. What happens if the valence band has holes and the conduction band has electrons? Normally, such a situation does not occur, but if electrons are injected into the conduction band and holes into the valence band (as happens for light-emitting devices), this could occur. *Under such conditions the electron-hole pairs could recombine and emit more photons than are absorbed.* Thus one must talk about the emission coefficient minus the absorption coefficient, or the gain of the material. If the gain is positive, an optical beam will grow as it moves through the material instead of decaying. In the simple parabolic bands we have the gain $g(\hbar\omega)$ given by the generalization of Eqn. 8.15 (gain = emission coefficient − absorption coefficient)

$$g(\hbar\omega) = \frac{\pi e^2 \hbar}{\epsilon_0 n_r c m_0^2 \hbar\omega} \mid \mathbf{a} \cdot \mathbf{p}_{if} \mid^2 N_{cv} \hbar\omega [f^e(E^e) - (1 - f^h(E^h))] \tag{8.28}$$

The term in the square brackets arises since the emission of photons is proportional to $f^e \cdot f^h$, while the absorption process is proportional to $(1 - f^e) \cdot (1 - f^h)$. The difference of these terms appears in Eqn. 8.28.

The energies E^e and E^h are (see Fig. 8.4) determined by noting the following:

$$\hbar\omega - E_g = \frac{\hbar^2 k^2}{2m_r^*} \tag{8.29}$$

$$E^e = E_c + \frac{\hbar^2 k^2}{2m_e^*} = E_c + \frac{m_r^*}{m_e^*}(\hbar\omega - E_g) \tag{8.30}$$

$$E^h = E_v - \frac{\hbar^2 k^2}{2m_h^*} = E_v - \frac{m_r^*}{m_h^*}(\hbar\omega - E_g) \tag{8.31}$$

If $f^e(E^e) = 0$ and $f^h(E^h) = 0$, i.e., if there are no electrons in the conduction band and no holes in the valence band, we see that the gain is simply $-\alpha(\hbar\omega)$, which we had discussed earlier. A positive value of gain occurs when

$$f^e(E^e) > 1 - f^h(E^h) \tag{8.32}$$

a condition that is called *inversion*. In this case the light wave passing in the material has the spatial dependence

$$I(z) = I_o \exp(gz) \tag{8.33}$$

which grows with distance instead of diminishing as it usually does if $g(\hbar\omega)$ is negative. The gain in the optical intensity is the basis for the semiconductor laser.

As the electrons and holes are "pumped" into the semiconductor they recombine through the process of spontaneous emission. This process does not require photons to be present for the photon emission process to occur. The spontaneous rate is given by (rate per unit volume) integrating Eqn. 8.19 over all energies after accounting for the occupation functions f^e and f^h. This gives, using the definition of τ_o from Eqns. 8.19 and 8.21,

$$R_{spon} = \frac{1}{\tau_o} \int d(\hbar\omega) N_{cv} \{f^e(E^e)\}\{f^h(E^h)\} \tag{8.34}$$

The spontaneous recombination rate is quite important for both electronic and optoelectronic devices. It is important to examine the rate for several important cases. We will give results for the electron hole recombination for the following cases:
i) **Minority Carrier Injection:** If $n \gg p$ and the sample is heavily doped n-type, we can assume that $f^e(E^e)$ is close to unity. We then have for the rate at which holes will recombine with electrons,

$$\begin{aligned} R_{spon} &\cong \frac{1}{\tau_o} \int d(\hbar\omega) N_{cv} f^h(E^h) \cong \frac{1}{\tau_o} \int d(\hbar\omega) N_h f^h(E^h) \left(\frac{m_r^*}{m_h^*}\right)^{3/2} \\ &\cong \frac{1}{\tau_o} \left(\frac{m_r^*}{m_h^*}\right)^{3/2} p \end{aligned} \tag{8.35}$$

Thus the recombination rate is proportional to the minority carrier density (holes in this case).
ii) **Strong Injection:** This case is important when a high density of both electrons and holes is injected. We can now assume that both f^e and f^h are sharp step functions and we get approximately

$$R_{spon} = \frac{n}{\tau_o} = \frac{p}{\tau_o} \tag{8.36}$$

iii) **Weak Injection:** In this case we can use the Boltzmann distribution to describe the Fermi functions. We get

$$R_{spon} = \frac{1}{2\tau_o} \left(\frac{2\pi\hbar^2 m_r^*}{k_B T m_e^* m_h^*}\right)^{3/2} np \tag{8.37}$$

If we write the total charge as equilibrium charge plus excess charge,

$$n = n_o + \Delta n; p = p_o + \Delta n \tag{8.38}$$

we have for the excess carrier recombination (note that at equilibrium the rates of recombination and generation are equal)

$$R_{spon} \cong \frac{1}{2\tau_o} \left(\frac{2\pi\hbar^2 m_r^*}{k_B T m_e^* m_h^*}\right)^{3/2} (\Delta n p_o + \Delta p n_o) \tag{8.39}$$

If $\Delta n = \Delta p$, we can define the rate of a single excess carrier recombination as $\frac{\Delta n}{\tau_r}$, where

$$\frac{1}{\tau_r} = \frac{1}{2\tau_o}\left(\frac{2\pi\hbar^2 m_r^*}{k_B T m_e^* m_h^*}\right)(n_o + p_o) \tag{8.40}$$

At low injection, τ_r is much larger than τ_o and is a result of the fact that at low injection, electrons have a low probability to find a hole with which to recombine.
iv) **Inversion Condition:** Another useful approximation occurs when the electron and hole densities are such that $f^e + f^h = 1$. *This is the condition for inversion when the emission and absorption coefficients become equal.* If we assume in this case $f^e \sim f^h = 1/2$, we get the approximate relation

$$R_{spon} \cong \frac{n}{4\tau_o} \cong \frac{p}{4\tau_o} \tag{8.41}$$

The recombination lifetime is approximately $4\tau_o$ in this case. This is a useful result to estimate the threshold current of semiconductor lasers.

The gain and recombination processes discussed here are extremely important in both electronic and optoelectronic devices that will be discussed later. We point out from the above discussion that the recombination time for a single excess carrier can be written, in many situations, in the form

$$\boxed{\tau_r = \frac{\Delta n}{R_{spon}}} \tag{8.42}$$

For minority carrier injection or strong injection $\tau_r \cong \tau_o$. In general, R_{spon} has a strong carrier density dependence as does τ_r. In Fig. 8.7 we show the radiative recombination time for GaAs. In the figure, this time is shown as a function of hole density for the case $n = p$ and also for the case where electrons are injected into a p-type material.

8.5.1 Application Example: Phosphors and Fluorescence

The optical processes we have described above are exploited in the design of semiconductor detectors, imagers, and light emitters. The general theory of electron–photon interactions is, of course, not just limited to semiconductors. Optical interactions occur in atoms, organic molecules, polymers, amorphous materials, etc. An important application of electron–photon interaction is light emission for display applications such as TV and computer screens. The screen is coated with special materials which have the property that when they are excited (say, by an electron beam) they emit light.

The general problem of light emission as a result of excitation can be represented by the three-level system shown in Fig. 8.8. Level M represents a metastable, or long-lived, state, while level E represents an excited state of the system. The level G denotes the ground state of the system. An excitation causes the electronic system to be excited to the level M where there are two distinct possibilities. Electrons

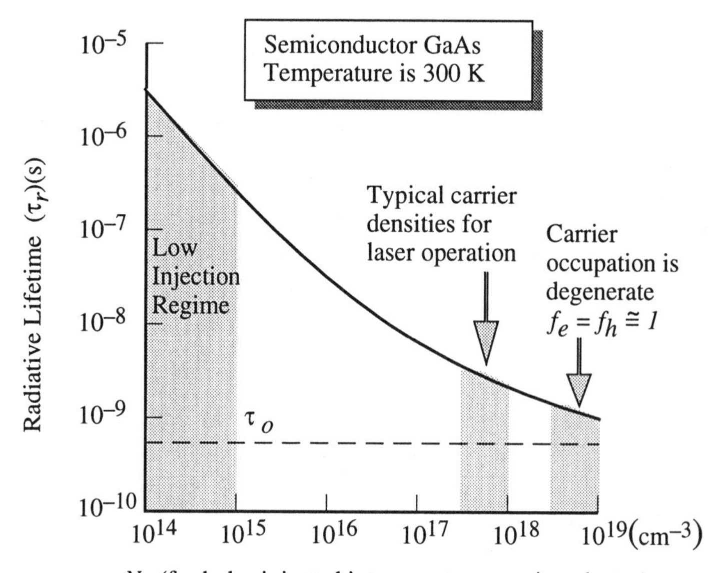

Figure 8.7: The radiative lifetime of electrons in GaAs at 300 K as a function of carrier density. The carrier density could be produced by injection ($n = p$) or by p-type doping in which case the lifetime is the minority carrier lifetime.

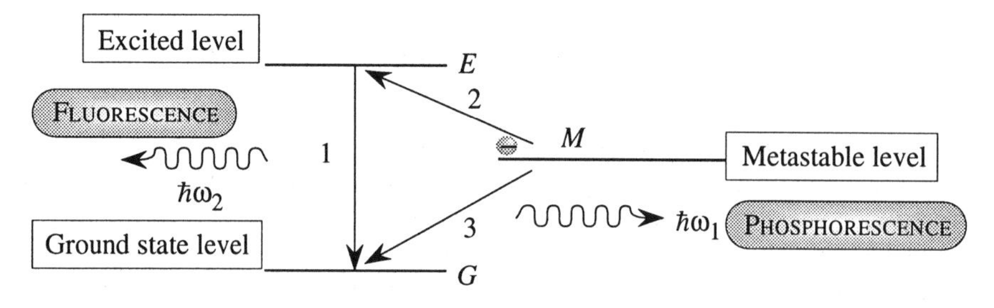

Figure 8.8: A general three-level system to represent light emission through excitation. Light can be emitted by a transition from $M \rightarrow G$ (process 3), and the process is usually known as phosphorescence. It can also be emitted by the process $M \rightarrow E \rightarrow G$, a process called delayed fluorescence.

excited to the state M can emit photons, as shown by process 3 of Fig. 8.8. In some materials, the transition $M \rightarrow G$ is forbidden due to certain selection rules. In this case, light emission can occur by the electron first being excited to level E due to phonon scattering and then through the $E \rightarrow G$ recombination.

The processes described by Fig. 8.8 are usually invoked for understanding "phosphors," a term used for light-emitting materials used in TV screens, cathode ray tubes, etc. The transition $M \rightarrow G$ mentioned above is called phosphorescence, while the transition $M \rightarrow E \rightarrow G$ is called fluorescence. As is clear form Fig. 8.8, the light emitted by fluorescence has a different wavelength from the light emitted by phosphorescence.

Phosphors in use in commercial applications range from organic materials used as dyes to inorganic materials used in the cathode ray tubes and TV screens. An interesting example of phosphors are the laundry brighteners that are based on colorless organic phenyl-based dyes. This phosphor is excited by ultraviolet light present in background lighting. Fluorescence occurs, emitting blue light, which compensates for the loss of blue light thorough absorption in textiles. As a result, the fabric appears "whiter."

In most inorganic phosphors, the light emission wavelength is determined by impurities that are introduced into the material. The impurity levels produce well-defined bandgap states. When electron-hole pairs are created through excitation, the carrier (electron or hole) is trapped at the defect level from where radiative recombination occurs. An important inorganic phosphor is based on the semiconductor ZnS which has a bandgap of ~3.8 eV. When this material is doped with impurities, levels are produced which can be exploited to produce green (copper-activated) or blue (silver-activated) colors. These materials are widely used in TV screen technology along with europium-activated yttrium ortho-vanadate for red color to produce color displays. In Table 8.1 we show the color response of some of the important phosphors used in modern display technology.

EXAMPLE 8.7 Using the Joyce-Dixon approximation, calculate the electron and hole densities needed to cause inversion at the bandedges of GaAs at 300 K. The electron and hole densities are equal.

The effective density of states N_c and N_v for GaAs at 300 K are 4.45×10^{17} cm^{-3} and 7.7×10^{18} cm^{-3}, respectively. The procedure for solving this problem is quite simple. We choose a value of $n(= p)$; use the Joyce-Dixon approximation to calculate E_{Fn} and E_{Fp}; calculate f^e and f^h at the bandedges and check if the inversion condition $f^e + f^h = 1$ is satisfied. The carrier density is increased in small steps until the condition is satisfied.

The above exercise gives a value of $n = p = 1.1 \times 10^{18}$ cm^{-3} for the inversion condition.

8.6 COLLISIONS AND BORN APPROXIMATION

In Fig. 8.9a we show a schematic of a collision of two particles mediated by some mutual interaction (e.g., Coulombic interaction, if the particles are charged). Intuitively, we can think of the collision process in the time domain where the particles start out at time t_i far from each other in regions where there is no interaction be-

MATERIAL (ACTIVATION)	EMISSION COLOR
Zinc Sulfide (silver)	Blue
Yttrium Silicate (cesium)	Purplish blue
Zinc sulfide (copper)	Green
Zinc orthosilicate (manganese)	Yellowish green
Gadolium oxysulfide (terbium)	Yellowish green
Zinc cadmium sulfide (silver)	Green
Yttrium oxysulfide (europium)	Red

Table 8.1: Characteristics of some phosphor materials used in display screen technology. The impurities used are specified in parenthesis.

tween them. As time passes, the particles come within a region where they interact with each other. Finally, at time t_f, the particles are once again so far from each other that there is no interaction.

A similar picture is shown in Fig. 8.9b, where a particle is scattering from a potential $U(r)$. At times less than t_i and greater than t_f, the particle is in a region where the potential has no influence. In between these times the particle feels the effect of the potential and scatters from it.

We assume that in such scattering processes there is a finite range of space over which the particles sense the scattering potentials. Away from this region they suffer no scattering and are described by a simple Schrödinger equation with known solutions. As shown schematically in Fig. 8.10, we can separate the Hamiltonian as

$$H = H_0 + U(r) \tag{8.43}$$

where H_0 is the simple Hamiltonian describing the particle with no scattering and $U(r)$ is the scattering potential. Since the scattering potential is finite only over a certain range of space, as shown in Fig. 8.10, outside this range the particle is represented by the solution of the equation

$$H_0 u_n = E_n u_n \tag{8.44}$$

The effect of the scattering potential is to cause transitions of a particle from one state, say u_m, to another. For example, if the Hamiltonian H_0 represents electrons in free space (or in a crystal) the problem of a particle coming from the left and scattering from the fixed potential is described by the electrons initially having a plane-wave-like behavior on the left of the potential. After scattering, there is a finite probability of finding the particle in other states.

The scattering rate from state u_i to state u_f is given by the Fermi golden

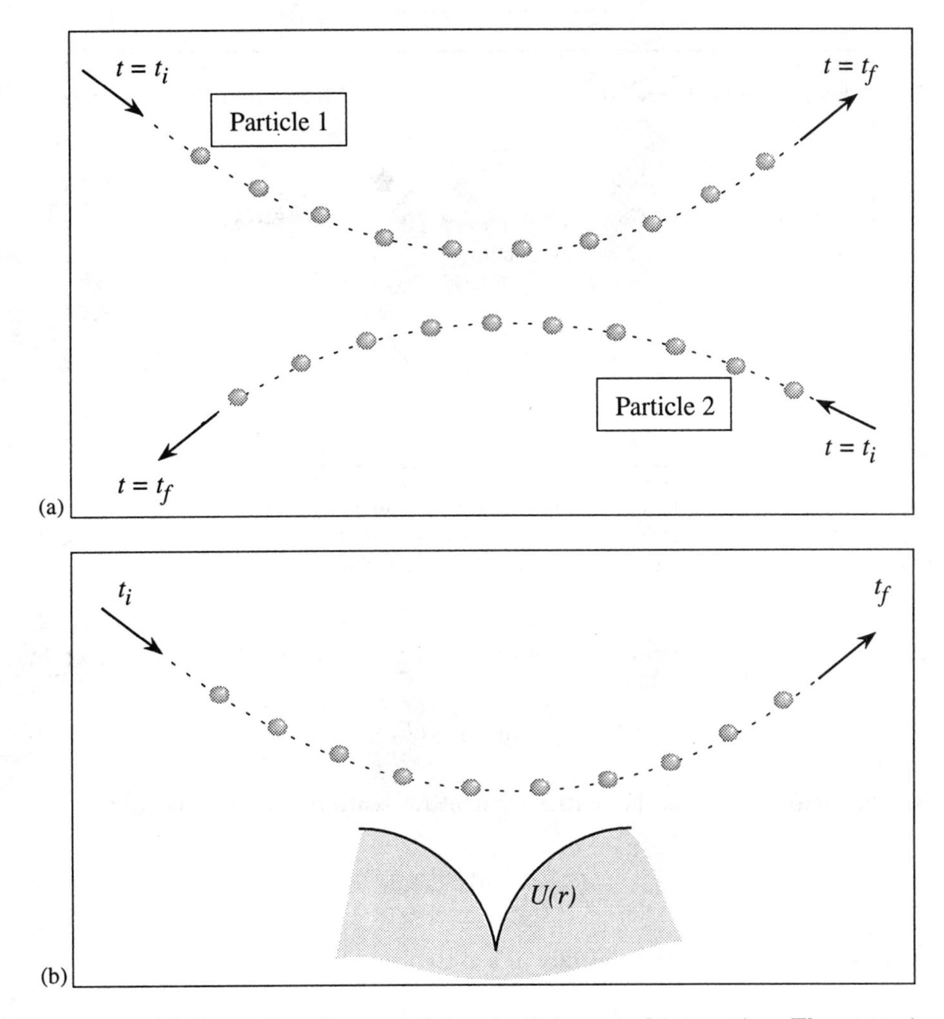

Figure 8.9: (a) Scattering of two particles via their mutual interaction. The scattering can be viewed as starting at time t_i and ending at time t_f.

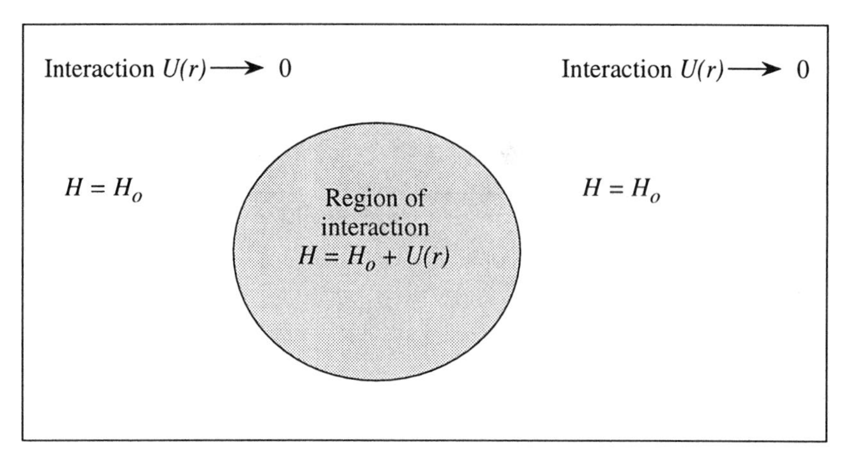

Figure 8.10: In the stationary state approach to scattering, we view the space as made up of regions where the scattering potential is zero and the eigensolutions are that of the simple Hamiltonian H_0, and a region where the scattering potential is non-zero.

rule with $\hbar\omega = 0$. The result is known as the *Born approximation* and is given by

$$W_{if} = \frac{2\pi}{\hbar} |M_{if}|^2 \delta(E_i - E_f) \tag{8.45}$$

Here the quantity M_{if} is the matrix element for scattering and is given by

$$M_{if} = \int u_f^* U(r) u_i d^3r \tag{8.46}$$

The rate at which a particle initally in a state u_i scatters out to any state is

$$W_i = \frac{2\pi}{\hbar} \sum_{\text{final states}} |M_{if}|^2 \delta(E_i - E_f) = \frac{2\pi}{\hbar} |M_{if}|^2 N(E_f) \tag{8.47}$$

where $N(E_f)$ is the density of states at energy $E_f(= E_i)$. We see that the total scattering rate is proportional to the density of states—the higher the number of states to scatter into, the stronger the scattering rate. In most scattering processes the spin of an electron is not affected by scattering. As a result, *the density of states to be used corresponds to a single spin value.* This is half of the total density of states.

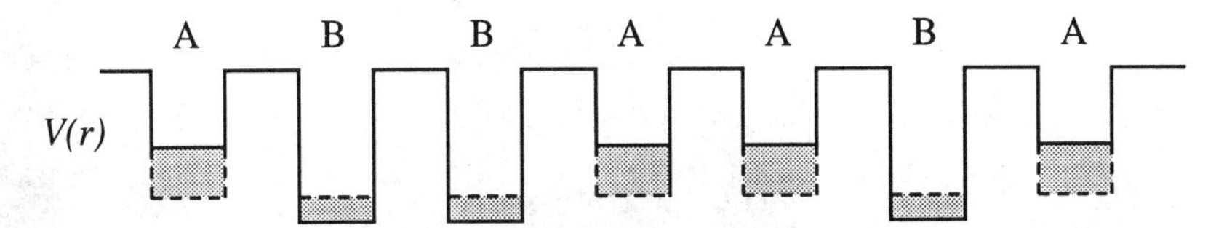

Figure 8.11: A schematic of the actual atomic potential (solid line) and the average virtual crystal potential (dashed line) of an $A - B$ alloy. The shaded area shows the difference between the real potential and the virtual crystal approximation.

8.6.1 Application Example: Alloy Scattering

An interesting scattering problem is the scattering of a particle from a spherically symmetric square well potential. An important application of such scattering is in the understanding of carrier transport in alloys. Alloys are used in a number of technologies and are based on a material synthesis process that can randomly mix two or more different materials. By using a mixture of materials it is possible to obtain new materials that have properties intermediate between those of the components. Ideal alloys represent a random arrangement of atoms on a lattice and, therefore, even if the lattice is periodic the potential seen by electrons is non-periodic. Thus when an electron is in an alloy, it sees a randomly varying potential. To simplify the problem the random potential is separated into a periodic potential and one with random fluctuations. We write

$$H = H_0 + H' \tag{8.48}$$

where H_0 results from the average potential of the alloy and H' arises from the difference of the real and average potentials.

Let us consider the problem of a perfectly random alloy where the smallest physical size over which the crystal potential fluctuates randomly is the unit cell. An electron moving in the alloy $A_x B_{1-x}$ will see a random potential, schematically shown in Fig. 8.11.

The average potential and the average bandstructure of the alloy are described to the lowest order by the *virtual crystal approximation*. In this approximation, the averaging of the atomic potentials

$$\{M\}_{\text{all}} = x\{M\}_{\text{A}} + (1-x)\{M\}_{\text{B}} \tag{8.49}$$

gives an average *periodic* potential represented by the dashed line in Fig. 8.11. An important approximation is now made. The difference between the real potential and the assumed virtual crystal potential is represented within each unit cell by a highly localized potential. For example, for A-type atoms, the difference is

$$\begin{aligned} E_{\text{all}} - E_{\text{A}} &= xE_{\text{A}} + (1-x)E_{\text{B}} - E_{\text{A}} \\ &= (1-x)\left[E_{\text{B}} - E_{\text{A}}\right] \\ &= (1-x)U_{\text{all}} \end{aligned} \tag{8.50}$$

Similarly, for the B atom, the difference is

$$E_{\text{B}} - E_{\text{all}} = x\left[E_{\text{B}} - E_{\text{A}}\right]$$

$$= x\, U_{\rm all} \tag{8.51}$$

The scattering potential is chosen to be of the form

$$\begin{aligned} U(\mathbf{r}) &= U_0 \quad \text{for } |\mathbf{r}| \leq r_0 \\ &= 0 \quad \text{for } |\mathbf{r}| > r_0 \end{aligned} \tag{8.52}$$

where r_0 is the interatomic distance. If we use the Born approximation to calculate the scattering rate, we have

$$W(\mathbf{k}) = \frac{2\pi}{\hbar} \sum_{\mathbf{k}'} \left| M_{\mathbf{k}\mathbf{k}'} \right|^2 \delta(E_{\mathbf{k}} - E_{\mathbf{k}'})$$

and

$$M_{\mathbf{k}\mathbf{k}'} = \int e^{i(\mathbf{k}-\mathbf{k}')\cdot\mathbf{r}}\, \Delta U(\mathbf{r})\, d^3r$$

We will now use the fact that the scattering potential only extends over a unit cell, and over this small distance

$$e^{i(\mathbf{k}-\mathbf{k}')\cdot\mathbf{r}} \approx 1 \tag{8.53}$$

Thus

$$M_{\mathbf{k}\mathbf{k}'} = \frac{4\pi}{3}\, r_0^3\, U_0 \tag{8.54}$$

and

$$\begin{aligned} W(\mathbf{k}) &= \frac{2\pi}{\hbar} \left(\frac{4\pi}{3}\, r_0^3\, U_0 \right)^2 \frac{1}{(2\pi)^3} \int \delta(E_{\mathbf{k}} - E_{\mathbf{k}'})\, d^3k' \\ &= \frac{2\pi}{\hbar} \left(\frac{4\pi}{3}\, r_0^3\, U_0 \right)^2 N(E_{\mathbf{k}}) \end{aligned} \tag{8.55}$$

Let us consider the face-centered cubic (fcc) lattice (the lattice for most semiconductors). We may write

$$r_0 = \frac{\sqrt{3}}{4} a$$

where a is the cube edge for the fcc lattice. This gives

$$\left(\frac{4\pi}{3} r_0^3 \right)^2 = \frac{3\pi^2}{16} V_0^2 \tag{8.56}$$

where $V_0 = a^3/4$ is the volume of the unit cell. We finally obtain for the scattering rate:

$$W(\mathbf{k}) = \frac{2\pi}{\hbar} \left(\frac{3\pi^2}{16} V_0^2 \right) U_0^2 N(E_{\mathbf{k}}) \tag{8.57}$$

We will now assume that all scattering centers scatter independently, so that we can simply sum the scattering rates. For A-type atoms, the scattering rate is (using $U_0 = (1-x)U_{all}$ from Eqn. 8.50)

$$W_{\rm A}(\mathbf{k}) = \frac{2\pi}{\hbar} \left(\frac{3\pi^2}{16} V_0^2 \right) (1-x)^2 U_{all}^2\, N(E_{\mathbf{k}}) \tag{8.58}$$

For B-type atoms, the rate is (using $U_0 = xU_{all}$ from Eqn. 8.51)

$$W_{\mathrm{A}}(\mathbf{k}) = \frac{2\pi}{\hbar}\left(\frac{3\pi^2}{16}V_0^2\right) x^2 U_{all}^2\, N(E_{\mathbf{k}}) \tag{8.59}$$

There are x/V_0 A-type atoms and $(1-x)/V_0$ B-type atoms in the unit volume, so that the total scattering rate is

$$\begin{aligned} W_{tot} &= \frac{2\pi}{\hbar}\left(\frac{3\pi^2}{16}V_0\right) U_{all}^2\, N(E_{\mathbf{k}}) \left[x\,(1-x)^2 + (1-x)\,x^2\right] \\ &= \frac{3\pi^3}{8\hbar} V_0 U_{all}^2\, N(E_{\mathbf{k}})\, x\,(1-x) \end{aligned} \tag{8.60}$$

8.6.2 Application Example: Screened Coulombic Scattering

As another example of the use of the Born approximation, we will examine the scattering of an electron from a charged particle. The scattering potential is Coulombic in nature. This scattering plays a very important role in many important applications. Problems that require an understanding of this scattering process include:

- *Scattering of α particles in matter*: When a thin film of metal is bombarded with α particles (He-nuclei), the properties of the outgoing particles are understood on the basis of Coulombic scattering.
- *Mobility in devices*: Semiconductor devices have regions that are doped with donors or acceptors. These dopants provide excess carriers in the conduction or the valence band. Without these carriers most devices will not function. When a dopant provides a free carrier, *the remaining ion provides a scattering center for the free carriers.* This causes scattering which is understood on the basis of electron-ion scattering. Additionally, at high densities one can have electron-electron scattering as well as electron-hole scattering, which is also understood for the general problem discussed in this subsection.

Before starting our study of scattering from a Coulombic interaction, it is important to note that in most materials there is a finite *mobile carrier density.* These carriers can adjust their spatial position in response to a potential and thus *screen* the potential. The screening is due to the dielectric response of the material and includes the effect that the background ions as well as the other free electrons have on the potential. A number of formalisms have been developed to describe the dielectric response function. We will use a form given by the Thomas-Fermi formalism.

Let us consider an electron scattering from a charged particle in a crystalline material. We will assume that the electron is described by the effective mass theory. We also asume that the density of free carriers is n_0. In the Thomas-Fermi formalism, the background free carriers modify their carrier concentration near the impurity so that when the scattering electron is far from the impurity it sees a potential much weaker than the Coulombic potential. Very close to the impurity the potential is not affected much by the screening.

The real-space behavior of the screened potential is given by

$$\phi_{\mathrm{tot}}(\mathbf{r}) = \frac{q}{4\pi\epsilon r}e^{-\lambda r} \tag{8.61}$$

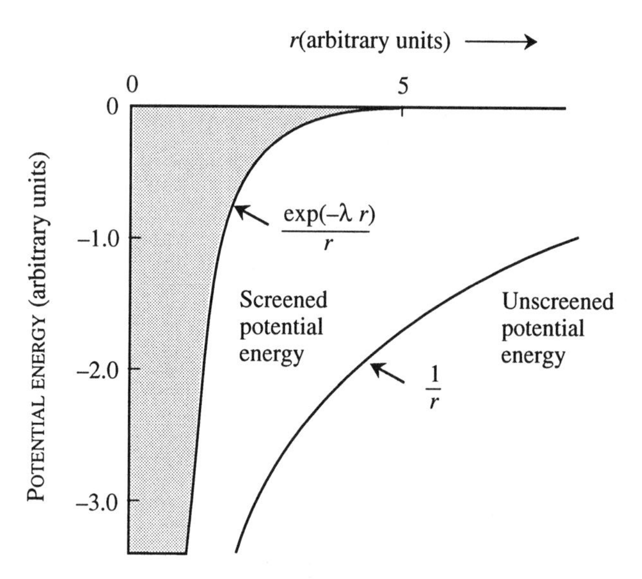

Figure 8.12: Comparison of screened and unscreened Coulomb potentials of a unit positive charge as seen by an electron. The screening length is λ^{-1}.

where q is the charge of the impurity and ϵ is the dielectric constant. The quantity λ, which represents the effect of the background free carriers is given for a non-degenerate carrier gas (i.e., a carrier distribution where the Fermi statistics is reasonably approximated by the Boltzmann statistics) as

$$\lambda^2 = \frac{n_0 e^2}{\epsilon k_B T} \tag{8.62}$$

When the free carrier density is high so that the carriers are degenerate,

$$\lambda^2 = \frac{3n_0 e^2}{2\epsilon E_F} \tag{8.63}$$

where E_F is the Fermi energy.

As noted, the effect of screening is to reduce the range of the potential from a $1/r$ variation to a $\exp(-\lambda r)/r$ variation. This is an extremely important effect and is shown schematically in Fig. 8.12.

We now calculate the matrix element for the screened Coulombic potential

$$U(\mathbf{r}) = \frac{Ze^2}{4\pi\epsilon} \frac{e^{-\lambda r}}{r} \tag{8.64}$$

where Ze is the charge of the impurity. We choose the initial normalized state to be $|\mathbf{k}\rangle = \exp(i\mathbf{k}\cdot\mathbf{r})/\sqrt{V}$ and the final state to be $|\mathbf{k}'\rangle = \exp(i\mathbf{k}'\cdot\mathbf{r})/\sqrt{V}$, where V is the volume of the crystal. The matrix element is then

$$M_{\mathbf{kk}'} = \frac{Ze^2}{4\pi\epsilon V} \int e^{-i(\mathbf{k}'-\mathbf{k})\cdot\mathbf{r}} \frac{e^{-\lambda r}}{r} r^2 dr \sin\theta' \, d\theta' \, d\phi'$$

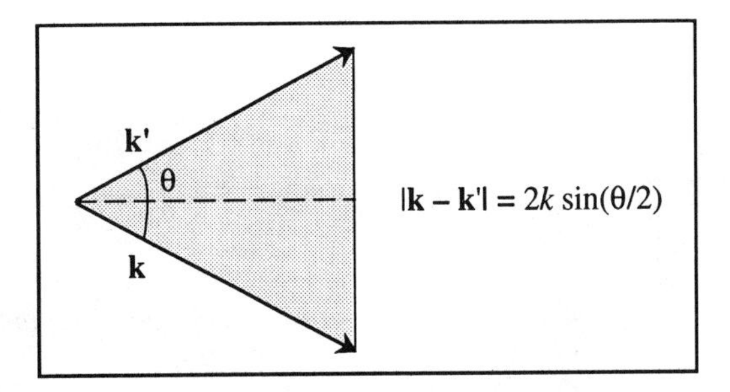

Figure 8.13: As a consequence of the elastic scattering, there is a simple relation between the magnitude of the scattered wavevector and the scattering angle q.

Carrying out the ϕ' integration which gives a factor of 2π, we have

$$\begin{aligned} M_{\mathbf{kk}'} &= \frac{Ze^2}{4\pi\epsilon V}2\pi\int_0^{\infty} r\,dr\int_{-1}^{1} d(\cos\theta')e^{-\lambda r}e^{-i|\mathbf{k}'-\mathbf{k}|r\cos\theta'} \\ &= \frac{Ze^2}{4\pi\epsilon V}2\pi\frac{2}{|\mathbf{k}'-\mathbf{k}|^2+\lambda^2} \end{aligned}$$

We note that $|\mathbf{k}'| = |\mathbf{k}|$ since the scattering is elastic. Then, as can be seen from Fig. 8.13,

$$\left|\mathbf{k}-\mathbf{k}'\right| = 2k\sin(\theta/2) \tag{8.65}$$

where θ is the polar scattering angle.

$$M_{\mathbf{kk}'} = \frac{Ze^2}{V\epsilon}\frac{1}{4k^2\sin^2(\theta/2)+\lambda^2}$$

The scattering rate is now given by the Born approximation:

$$W(\boldsymbol{k},\boldsymbol{k}') = \frac{2\pi}{\hbar}\left(\frac{Ze^2}{V\epsilon}\right)^2\frac{\delta\left(E_{\mathbf{k}}-E_{\mathbf{k}'}\right)}{\left(4k^2\sin^2(\theta/2)+\lambda^2\right)^2} \tag{8.66}$$

One can see that in the two extremes of no screening ($\lambda \to 0$) and strong screening ($\lambda \to \infty$), the rate becomes, respectively,

$$W(\boldsymbol{k},\boldsymbol{k}') = \frac{2\pi}{\hbar}\left(\frac{Ze^2}{V\epsilon}\right)^2\frac{\delta\left(E_{\mathbf{k}}-E_{\mathbf{k}'}\right)}{16k^4\sin^4(\theta/2)} \tag{8.67}$$

and

$$W(\boldsymbol{k},\boldsymbol{k}') = \frac{2\pi}{\hbar}\left(\frac{Ze^2}{V\epsilon}\right)^2\frac{\delta\left(E_{\mathbf{k}}-E_{\mathbf{k}'}\right)}{\lambda^4} \tag{8.68}$$

The angular dependence of the scattering process is very important. One can intuitively see that scattering that produces a large angle scattering is much more effective in altering the motion of electrons than small-angle scattering. In fact, since the scattering is elastic, a forward-angle scattering (scattering angle is

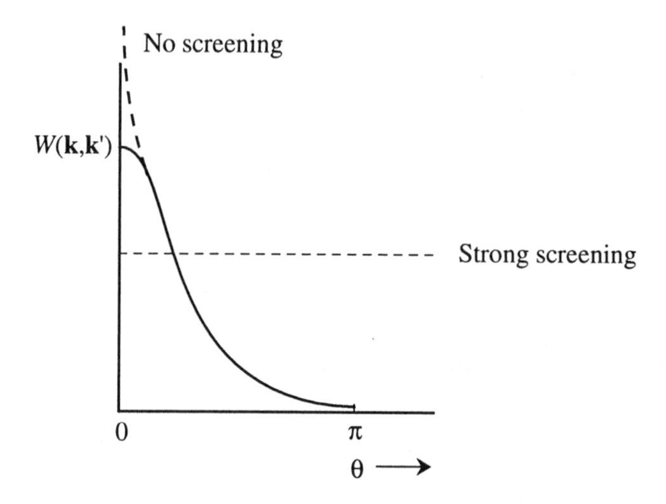

Figure 8.14: Angular dependence of the scattering by ionized impurities. The scattering has a strong forward angle preference.

zero) has no effect on the motion of the initial electron. Thus it is important to examine the angular dependence of $W(\mathbf{k}, \mathbf{k}')$. This is plotted as a function of the scattering angle θ in Fig. 8.14. The ionized impurity scattering has a strong forward angle bias, as can be seen.

8.7 SCATTERING AND TRANSPORT THEORY

Transport of electrons (holes) in materials is the basis of a large number of devices. Semiconductor transistors, diodes, and lasers depend upon the movement of electrons and holes from one region to another. Current flow in metals is exploited in a wide range of technologies, ranging from electric power grids to interconnects on an integrated chip. Transport of charged mobile carriers in materials is described by phenomenological laws known as Ohm's law and Fick's law. Ohm's law describes the response of charged carriers to an electric field, while Fick's law tells us how particles move from a region of high concentration to a region of low concentration. Quantum mechanics and our understanding of band theory allows us to understand the origin of such phenomenological laws.

Let us express Ohm's law in some other equivalent forms. If A is the area of the sample and L is its length, the resistance of the sample is

$$R = \rho \frac{L}{A} = \frac{1}{\sigma} \frac{L}{A} \tag{8.69}$$

where ρ and σ are the resistivity and conductivity of the material, respectively. We may write the current as

$$I = JA \tag{8.70}$$

where J is the current density. We may also write the electric field applied to the sample as

$$E = \frac{V}{L} \tag{8.71}$$

Ohm's law now becomes

$$J = \sigma E \tag{8.72}$$

Let us proceed further along this path. We have, by the definition of current density,

$$J = nev \tag{8.73}$$

where n is the density of the carriers carrying current, e is their charge, and v is the average velocity with which the carriers are moving in the direction of the field. We may also define a relation between velocity of the carriers and the applied field through a definition

$$v = \mu E \tag{8.74}$$

where μ is called the mobility of the carriers. With this definition we get

$$\sigma = ne\mu \tag{8.75}$$

All of the equations we have given above are just different ways of expressing Ohm's law. Even though we have introduced carrier density and charge and mobility, *we have not derived Ohm's law.* The first attempt at deriving Ohm's law was due to Drude at the turn of the twentieth century.

In the case of metals, we have seen that the quantity n above is essentially the density of electrons in the outermost partially filled band. In the case of semiconductors, we have the modified equation

$$\sigma = ne\mu_n + pe\mu_p \tag{8.76}$$

where n is the electron density in the conduction band and p is the hole density in the valence band. The mobilities of electrons is μ_n and that of holes is μ_h.

Current can also flow if there is a concentration gradient in particle density. This current is given by

$$\begin{aligned} J_{tot}(diff) &= J_n(diff) + J_p(diff) \\ &= eD_n \frac{dn}{dx} - eD_p \frac{dp}{dx} \end{aligned} \tag{8.77}$$

where D_n and D_p are the electron and hole diffusion coefficients. The diffusion coefficients are related to the mobilites by Einstein relation

$$D = \frac{\mu k_B T}{e} \tag{8.78}$$

The perturbation theory discussed in this chapter, coupled with band theory, allows us to calculate mobilites of materials. It is important to note that the observation that the average velocity of electrons is proportional to the electric field and has no dependence on the time the field is on or the distance the electrons have traveled shows that electrons must suffer scattering as they move in materials. We also know from the Bloch theorem that in a perfectly crystalline material electrons behave just as they do in free space and suffer no scattering. However, a real material is never perfect and electrons do suffer scattering. As we have noted in the previous section, impurity scattering and alloy scattering can cause scattering. So can lattice vibrations in a crystal. At a finite temperature, the atoms in a crystal vibrate due to their thermal energies. This vibration causes local strain waves in the crystal, which result in variations in the bandedge positions of the semiconductor. The electrons scatter from these vibrations. As the temperature of the crystal is increased, the amplitude of vibration increases, causing an enhancement of the scattering rates. In general, as the electron is moving in the semiconductor, it suffers scattering from all the various scattering processes.

If one considers a beam of electrons moving initially with the same momentum, then due to the scattering processes, the momentum and energy will gradually lose coherence with the initial state values. The average time it takes to lose coherence or memory of the initial state properties is called the "relaxation time" or scattering time. Thus one can define a relaxation time for momentum (τ) or velocity, as shown in Fig. 8.15.

The scattering process and the relaxation times are related to the microscopic properties of the electrons in the semiconductor and the scattering potential of the imperfection, as discussed above. The transport of the electron is schematically shown in Fig. 8.16. The electron suffers collisions which change its momentum. *In between collisions, the electron moves under the electric field as a "free particle," obeying Newton's equation of motion with free electron mass replaced by its effective mass.*

When an electron distribution is subjected to an electric field, the electrons tend to move in the field direction (opposite to the field **E**) and gain velocity from the field. However, because of imperfections, they scatter in random directions. A steady state is established in which the electrons have some net drift velocity in the field direction.

Let us develop a very simple model for electron mobility or conductivity in materials. We will then discuss the quantum interpretation and how one goes about calculating mobility. We start with the following simple assumptions:
i) The electrons in the semiconductor do not interact with each other. This approximation is called the independent electron approximation.
ii) Electrons suffer collisions from various scattering sources and the time τ describes the mean time between successive collisions.
iii) The electrons move according to the free electron equation

$$\frac{\hbar dk}{dt} = \mathbf{F}_{ext} \tag{8.79}$$

in between collisions. After a collision, the electrons lose all their excess energy (on the average) so that the electron gas is essentially at thermal equilibrium. This assumption is really valid only at very low electric fields.

According to these assumptions, immediately after a collision the electron velocity is the same as that given by the thermal equilibrium conditions. This average velocity is thus zero, after collisions. The electron gains a velocity in between

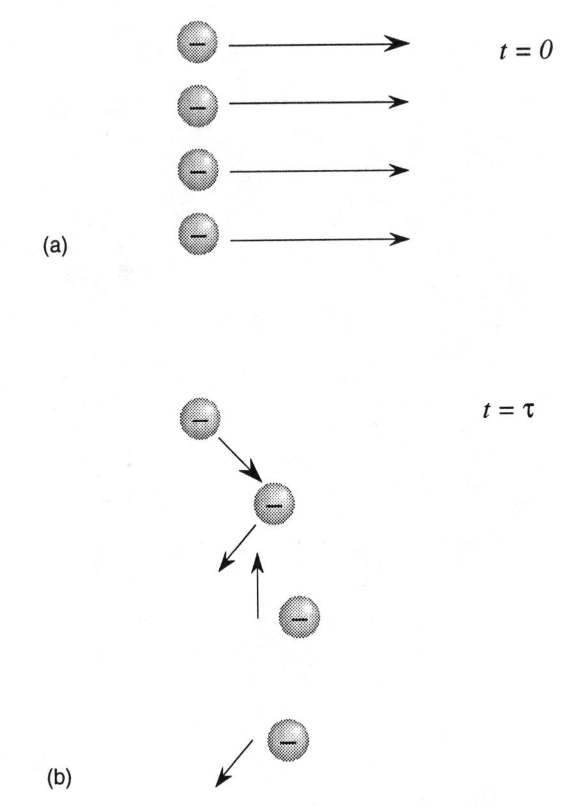

Figure 8.15: A schematic of the effect of scattering on electron velocities. (a) A beam of electrons with constant velocity is considered at time $t = 0$; the arrow is the vector representing the velocity. (b) After a relaxation time (scattering time) τ_{sc}, the electrons have lost memory of their velocities.

collisions, i.e., only for the time τ.

This average velocity gain is then that of an electron with mass m^* travelling in a field $\boldsymbol{E}$ for a time τ:

$$\boldsymbol{v}_{avg} = -\frac{e\boldsymbol{E}\tau}{m^*} = \boldsymbol{v}_d \tag{8.80}$$

where $\boldsymbol{v}_d$ is the drift velocity. The current density is now

$$\mathbf{J} = -ne\mathbf{v}_d = \frac{ne^2\tau}{m^*}\boldsymbol{E} \tag{8.81}$$

Comparing this with the Ohm's law result for conductivity σ

$$\mathbf{J} = \sigma\boldsymbol{E}$$

we have

$$\boxed{\sigma = \frac{ne^2\tau}{m^*}} \tag{8.82}$$

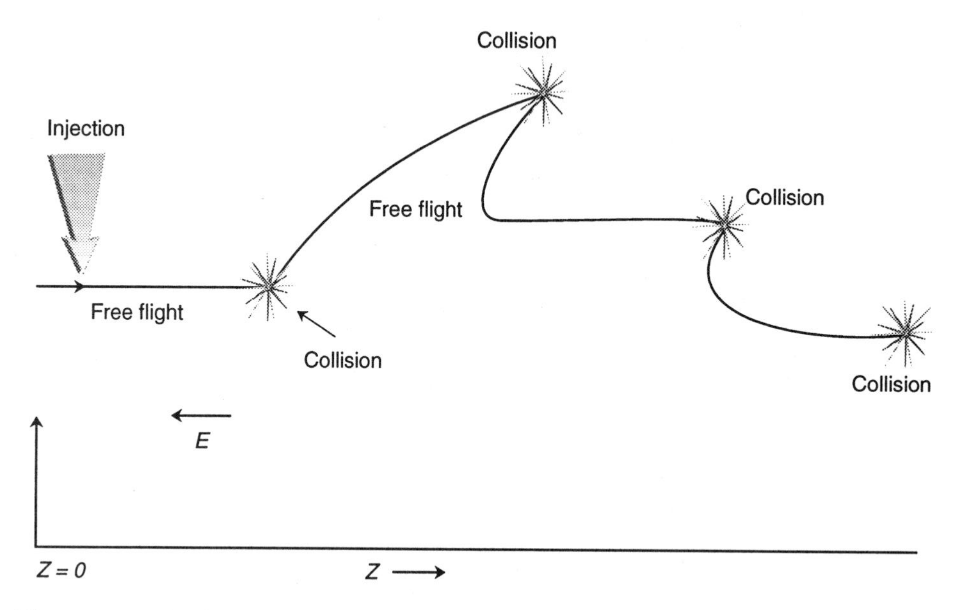

Figure 8.16: A schematic view of an electron moving in a semiconductor under an applied field. The flight of the electron is made up of a series of free flights followed by scattering.

The resistivity of the semiconductor is simply the inverse of the conductivity. From the definition of mobility μ, for electrons

$$\mathbf{v}_d = -\mu \mathbf{E}$$

Note that the electrons move in a direction opposite to the field, while the holes move in the same direction. From Eqns. 8.82 and the definition of mobility, we have

$$\boxed{\mu = \frac{e\tau}{m^*}} \tag{8.83}$$

If both electrons and holes are present, the conductivity of the material becomes

$$\sigma = ne\mu_n + pe\mu_p$$

where μ_n and μ_p are the electron and hole mobilities and n and p are their densities.

An important question that arises is the following: How does one calculate the proper scattering time τ to be used in the mobility or conductivity expression? In the previous section we have seen how one calculates the scattering rate W_{if} from an initial state to a final state using the Fermi golden rule or Born approximation. Clearly, τ is related to W_{if}. But what is the precise relation? This relationship is given by solving the Boltzmann transport equation. The problem is addressed in Appendix B and we will present the outcome here.

There are three important issues in relating the results of transport calculations to measurements made in the lab. Consider an electron initially in a state described by momentum $\hbar \boldsymbol{k}$ scattering into a state with momentum $\hbar \boldsymbol{k}^{'}$. Let us say that we know how to calculate $W(\boldsymbol{k}, \boldsymbol{k}^{'})$. We have carried out such calculations for charged impurity scattering and alloy scattering. The three questions that need to be addressed are: i) How is τ for an electron related to the scattering rate $W(\boldsymbol{k}, \boldsymbol{k}^{'})$? ii) Since electrons are distributed with different energies, what is the proper way to average the scattering time? iii) If there are several different sources of scattering, what is the procedure for calculating their total effect?

Relation Between τ and $W(\boldsymbol{k}, \boldsymbol{k}^{'})$

In general, the relations between τ to be used for mobility, and $W(\boldsymbol{k}, \boldsymbol{k}^{'})$ calculated by the Fermi golden rule (or Born approximation) is quite complicated. However, if the scattering is elastic, i.e., no energy is lost in the scattering process, the relation becomes quite simple. We find that if the angle between $\boldsymbol{k}$ and $\boldsymbol{k}^{'}$ is θ, we have

$$\frac{1}{\tau} = \int W(\boldsymbol{k}, \boldsymbol{k}^{'})(1 - \cos\theta)\frac{d^3 \boldsymbol{k}^{'}}{(2\pi)^3} \tag{8.84}$$

The factor $(2\pi^3)$ in the denominator comes from the definition of the density of states in a k-space volume $d^3\boldsymbol{k}'$. From this expression we see that elastic scattering processes which do not change the angle of a particle are ineffective in transport. Scattering where $\theta = 0$ is called forward-angle scattering, while if $\theta = \pi$ it is called back-scattering.

In the next section we will apply the expression given above to ionized impurity and alloy scattering. For inelastic scattering processes, there is no simple relation between τ and $W(\boldsymbol{k}, \boldsymbol{k}^{'})$ and the problem has to be solved numerically as discussed in Appendix B. However, an approximate value of τ can be obtained by taking the inverse of $W(k)$, i.e., of the integral of $W(\boldsymbol{k}, \boldsymbol{k}^{'})$ over all final states.

Averaging Over Electron Energies

The expression given above for the scattering time τ will, in general, depend on the energy of the electron. Since the electrons are distributed with different energies, how does one find the averaged τ to be used in transport? Time averaging is derived in Appendix B and is given by

$$\langle\langle \tau \rangle\rangle = \frac{\int \tau E f^{\circ}(E) dE}{\int E f^{\circ}(E) dE} \tag{8.85}$$

where $f^{\circ}(E)$ is the Fermi-Dirac distribution function. For the case where the value of $f^{\circ}(E)$ is small, i.e., we can use the Boltzmann distribution function, the averaging essentially means that $\tau(E)$ is approximately replaced by $\tau(k_B T)$, i.e., the average energy of the electrons is taken as $\sim k_B T$.

Effect of Various Scattering Processes

In general, as electrons move in a semiconductor they will suffer from a number of distinct scattering mechanisms. These may involve ionized impurities, alloy disorder, lattice vibrations, etc. These scattering processes can be assumed to be independent and, to a reasonable approximation, the following approximate rule (called Mathieson's rule) can be applied

$$\frac{1}{\mu_{tot}} = \sum_i \frac{1}{\mu_i} \tag{8.86}$$

where μ_i is the mobility limited by the scattering process I.

In modern semiconductor devices, scattering is an integral part of device operation. However, as device dimensions shrink, it will be possible for electrons to move without scattering. In Appendix C we discuss devices which exploit transport of particles without scattering.

EXAMPLE 8.8 The mobility of electrons in pure GaAs at 300 K is 8500 $\text{cm}^2/\text{V.s}$. Calculate the relaxation time. If the GaAs sample is doped at $N_d = 10^{17}\ \text{cm}^{-3}$, the mobility decreases to 5000 $\text{cm}^2/\text{V.s}$. Calculate the relaxation time due to ionized impurity scattering.

The relaxation time is related to the mobility by

$$\begin{aligned}\tau_{sc}^{(1)} &= \frac{m^*\mu}{e} = \frac{(0.067 \times 0.91 \times 10^{-30}\ \text{kg})(8500 \times 10^{-4}\ \text{m}^2/\text{V.s})}{1.6 \times 10^{-19}\ \text{C}} \\ &= 3.24 \times 10^{-13}\ \text{s}\end{aligned}$$

If the ionized impurities are present, the time is

$$\tau_{sc}^{(2)} = \frac{m^*\mu}{e} = 1.9 \times 10^{-13}\ \text{s}$$

According to Mathieson's rule, the impurity related time τ_{imp} is given by

$$\frac{1}{\tau_{sc}^{(2)}} = \frac{1}{\tau_{sc}^{(1)}} + \frac{1}{\tau_{sc}^{(imp)}}$$

which gives

$$\tau_{sc}^{(imp)} = 4.6 \times 10^{-13}\ \text{s}$$

EXAMPLE 8.9 The mobility of electrons in pure silicon is 1500 $\text{cm}^2/\text{V.s}$. Calculate the scattering time using the conductivity effective mass. The conductivity mass is given by

$$\begin{aligned}m_\sigma^* &= 3\left(\frac{2}{m_t^*} + \frac{1}{m_\ell^*}\right)^{-1} \\ &= 3\left(\frac{2}{0.19m_o} + \frac{1}{0.98m_o}\right)^{-1} = 0.26m_o\end{aligned}$$

The scattering time is then

$$\begin{aligned}\tau_{sc} &= \frac{\mu m_\sigma^*}{e} = \frac{(0.26 \times 0.91 \times 10^{-30})(1500 \times 10^{-4})}{1.6 \times 10^{-19}} \\ &= 2.2 \times 10^{-13}\ \text{s}\end{aligned}$$

8.8 TRANSPORT: VELOCITY FIELD RELATION

In the previous section we have discussed the formal theory behind electron (hole) transport in materials. In this section we will apply this formalism to some important scattering processes. Before doing so it is important to note that the discussion of the previous section (i.e., Ohm's law, mobility, conductivity, etc.) is applicable to cases where the applied electric field is small (typically less than 1.0 kV/cm). In this case, the velocity is proportional to the electric field or mobility is independent of the field.

In many problems of interest the fields are indeed small and the simple linear relation between velocity and field is accurate. However, in many semiconductor devices, particularly modern devices being used in digital and microwave applications, electrons see very high electric fields. Typical electric fields in MOSFETs used in modern microprocessors are $\sim$ 100–200 kV/cm. Under such conditions the response of electron velocity to the field is not linear. The electrons gain a lot of energy from the applied electric field and the scattering rate increases to very high values. Thus τ becomes strongly dependent on the electric field, and the mobility, instead of being independent of field, starts to decrease with field. We will briefly examine this regime of transport in this section.

8.8.1 Ionized Impurity Limited Mobility

Based on the brief discussion of transport theory in the previous subsection, let us evaluate the relaxation time weighted with the $(1-\cos\theta)$ factor (this time is known as the *momentum relaxation time*), which is used to obtain the low field carrier mobility.

$$
\begin{aligned}
\frac{1}{\tau} &= \frac{V}{(2\pi)^3}\int (1-\cos\theta)\, W(\boldsymbol{k},\boldsymbol{k}')\, d^3k' \\
&= \frac{2\pi}{\hbar}\left(\frac{Ze^2}{\epsilon}\right)^2 \frac{1}{V^2}\frac{V}{(2\pi)^3} \\
&\quad \times \int (1-\cos\theta)\, \frac{\delta\left(E_{\mathbf{k}}-E_{\mathbf{k}'}\right)}{\left(4k^2\sin^2(\theta/2)+\lambda^2\right)^2} k'^2\, dk' \sin\theta\, d\theta\, d\phi \\
&= \frac{1}{2\hbar}\left(\frac{Ze^2}{\epsilon}\right)^2 \frac{1}{V} \\
&\quad \times \int (1-\cos\theta)\left(4k^2\sin^2(\theta/2)+\lambda^2\right)^2 N(E_{\mathbf{k}})\, dE_{\mathbf{k}}\, d(\cos\theta)\, d\phi \\
&= \frac{1}{\hbar}\left(\frac{Ze^2}{\epsilon}\right)^2 \frac{1}{V}\frac{N(E_{\mathbf{k}})}{32k^4} \\
&\quad \times \int (1-\cos\theta)\, \frac{1}{\left[\sin^2(\theta/2)+\left(\frac{\lambda}{2k}\right)^2\right]^2}\, d(\cos\theta)\, d\phi
\end{aligned}
$$

$$= F \int (1 - \cos\theta) \frac{1}{\left[\sin^2(\theta/2) + \left(\frac{\lambda}{2k}\right)^2\right]^2} d(\cos\theta)\, d\phi \tag{8.87}$$

with

$$F = \frac{1}{\hbar}\left(\frac{Ze^2}{\epsilon}\right)^2 \frac{1}{V}\frac{N(E_{\mathbf{k}})}{32k^4} \tag{8.88}$$

Let $z = \cos\theta$, so $\sin^2(\theta/2) = (1 - z)/2$.

$$\frac{1}{\tau} = 8\pi F \int_{-1}^{1} \frac{(1-z)\, dz}{\left(1 + 2\left(\frac{\lambda}{2k}\right)^2 - z\right)^2}$$

$$= 8\pi F \int_{-1}^{1} \frac{dz}{\left(1 + 2\left(\frac{\lambda}{2k}\right)^2 - z\right)^2} - \int_{-1}^{1} \frac{z\, dz}{\left(1 + 2\left(\frac{\lambda}{2k}\right)^2 - z\right)^2} \tag{8.89}$$

Let $a = 2(\lambda/2k)^2$ and $w = 1 - z + a$.

$$\frac{1}{\tau} = 8\pi F \int_a^{2+a} \frac{(w-a)}{w^2} dw$$

$$= 8\pi F \int_a^{2+a} \left(\frac{1}{w} - \frac{a}{w^2}\right) dw$$

$$= 8\pi F \left[\left(\ln|w| + \frac{a}{w}\right)\Big|_a^{2+a}\right]$$

$$= 8\pi F \left[\ln\left|\frac{2+a}{a}\right| - \frac{1}{1 + a/2}\right]$$

$$= 8\pi F \left[\ln\left(1 + \left(\frac{2k}{\lambda}\right)^2\right) - \frac{1}{1 + \left(\frac{\lambda}{2k}\right)^2}\right] \tag{8.90}$$

Finally

$$\frac{1}{\tau} = \frac{\pi}{4\hbar}\left(\frac{Ze^2}{\epsilon}\right)^2 \frac{N(E_{\mathbf{k}})}{Vk^4}\left[\ln\left(1 + \left(\frac{2k}{\lambda}\right)^2\right) - \frac{1}{1 + (\lambda/2k)^2}\right]$$

$$N(E_{\mathbf{k}}) = \frac{m^{*3/2}E^{1/2}}{\sqrt{2}\pi^2\hbar^3} \tag{8.91}$$

Note that the spin degeneracy is ignored, since the ionized impurity scattering cannot alter the spin of the electron. In terms of the electron energy, $E_{\mathbf{k}}$, we have

$$\frac{1}{\tau} = \frac{1}{V16\sqrt{2}\pi}\left(\frac{Ze^2}{\epsilon}\right)^2 \frac{1}{m^{*1/2}E_{\mathbf{k}}^{3/2}}$$
$$\times \left[\ln\left(1 + \left(\frac{8m^* E_{\mathbf{k}}}{\hbar^2\lambda^2}\right)\right) - \frac{1}{1 + (\hbar^2\lambda^2/8m^* E_{\mathbf{k}})}\right] \tag{8.92}$$

To calculate the mobility limited by ionized impurity scattering, we have to find the ensemble averaged τ. To a good approximation, the effect of this averaging is essentially to replace $E_{\mathbf{k}}$ by k_BT in the expression for $1/\tau$. A careful evaluation of the average $\langle\langle\tau\rangle\rangle$ gives (see Eqn. 8.85)

$$\frac{1}{\langle\langle\tau\rangle\rangle} = \frac{1}{V128\sqrt{2\pi}}\left(\frac{Ze^2}{\epsilon}\right)^2 \frac{1}{m^{*1/2}\left(k_BT\right)^{3/2}} \times \left[\ln\left(1+\left(\frac{24m^*k_BT}{\hbar^2\lambda^2}\right)\right) - \frac{1}{1+\left(\frac{\hbar^2\lambda^2}{24m^*k_BT}\right)}\right] \quad (8.93)$$

If there are N_i impurities per unit volume, and if we assume that they scatter electrons independently, the total relaxation time is simply obtained by multiplying the above results by N_iV,

$$\frac{1}{\langle\langle\tau\rangle\rangle} = \frac{N_i}{128\sqrt{2\pi}}\left(\frac{Ze^2}{\epsilon}\right)^2 \frac{1}{m^{*1/2}\left(k_BT\right)^{3/2}} \times \left[\ln\left(1+\left(\frac{24m^*k_BT}{\hbar^2\lambda^2}\right)\right) - \frac{1}{1+\left(\frac{\hbar^2\lambda^2}{24m^*k_BT}\right)}\right] \quad (8.94)$$

The mobility is then

$$\mu = \frac{e\langle\langle\tau\rangle\rangle}{m^*}$$

Mobility limited by ionized impurity scattering has the special $\mu \sim T^{3/2}$ behavior that is represented in Eqn. 8.94. This temperature dependence (the actual temperature dependence is more complex due to the other T-dependent terms present) is a special signature of the ionized impurity scattering. One can understand this behavior physically by realizing that at higher temperatures the electrons are traveling faster and are less affected by the ionized impurities.

Ionized impurity scattering plays a very central role in controlling the mobility of carriers in semiconductor devices. This is especially true at low temperatures where the other scattering processes (due to lattice vibrations) are weak. To avoid impurity scattering, the concept of *modulation doping* has been developed. In this approach, the device is made from two semiconductors—a large bandgap *barrier* layer and a smaller bandgap *well* layer. The barrier layer is doped so that the free carriers spill over into the well region where they are physically separated from the dopants. This essentially eliminates ionized impurity scattering. Figure 8.17 compares the mobilities of conventionally doped and modulation doped GaAs channels. As can be seen, there is a marked improvement in the mobility, especially at low temperatures. Modulation doping forms the basis of the highest performance semiconductor devices in terms of speed and noise.

In Fig. 8.18 we show how the mobility in Ge, Si and GaAs varies as a function of doping density. The mobility shown includes the effects of lattice scattering, as well as ionized impurity scattering.

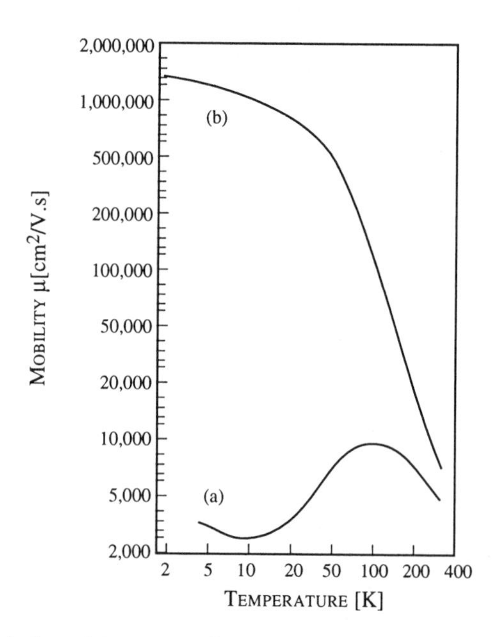

Figure 8.17: A typical plot of electron mobility as a function of temperature in a uniformly doped GaAs with $N_D = 10^{17}$ cm^{-3}. The mobility drops at low temperature due to ionized impurity scattering, becoming very strong. In contrast, the curve *b* shows a typical plot of mobility in a modulation-doped structure where ionized impurity is essentially eliminated.

8.8.2 Alloy Scattering Limited Mobility

The ensemble averaged relaxation time for the alloy scattering is quite simple (see Eqns. 8.60 and 8.85):

$$\frac{1}{\langle\langle\tau\rangle\rangle} = \frac{3\pi^3}{8\hbar} V_0 U_{all}^2 x(1-x) \frac{m^{*3/2}(k_B T)^{1/2}}{\sqrt{2}\pi^2\hbar^3} \frac{1}{0.75} \tag{8.95}$$

according to which the mobility due to alloy scattering is

$$\mu_{all} \propto T^{-1/2} \tag{8.96}$$

Thus in 3D systems the mobility decreases with temperature. This temperature dependence should be contrasted to the situation for the ionized impurity scattering.

8.8.3 Lattice Vibration Limited Mobility

In Chapter 2 we discussed the structural nature of crystalline materials. In that discussion we assumed that the atoms are fixed in space. In reality, the atoms vibrate, the amplitude of vibration increasing with temperature. In quantum mechanics these vibrations are represented by particles called *phonons*.

In Section 5.3.1 we discussed the vibrations of atoms in a crystalline structure. These lattice vibrations satisfy an equation of motion similar to that of masses

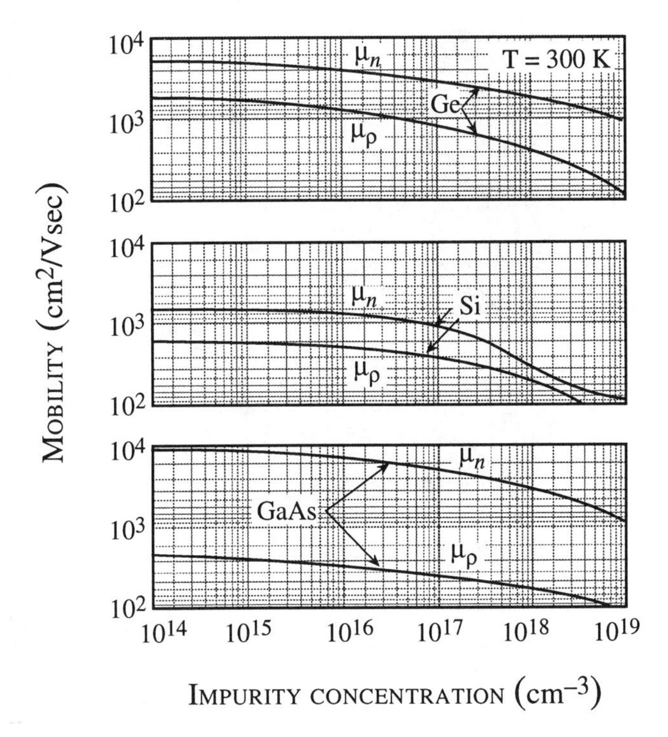

Figure 8.18: Drift mobility of Ge, Si, and GaAs as 300 K versus impurity concentration. (Reproduced with permission, S. M. Sze, *Physics of Semiconductor Devices,* 2nd ed., John Wiley and Sons, New York (1981).)

coupled to each other by springs. The properties of the lattice vibrations are represented by the relation between the vibration, amplitude, u, frequency, ω, and the wavevector, q. The vibration of a particular atom is given by

$$u_i(q) = u_{oi} \exp i(q \cdot r - \omega t) \tag{8.97}$$

which has the usual plane wave form that all solutions of periodic structures have. Recall that in a semiconductor there are two kinds of atoms in a basis. This results in an ω vs. k relation shown in Fig. 8.19. This relation, which is for GaAs, is typical of materials with two distinct atoms on the basis. One notices two kinds of lattice vibrations, denoted *acoustical* and *optical.* Additionally, there are two transverse and one longitudinal modes of vibration for each kind of vibration. The acoustical branch can be characterized by vibrations where the two atoms in the basis of a unit cell (i.e., Ga and As for GaAs) vibrate with the same sign of the amplitude as shown in Fig. 8.19b. In optical vibrations, the two atoms with opposing amplitudes are shown.

While the dispersion relations represent the allowed lattice vibration modes, an important question is: How many such modes are actually being excited at a given temperature? As noted in Section 5.3.1, in quantum mechanics, the modes are called phonons and the number of phonons with frequency ω are given by (phonons are

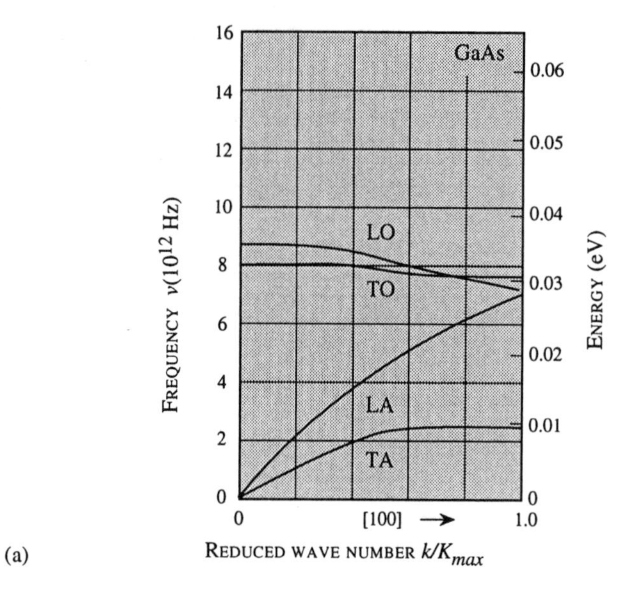

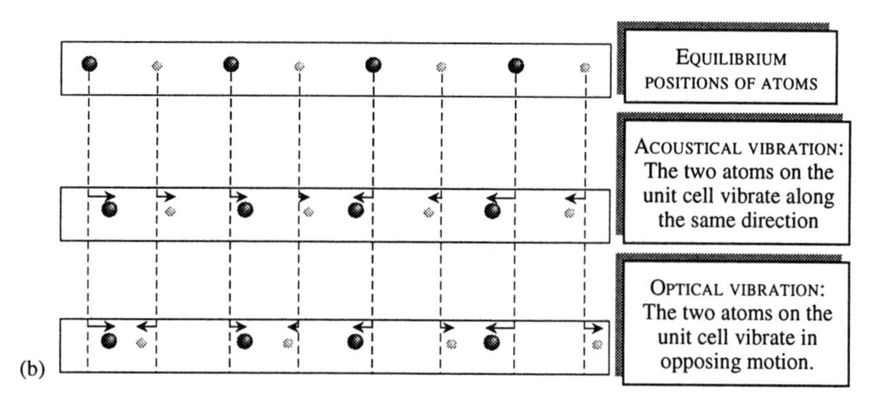

Figure 8.19: (a) Typical dispersion relations of a semiconductor (GaAs in this case). (b) The displacement of atoms in the optical and acoustic branches of the vibrations are shown. The motion of the atoms is shown for small k vibrations.

bosons and obey Bose-Einstein statistics)

$$n_\omega = \frac{1}{\exp\left(\frac{\hbar\omega}{k_BT}\right) - 1} \tag{8.98}$$

The lattice vibration problem is mathematically similar to the harmonic oscillator problem. The quantum mechanics of the harmonic oscillator problem tells us that the energy in the mode frequency ω is

$$E_\omega = (n_\omega + \frac{1}{2})\hbar\omega \tag{8.99}$$

Note that even if there are no phonons in a particular mode, there is a finite "zero

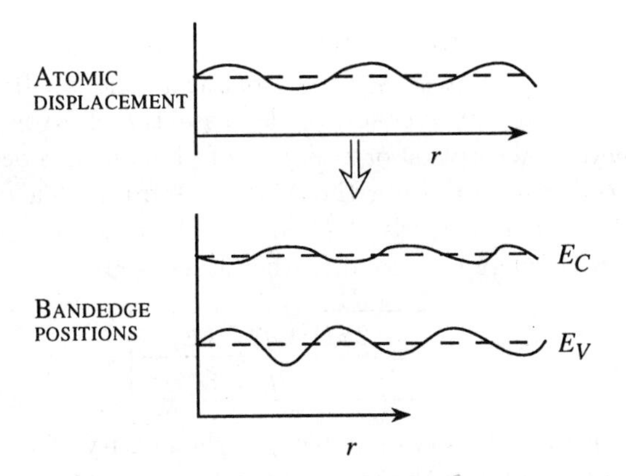

Figure 8.20: A schematic showing the effect of atomic displacement on bandedge energy levels in real space. The lattice vibrations cause spatial and time-dependent variations in the bandedges from which the electrons scatter.

point" energy $\frac{1}{2}\hbar\omega$ in the mode.

The vibrations of the atoms produce potential disturbances which result in the scattering of electrons. The phonon scattering is a key source of scattering and limits the performance of both electronic and optoelectronic devices. A schematic of the potential disturbance created by the vibrating atoms is shown in Fig. 8.20. In a simple physical picture, one can imagine the lattice vibrations causing spatial and temporal fluctuations in the conduction and valence band energies. The electrons (holes) then scatter from these disturbances. The acoustic phonons produce a strain field in the crystal, and the electrons see a disturbance which produces a potential of the form (this is a time-dependent perturbation)

$$V_{AP} = D\frac{\partial u}{\partial x} \tag{8.100}$$

where D is called a deformation potential (units are eV) and $\frac{\partial u}{\partial x}$ is the amplitude gradient of the atomic vibrations.

The optical phonons produce a potential disturbance which is proportional to the atomic vibration amplitude, since in the optical vibrations the two vibrations of the atoms in the basis oppose each other:

$$V_{op} = D_o u \tag{8.101}$$

where D_o (units are eV/cm) is the optical deformation potential.

In compound semiconductors where the two atoms in the basis of the crystal structure are different, an extremely important scattering potential arises additionally from optical phonons. Since the two atoms are different, there is an effective positive and negative charge of magnitude e^* on each atom. When optical vibrations take place, the effective dipole in the unit cell vibrates, causing polarization fields from which the electron scatters. This scattering, called polar optical phonon scattering, has a scattering potential of the form

$$V_{po} \sim e^* u \tag{8.102}$$

In acoustic phonon scattering, the electron energy does not change much after scattering (since $\omega \to 0$ at $q \to 0$ for acoustic phonons). However, for optical phonons, the scattering can increase or decrease the electron energy by $\hbar\omega_0$, depending upon whether absorption or emission of phonons has occurred.

Using first-order perturbation theory, the Fermi golden rule gives us the scattering rate for the different kinds of phonons. The acoustic phonon scattering rate for an electron with energy E_k to any other state is given by

$$\boxed{W_{ac}(E_k) = \frac{2\pi D^2 k_B T N(E_k)}{\hbar \rho v_s^2}} \tag{8.103}$$

where $N(E_k)$ is the electron density of states, ρ is the density of the semiconductor, v_s is the sound velocity, and T is the temperature.

For optical phonon scattering, the phonon energies are not negligible. As a result, the processes of phonon absorption and emission give distinct scattering rates. Using the Fermi golden rule, one finds that the rates are

$$W_{abs}(E_k) = \frac{\pi D_0^2}{\rho\omega_0}\left[n(\omega_0)N(E_k + \hbar\omega_0)\right] \tag{8.104}$$

$$W_{em}(E_k) = \frac{\pi D_0^2}{\rho\omega_0}\left[(n(\omega_0)+1)\,N(E_k - \hbar\omega_0)\right] \tag{8.105}$$

where ω_0 is the optical phonon energy, $n(\omega_0)$ is the phonon occupation number, and $N(E)$ is the density of states.

In Fig. 8.21, we show the mobility of electrons in silicon for a pure ($N_d <$ 10^{12} cm^{-3}) and doped sample. As the temperature increases, the mobility in the pure sample decreases, since lattice vibration scattering increases. In the doped sample, at low temperatures, the mobility is dominated by ionized impurity scattering and, as a result, the mobility shows a hump.

In metals the electric field is quite small under most conditions and the conductivity of the material is $ne\mu$, where n is the conduction electron density. In Table 8.2 we show the resistivities of some metals. The mobility of most metals is in the range of ~100 cm^2/V.s. The mobilities in semiconductors are much higher than those in metals (though the conductivity is much smaller). Table 8.3 lists mobilities in some semiconductors.

8.8.4 High Field Transport

In most electronic devices a significant portion of the electronic transport occurs under strong electric fields. This is especially true of field effect transistors. At such high fields (F ~1 - 100 kV/cm) the electrons get "hot" and their temperature (electron temperature defined through their average energy) can be much higher than the lattice temperature. The extra energy comes due to the strong electric fields. The drift velocities are also quite high. The description of electrons at such high electric fields is quite complex and requires either numerical techniques or computer simulations. In particular, the details of the bandstructure become very

Element	ρ(μΩ-em)
Na	4.2
Cu	1.56
Ag	1.51
Au	2.04
Fe	8.9
Zn	5.5
Cd	6.8
Al	2.45
In	8.0
Sn	10.6

Table 8.2: Resistivities of some materials at 273 K in micro ohm centimeters.

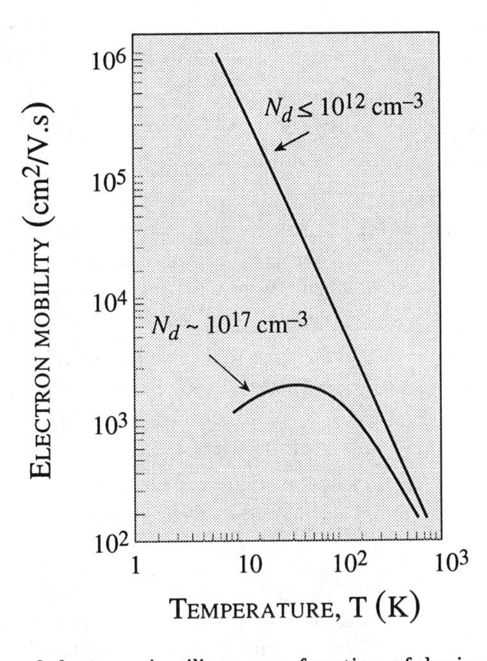

Figure 8.21: Mobility of electrons in silicon as a function of doping and temperature.

	Bandgap (eV)	Mobility at 300 K (cm^2/V-s)	
Semiconductor	300 K	Electrons	Holes
C	5.47	800	1200
Ge	0.66	3900	1900
Si	1.12	1500	450
α-SiC	2.996	400	50
GaSb	0.72	5000	850
GaAs	1.42	8500	400
GaP	2.26	110	75
InAs	0.36	33000	460
InP	1.35	4600	150
CdTe	1.56	1050	100

Table 8.3: Bandgap, electron, and hole mobilities of some semiconductors. Notice that narrow bandgap materials have superior mobility, since the effective mass is smaller.

important. We will consider two generic cases: Si and GaAs, which are representative of transport in indirect and direct bandgap materials.

High Field Transport in Si

As discussed in Chapter 4, the bottom of conduction band in Si has six equivalent valleys. Up to electric fields of ~100 kV/cm, the electrons remain in these valleys. As the field is increased, the velocity of the electrons increases steadily. However, gradually, as the electrons get *hotter* i.e., have higher kinetic energy, the scattering rates start to increase as well. *The velocity eventually saturates as shown in Fig. 8.22 at a value of* $\sim 1.2 \times 10^7$ cm/s.

The monotonic increase and then saturation of velocity with electric field is generic of all indirect bandgap semiconductors (AlAs, Ge, etc.) where electrons remain in the same lower valleys with high effective mass.

High Field Transport in GaAs

Electron transport in GaAs has some very remarkable features which are exploited for microwave applications. These features are related to the negative differential resistance in the v–E relationships. In GaAs and other direct bandgap semiconductors, the conduction band minima are made up of a single valley with a very light

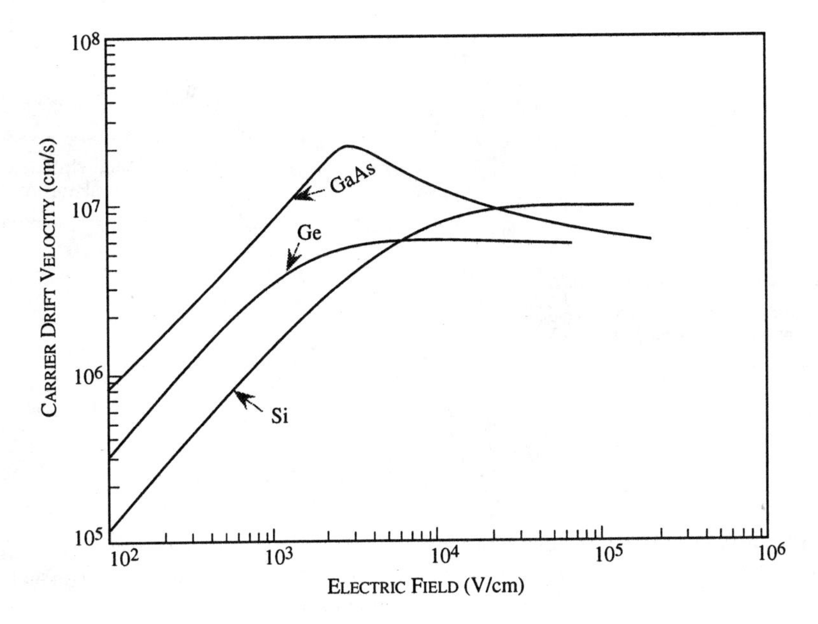

Figure 8.22: Velocity–field relation for electrons in Si, Ge, and GaAs.

mass $m* = 0.067m_o$ (the so-called Γ- valley). Slightly higher up in energy, separated by ~0.3 eV, is a heavy mass valley (the L-valley) with $m* = 0.22m_o$. At electric fields below ~3 kV/cm, the electrons are primarily in the low mass valley and have a high mobility. The peak velocity in pure GaAs can reach a value of ~ 2×10^7 cm/s at ~2.5 kV/cm, as shown in Fig. 8.22. As the field increases, the fraction of electrons in the upper valley starts to increase. Since the upper valley electrons are heavy, the *electron velocity decreases even though the field is increasing.* Eventually, when most of the electrons are in the upper valley, the velocity saturates at ~ 10^7 cms^{-1}. The results described for GaAs are typical of electron transport in most direct bandgap semiconductors.

8.8.5 Very High Field Transport: Avalanche Breakdown

When the electric field across an insulator or semiconductor becomes extremely high, the material suffers a "breakdown" in which the current has a "runaway" behavior. The breakdown occurs due to carrier multiplication, which means that the number of electrons in the conduction band and holes in the valence band that can participate in current flow, increases. Of course, the total number of elections is always conserved.

In the avalanche process (also known as the impact ionization process) shown schematically in Fig. 8.23, an electron which is "very hot" scatters with an electron in the valence band via the Coulombic interaction and knocks it out into the conduction band. The initial electron must provide enough energy to bring the valence band electron up into the conduction band. Thus the *initial electron should have energy slightly larger than the bandgap* (measured from the conduction band

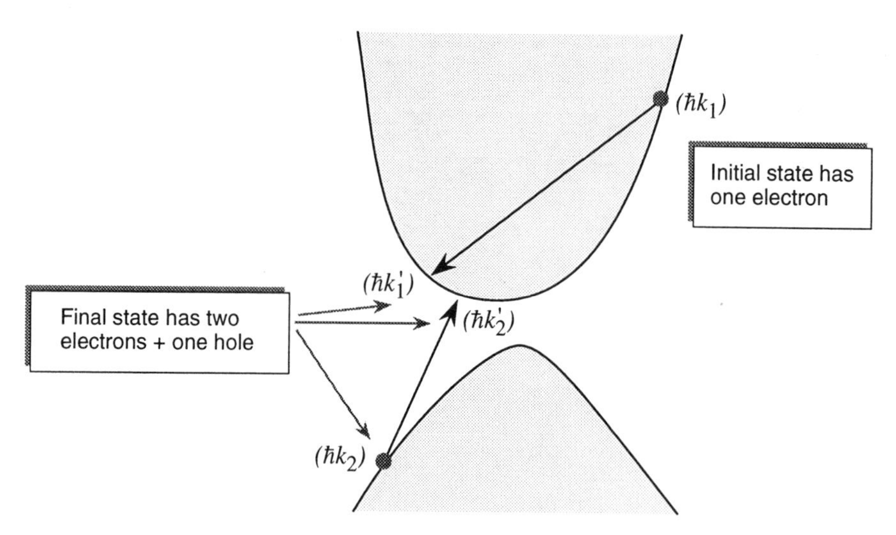

Figure 8.23: The impact ionization process, where a high-energy electron scatters from a valence band electron, producing two conduction band electrons and a hole. Hot holes can undergo a similar process. Both energy and momentum conservation are required in such a process. Thus the initiating electron must have an energy above a minimum threshold.

minimum). In the final state we now have two electrons in the conduction band and one hole in the valence band. Thus the number of current-carrying charges have multiplied, and the process is often called avalanching. Note that the same could happen to "hot holes," which could then trigger the avalanche.

In order for the impact ionization process to begin, the initiating electron with momentum $\hbar k_1$ must have enough energy to knock an electron from the valence band to the conduction band. Additionally, momentum must be conserved in the scattering. Thus the initial electron must overcome a threshold energy before the carrier multiplication can begin.

For the purpose of device applications, the current in the device, once avalanching starts, is given by

$$\frac{dI(t)}{dt} = \alpha_t I \quad \text{or} \quad \frac{dI(z)}{dz} = \alpha_z I$$

Where I is the current and α_t or α_z represent the average rate of ionization per unit time or distance, respectively.

The coefficients α_z (or simply α_{imp}) for electrons and β_z (or β_{imp}) for holes strongly depend upon the bandgap of the material. This is because, as discussed above, the process can only start if the initial electron has a kinetic energy equal to a certain threshold (roughly equal to the bandgap). This is achieved for lower electric fields in narrow gap materials.

A critical breakdown field F_{crit} is defined where α_{imp} or β_{imp} approaches 10^4 cm^{-1}. When α_{imp} (β_{imp}) approaches 10^4 cm^{-1}, there is about one impact ionization when a carrier travels a distance of one micrometer. The avalanche process places an important limitation on the power output of devices. Once the process

BREAKDOWN ELECTRIC FIELDS IN SEMICONDUCTORS

Material	Bandgap (eV)	Breakdown Electric Field (V/cm)
GaAs	1.43	4×10^5
Ge	0.664	10^5
InP	1.34	5×10^5
Si	1.1	3×10^5
$In_{0.53}Ga_{0.47}As$	0.8	2×10^5
C	5.5	$\sim 10^7$
SiC	2.9	$\sim 3 \times 10^6$
SiO_2	9	$\sim 10^7$
Si_3N_4	5	$\sim 10^7$

Table 8.4: Breakdown electric fields in some semiconductors.

starts, the current rapidly increases, due to carrier multiplication and the control over the device is lost. The push for high-power devices is one of the reasons for research in large gap semiconductor devices. It must be noted that in certain devices, such as avalanche photodetectors, the process is exploited for high gain detection. The process is also exploited in special microwave devices. Table 8.4 lists the critical field for breakdown of some materials.

EXAMPLE 8.10 If the measured room temperature mobility of electrons in GaAs doped n-type at 5×10^{17} cm^{-3} is 3500 cm$^2V^{-1}$ s^{-1} calculate the relaxation time for phonon scattering.

To find the contribution of the phonon scattering, we use the Mathieson rule for adding mobilities. Note that this rule is not exact, since, in reality, various scattering processes influence each other. However, it can be used to get an approximate idea of the various contributions to scattering. We have

$$\frac{1}{\mu_{total}} = \frac{1}{\mu_{imp}} + \frac{1}{\mu_{phonon}}$$

or

$$\frac{1}{\tau_{total}} = \frac{1}{\tau_{imp}} + \frac{1}{\tau_{phonon}}$$

The ionized impurity scattering time is found to be 4.0×10^{-13} s. Using the values given for the total mobility we get for the total scattering time 1.33×10^{-13} s. This gives us

$$\begin{aligned}\frac{1}{\tau_{phonon}} &= \left(\frac{1}{1.33 \times 10^{-13}\ \text{s}} - \frac{1}{4.0 \times 10^{-13}\ \text{s}}\right)^{-1} \\ &= 5 \times 10^{12}\ \text{s}^{-1}\end{aligned}$$

This gives a phonon scattering time of 2×10^{-13} s.

EXAMPLE 8.11 Calculate the alloy scattering limited mobility in $Al_{0.3}Ga_{0.7}As$ at 77 K and 300 K. Assume that the alloy scattering potential is 1.0 eV. The relaxation time at 300 K is ($m^* = 0.07m_0$)

$$\begin{aligned}\frac{1}{\ll \tau \gg} &= \frac{3\pi V_0 (U_{all})^2 x(1-x) m^{*3/2} (k_B T)^{1/2}}{8\sqrt{2}\hbar^4 (0.75)} \\ &= 2.1 \times 10^{12}\ \text{s}\end{aligned}$$

Here we have used $x = 0.3$, $V_0 = \frac{a^3}{4}$ with $a = 5.65$ Å.

The value of $\ll \tau \gg$ is 4.77×10^{-13} s. The mobility is then

$$\mu_{all}(300\ \text{K}) = 1.2 \times 10^4\ \text{cm}^2/V.s$$

The mobility goes as $T^{-1/2}$ which gives

$$\mu_{all}(77K) = 2.36 \times 10^4\ \text{cm}^2/\text{V.s}$$

8.9 CHAPTER SUMMARY

Tables 8.5 and 8.6 summarize the topics covered in this chapter.

8.10 PROBLEMS

Problem 8.1 The performance of a silicon detector is poor for photons of wavelength 1.0 μm, but are quite good for photons of wavelength shorter than 0.4 μm. Explain this observation by examining the bandstructure of silicon.

Problem 8.2 In long-distance fiber optics communication, it is important that photons with energy with low absorption in the fiber be used. The lowest absorption for silica fibers occurs at 1.55 μm. Find a semiconductor alloy using InAs, GaAs, and InP that has a bandgap corresponding to this wavelength (semiconductor lasers emit at close to bandgap energy). Also find an alloy with a bandgap corresponding to 1.3 μm, the wavelength at which fiber dispersion is minimum.

Problem 8.3 Calculate and plot the absorption coefficients for $In_{0.53}Ga_{0.47}As$ and InP. Assume that the momentum matrix element values are the same as those for GaAs.

Problem 8.4 Calculate the rate (per second) at which photons with energy 1.6 eV are absorbed in GaAs ($E_g = 1.43$ eV).

TOPICS STUDIED	KEY OBSERVATIONS
Fermi golden rule	The Fermi golden rule gives us the scattering rate of electrons from one state to another in the presence of a time-dependent perturbation. If the time-dependence has a harmonic frequency ω, the scattering could be one in which the electron gains an energy $\hbar\omega$ (absorption) or loses an energy $\hbar\omega$ (emission). Scattering rate is proportional to the density of states at the final energy and the square of the matrix element for scattering.
Optical absorption in semiconductors	• A light particle (photon) can cause an electron in the valence band to go into the conduction band. The phonon is absorbed and the process is defined by an absorption coefficient $\alpha(\hbar\omega \geq E_g)$. • The absorption process is strong in direct-gap semiconductors due to momentum conservation. Thus the optical transitions are vertical in k-space.
Optical emission in semiconductors	An electron with a certain momentum can recombine with a hole with the same momentum to emit a photon. The emission rate is the radiative lifetime r_0 and has a value of ~0.6 ns for GaAs.

Table 8.5: Summary table.

Problem 8.5 A light beam of intensity 1 Wcm^{-2} impinges upon a GaAs sample. Calculate the *e-h* pair generation rate at $z = 0$ and $z = 10$ μm. The energy of photons in the beam is 1.6 eV. The bandgap of GaAs is 1.43 eV.
Problem 8.6 Calculate the electron densities at which the quasi-Fermi level for electrons just enters the conduction band in Si and GaAs at 77 K and 300 K.
Problem 8.7 Calculate the hole densities at which the hole quasi-Fermi level enters the valence band in Si and GaAs at 77 K and 300 K.
Problem 8.8 Assume that equal densities of electrons and holes are injected into GaAs. Calculate the electron or hole density at which $f^e(E^e = E_c) + f^h(E^h = E_v) = 1$. Calculate the densities at 77 K and 300 K.
Problem 8.9 In a p-type GaAs doped at $N_a = 10^{18}$ cm^{-3}, electrons are injected to produce a constant electron density of 10^{15} cm^{-3}. Calculate the rate of photon emission, assuming that all electron–hole recombination results in photon emission. What is the optical power output if the device volume is 10^{-7} cm^3?

TOPICS STUDIED	KEY OBSERVATIONS
Quasi-Fermi levels	For non-equilibrium electron and hole concentrations, the occupations of the carriers are given by electron and hole Fermi functions independent of each other. In equilibrium, the same Fermi function describes both electrons and holes.
Electron-hole recombination of excess carriers	• Electrons and holes can recombine by emission of photons. The process depends upon the electron finding a hole and has a strong dependence on charge density.
Scattering of electrons from fixed potentials	In the presence of potentials which act as perturbations, electrons can scatter from one state to another. The momentum direction of the electron can change as a result of this scattering.
Scattering time and mobility	• Due to imperfections, electrons scatter as they move through the crystal. • The mobility of electrons (holes) is proportional to the scattering time.
Low field transport	Carrier mobility is constant and velocity increases linearly with applied electric field (for fields $\leq$ 1 kV/cm).
High field transport	• Carrier velocity tends to saturate and mobility = v/F starts to decrease. • GaAs and other direct-gap semiconductors show regions where velocity decreases as field increases, i.e., they have negative resistance regions.
Very high field transport	• Electrons and holes have so much energy that they can cause impact ionization or carrier multiplication. At a critical field F_{crit} the coefficients approach 10^4 cm^{-1}.

Table 8.6: Summary table.

Problem 8.10 Electrons and holes are injected into a GaAs device where they recombine to produce photons. The volume of the active region of the device where recombination occurs is 10^{-6} cm^2 and the temperature is 300 K. Calculate the output power for the cases: i) $n = p = 10^{15}$ cm^{-3}; and ii) $n = p = 10^{18}$ cm^{-3}.
Problem 8.11 The photon number in a GaAs laser diode is found to be 100. Calculate the electron–hole recombination time for carriers corresponding to the lasing energy.
Problem 8.12 Calculate the gain versus energy curves for GaAs at injection densities of:

$$(1)\ n = p = 0$$
$$(2)\ n = p = 5 \times 10^{16} \text{ cm}^{-3}$$
$$(3)\ n = p = 1.0 \times 10^{17} \text{ cm}^{-3}$$
$$(4)\ n = p = 1.0 \times 10^{18} \text{ cm}^{-3}$$
$$(5)\ n = p = 2.0 \times 10^{18} \text{ cm}^{-3}$$

Plot the results for $\hbar\omega$ going from E_g to $E_g + 0.3$ eV. Calculate the results for 77 K and 300 K.
Problem 8.13 Assume that a GaAs laser lases when the injection density $n = p$ reaches 1.2 times the density at which the hole quasi-Fermi level enters the valence band. Calculate the threshold carrier density and threshold current density for the laser at 77 K and at 300 K.

The width of the active region of the laser is

$$d_{las} = 500 \text{ Å}$$

The threshold current density is given by

$$J_{th} = e\frac{n_{th} d_{las}}{\tau_r}$$

where n_{th} is the threshold density and τ_r is the e-h recombination rate.
Problem 8.14 Calculate the alloy scattering rate as a function of energy for electrons in $Al_{0.3}Ga_{0.7}As$. Assume that the alloy scattering potential is 1.0 eV. Plot the results from $E = 0$ to $E = 0.3$ eV.
Problem 8.15 Plot the scattering rate (in arbitrary units) versus scattering angle for electrons with energy equal to $3/2k_BT$ in Si doped at i) 10^{17} cm^{-3} and ii) 10^{18} cm$-$3. Calculate the results at 77 K and 300 K.
Problem 8.16 Calculate the alloy scattering limited mobility in $Al_{0.3}Ga_{0.7}As$ at 77 K and 300 K. Assume that the alloy scattering potential is 1.0 eV.
Problem 8.17 If the measured room temperature mobility of electrons in GaAs doped n-type at 5×10^{17} cm^{-3} is 3500 cm^2 V^{-1}s^{-1}, calculate the relaxation time for phonon scattering.
Problem 8.18 Consider a silicon sample in which there are 10^{13} defects. The defect potential is similar to the alloy potential fluctuation discussed in the text, i.e.,

$$V(r) = -1.0 \text{ eV} \quad \text{if} \quad r \leq 10 \text{ Å}$$
$$= 0 \text{ otherwise}$$

Calculate the electron mobility (use $m^* = 0.26\ m_0$) limited by these defects at room temperature. How important are these defects in controlling mobility in silicon at room temperature?

Problem 8.19 Show that the average energy gained between collisions is

$$\delta E_{av} = \frac{1}{2} m^* (\mu E)^2$$

where E is the applied electric field. If the "optical phonon" energies in GaAs and Si are 36 meV and 47 meV, and the mobilities are 8000 cm^2/V.s and 1400 cm^2/V.s, respectively, what are the electric fields at which optical phonon emission can start?

Problem 8.20 The electron mobility of Si at 300 K is 1400 cm^2/V.s. Calculate the mean free path and the energy gained in a mean free path at an electric field of 1 kV/cm. Assume that the mean free path = $v_{th} \cdot \tau_{sc}$, where v_{th} is the thermal velocity of the electron.

Problem 8.21 The mobility of electrons in the material InAs is $\sim$ 35,000 cm^2/V.s at 300 K compared to a mobility of 1400 cm^2/V.s for silicon. Calculate the scattering times in the two semiconductors.

Problem 8.22 Using the resistivities of metals given in this chapter, calculate the 273 K scattering time and mobilities of electron in: i) Cu; ii) Ag; iii) Au; iv) Fe; and v) Al. Effective mass of electrons is m_0.

Problem 8.23 Calculate the change in conductivity of Al if 10^{21} electrons are added to the conduction band by "n-type doping."

Problem 8.24 Plot the room temperature conductivities of Si and GaAs as the Fermi level moves from the valence bandedge to the conduction bandedge. Such a change in the position of the Fermi level can be made by a properly applied potential. You can use the mobility data provided in this chapter.

Problem 8.25 Use the velocity–field relations for Si and GaAs to calculate the transit time of electrons in a 1.0 μm region for a field of 1 kV/cm and 50 kV/cm.

Problem 8.26 The velocity of electrons in silicon remains $\sim 1 \times 10^7$ cm s^{-1} between 50 kV/cm and 200 kV/cm. Estimate the scattering times at these two electric fields.

Problem 8.27 The power output of a device depends upon a maximum voltage that the device can tolerate before impact ionization generated carriers become significant (say, 10% excess carriers). Consider a device of length L over which a potential V drops uniformly. What is the maximum voltage that can be tolerated by an Si and a diamond device for $L = 2\ \mu$m and $L = 0.5\ \mu$m?

CHAPTER 9

SPECIAL THEORY OF RELATIVITY

CHAPTER AT A GLANCE

9.1 INTRODUCTION

So far in this text we have examined what is known as non-relativistic modern physics. In this physics we have seen what happens when the wave nature of particles becomes important, but we have assumed that the velocity of the particles is small compared with the velocity of light. Of course, we have discussed that light travels at a speed of 2.998×10^8 m/s, but we have not addressed several important questions. Questions such as: i) Can any object travel at a speed greater than the speed of light? ii) Does the speed of light change if we measure it in a reference frame which is itself moving? According to Newtonian physics, the answers to both these questions should be "yes." However, nature once again defies our intuition, and experiments suggest that there is something special about light.

We saw in Chapter 2 how certain experiments seem to fly against our "natural expectation" based on classical physics. In these experiments, particles seem to behave like waves and waves seem to behave as particles. This dilemma in our understanding was resolved through the revolutionary theory of quantum mechanics. Interestingly, around the same time when the experiments which led to quantum mechanics were being carried out, other experiments were being carried out that challenged classical physics on another front—they address the question: What happens when we deal with speeds approaching the speed of light? The outcome of these experiments was very counter-intuitive and eventually led to the special theory of relativity proposed by Albert Einstein.

The special theory of relativity was as revolutionary as quantum mechanics. And just as quantum mechanics reduces to Newtonian physics under "normal" conditions (particle mass is large, potential energy variation is slow, etc.), relativistic physics reduces to classical physics when speed of particles is slow. In this chapter we will take a look at the remarkable development known as the special theory of relativity.

9.2 CLASSICAL MECHANICS IN MOVING REFERENCE FRAMES

Classical mechanics is governed by Newton's equation of motion

$$m\frac{d\boldsymbol{r}}{dt} = \boldsymbol{F} \tag{9.1}$$

This equation tells us how a particle's trajectory will be in space and time. What happens if the particle is observed by observers who are themselves moving with respect to each other? Let us define this question a little more rigorously. To do so we will introduce some terminology.

Inertial Frames

Let us consider two observers O and O′ at rest in two frames of reference, as shown in Fig. 9.1. Let us say that each observer finds that a particle at rest or at a constant velocity stays in that state forever, unless a force is applied to it. Such frames of

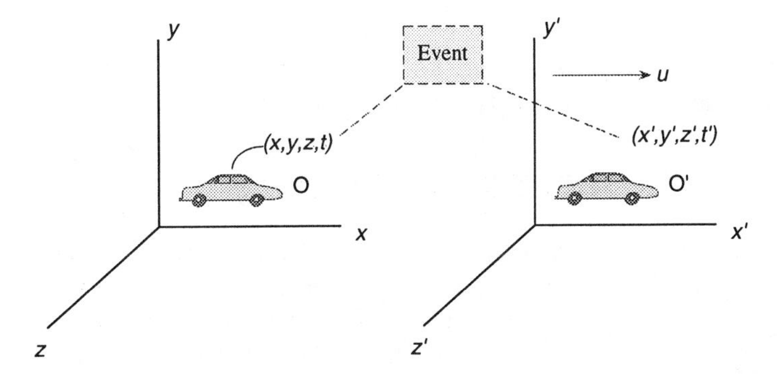

Figure 9.1: Two observers, O and O′ observe the coordinates of the same event. O′ is moving relative to O, as shown.

reference are called *inertial frames.* Inertial frames differ from each other by the fact that they have a constant velocity with respect to each other.

According to our definition, a child observing the slide on a playground while sitting on a twirling merry-go-round is not on an inertial frame. In principle, an observer at rest on Earth is not on an inertial frame, since Earth is spinning on its axis. However, for many purposes, the acceleration seen by an object on Earth is so small that one may regard Earth as an inertial frame.

Galilean Transformation

Let us say that two observers in two inertial frames observe an event. The coordinates of the event are $\{x, y, z, t\}$ and $\{x', y', z', t'\}$, as shown in Fig. 9.1. According to classical physics, the time of observation is the same for all observers, i.e., $t = t'$. For simplicity, let us assume that the origins of the two coordinate systems coincide at $t = 0$, and the frame O′ is moving at a velocity $\boldsymbol{u}$ with respect to O in the x-direction. We now have

$$\begin{aligned} x' &= x - ut \\ y' &= y \\ z' &= z \\ x' &= x - ut \end{aligned} \tag{9.2}$$

We can now find the relation between the velocities observed by the two observers by taking the derivative of the above equations

$$\begin{aligned} v_x' &= v_x - u \\ v_y' &= v_y \\ v_z' &= v_z \end{aligned} \tag{9.3}$$

The relation between the accelerations observed are, similarly (u is a constant),

$$\begin{aligned} a_x' &= a_x \\ a_y' &= a_y \\ a_z' &= a_z \end{aligned} \tag{9.4}$$

Thus the accelerations observed by the two observers are the same and Newton's law $F = ma$ is valid in both frames. Also the "obvious" fact the mass of a particle is the same in all frames is an integral part of classical mechanics.

9.3 LIGHT IN DIFFERENT FRAMES

The impetus for the special theory of relativity arose from Maxwell's theory of electromagnetism. This theory accounted for all observed phenomena involving electricity and magnetism and predicted that electromagnetic disturbances propagate at waves with a definite speed, c, in a vacuum. The presence of a well-defined speed, c, posed a question for classical physics. It suggested that there is a preferred system in which the velocity of light is c. In other systems moving with a speed u with respect to this preferred system the velocity would be $c+u$. What was this preferred system? It was assumed that this system was one in which ether, the medium of propagation of light, was at rest.

In the previous section, we have seen that if an inertial frame moves with a constant velocity with respect to another frame, observers on the two frames will measure different velocities of a particle. It appears correct to assume that this holds for the velocity of light as well. The velocity of light is measured on Earth as having a value of 2.998×10^8 m/s. Consider an observer moving at a velocity u with respect to Earth. Will he/she measure a velocity of light $c+u$? A hundred years ago it was not possible to find such an inertial frame, so this velocity transformation could not be checked for light. However, it was possible to devise a clever experiment which tests the Galilean transformation for velocities as applied to electromagnetic radiation. This experiment was carried out by Albert A. Michelson and his associate, E. W. Morley.

9.3.1 Michelson-Morley Experiment

Maxwell equations describe electromagnetic radiation as waves which travel at a speed $c(= 2.99 \times 10^8$ m/s). What is the medium through which these waves travel? We know now that electromagnetic waves do not require a medium through which to travel, but in the nineteenth century, it was believed that for a wave to propagate, one needed a medium. Scientists had predicted that light propagated through an invisible, massless medium called *ether*. It was postulated that it was not possible to detect the presence of ether by any mechanical means. However, it appeared reasonable to detect the presence of ether by examining the speed of light on Earth as Earth moved through ether in its motion around the sun. Earth moves around the sun at a speed of 18 miles per second, a speed which can be exploited to detect motion through ether by using a clever experiment.

In Fig. 9.2, we show the arrangement used by Michelson and Morley in their famous experiment. The apparatus consists of a light source S, which generates a monochromatic optical beam. The beam can be split by a beam splitter B into two beams traveling perpendicular to each other. The beams are reflected back from mirrors C and E to the splitter B. They pass through the splitter and recombine as two waves D and F.

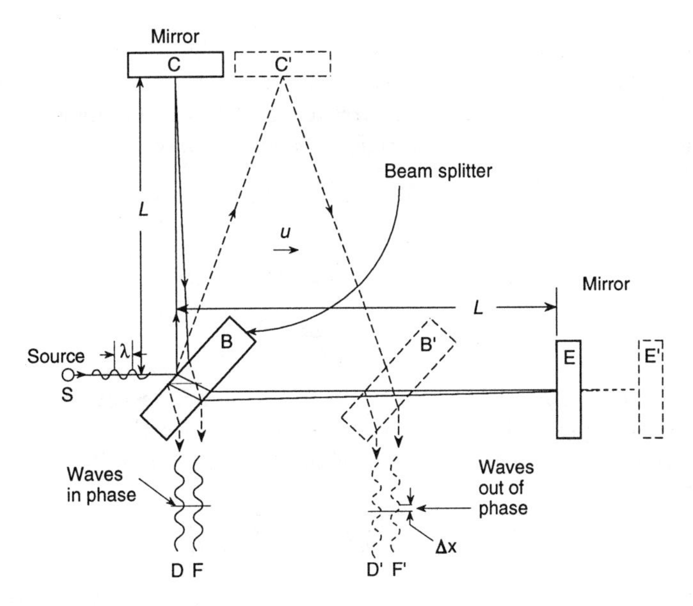

Figure 9.2: A schematic of the Michelson-Morley apparatus. The apparatus is an interferometer capable of detecting small variations in optical paths of two beams.

The beams D and F can interfere with each other constructively or destructively, depending upon the time it takes them to travel their respective paths. If, for example, the length BC and BE are exactly the same and the apparatus is at rest with respect to ether, the beams will interfere constructively. However, if the apparatus is moving with a velocity u, there should be a phase difference between the two beams, resulting in a loss of constructive interference. We can see this as follows.

First, let us calculate the time required for the light to go from B to E and back. Let us say that the time for light to go from plate B to mirror E is t_1, and the time for the return is t_2. Now, while the light is on its way from B to the mirror, the apparatus moves a distance ut_1, so the light must traverse a distance $L + ut_1$, at the speed c. We can also express this distance as ct_1, so we have

$$ct_1 = L + ut_1, \text{ or } t_1 = L/(c - u) \tag{9.5}$$

This result is also obvious from the point of view that the velocity of light relative to the apparatus is $c - u$, so the time is the length L divided by $c - u$. In a like manner, the time t_2 can be calculated. During this time the plate B advances a distance ut_2, so the return distance of the light is $L - ut_2$. Then we have

$$ct_2 = L - ut_2, \text{ or } t_2 = L/(c + u)$$

The total time is

$$t_1 + t_2 = 2Lc/(c^2 - u^2) \tag{9.6}$$

We write this as

$$t_1 + t_2 = \frac{2L/c}{1 - u^2/c^2} \tag{9.7}$$

Our second calculation will be of the time t_3 for the light to go from B to the mirror C. As before, during time t_3, the mirror C moves to the right a distance ut_3 to the position C'; in the same time, the light travels a distance ct_3 along the hypotenuse of a triangle, which is BC'. For this right triangle, we have

$$(ct_3)^2 = L^2 + (ut_3)^2$$

or

$$L^2 = c^2t_3^2 - u^2t_3^2 = (c^2 - u^2)t_3^2$$

from which we get

$$t_3 = L/\sqrt{c^2 - u^2} \tag{9.8}$$

For the return trip from C' the distance is the same, as can be seen from the symmetry of the figure; therefore the return time is also the same, and the total time is $2t_3$. With a little rearrangement of the form we can write

$$2t_3 = \frac{2L}{\sqrt{c^2 - u^2}} = \frac{2L/c}{\sqrt{1 - u^2/c^2}} \tag{9.9}$$

We are now able to compare the times taken by the two beams of light. In expressions 9.7 and 9.9 the numerators are identical, and represent the time that would be taken if the apparatus were at rest. In the denominators, the term u^2/c^2 will be small, unless u is comparable in size to c. The denominators represent the modifications in the times caused by the motion of the apparatus. And behold, these modifications are *not the same*—the time to go to C and back is a little less than the time to E and back, even though the mirrors are equidistant from B, and all we have to do is to measure that difference with precision.

In a real apparatus, it is not possible to make the lengths BC and BE exactly equal. To avoid this problem the fringes are observed in one setting and then the apparatus rotated by 90°. This makes BC the line of motion and BE the line perpendicular to the motion. One may also argue that the ether itself may be moving along the direction of Earth. To address this doubt, the experiment has been done with a six month interval so that the direction of Earth's motion is reversed.

Michelson and Morley carried out their experiment with utmost care and precision. To their and other's surprise, there was no phase difference between the light traveling along the arm parallel to Earth's motion and along the arm perpendicular to it. This result was extremely puzzling. The null results seemed to suggest that the Galilean transformation of velocities (an important pillar of classical mechanics) was not applicable to light.

9.3.2 Lorentz's Proposal

In order to reconcile the null results of Michelson and Morley's experiment, Lorentz put forth an idea known as Lorentz transformations. He suggested that the length of an object changes when the object is moving. The length change only occurs in the direction of motion. If L_0 is the length at rest, the length when the object moves with a speed u parallel to the length is

$$L_{\|} = L_0\sqrt{1 - u^2/c^2} \tag{9.10}$$

When this transformation is applied to the lengths shown in Fig. 9,2, the length from B to C does not change and the time $2t_3$ does not change. However, the length from B to E is shortened and we have

$$t_1 + t_2 = \frac{2L}{c}\frac{\sqrt{1-u^2/c^2}}{1-u^2/c^2} = \frac{2L}{c}\frac{1}{\sqrt{1-u^2/c^2}} = 2t_3 \tag{9.11}$$

This length contraction would thus explain the null result observed by Michelson and Morley.

The contraction of length along the direction of motion would also imply that the time duration of an event would be different if the measurement was made at rest or by an observer in the moving frame. For example, if the measurement of the time it takes for light to go from the beam splitter B to the mirror E and back were to be measured in the moving frame, it would just be

$$t_0 = \frac{2L}{c} \tag{9.12}$$

However, for a person at rest it would be given by Eqn. 9.11

$$t_{\parallel} = \frac{2L}{c}\frac{1}{\sqrt{1-u^2/c^2}} = \frac{t_0}{\sqrt{1-u^2/c^2}}$$

This suggests that the time would pass at a slower rate in a moving system than in a frame where the time is measured at rest. The implications of time slowing down in moving systems are very profound and we will discuss them later.

Lorentz's suggestions, while able to explain the Michelson-Morley experiment, seemed overly artificial. They appeared to have been "concocted" just to explain an experiment. It was Einstein who put forward a consistent theory which allowed Michelson-Morley and other experiments to be interpreted coherently.

EXAMPLE 9.1 Consider the Michelson-Morley experiment carried out with an interferometer with arms 10 m long. Calculate the phase change one would expect if Earth were indeed moving through ether at 28.8 km/s and classical mechanics were true. Assume that light of wavelength 5000 Å is used for the experiment.

The time difference between the two flights of light is

$$\begin{aligned}\Delta t &= \frac{2L}{c}\frac{1}{1-u^2/c^2} - \frac{2L}{c}\frac{1}{\sqrt{1-u^2/c^2}} \\ &\cong \frac{2L}{c}\frac{u^2}{2c^2} \\ &= 3.07 \times 10^{-16} \text{ s}\end{aligned}$$

The phase difference is given by

$$\begin{aligned}\Delta\phi &= \frac{2\pi c}{\lambda}\Delta T \\ &= 1.15 \text{ rad}\end{aligned}$$

This kind of phase change would certainly be detected if it were present.

9.4 EINSTEIN'S THEORY

In 1905, Einstein based his special theory of relativity on two postulates. These are:

• *The laws of physics are the same in all inertial frames.* The laws include not only the laws of mechanics, but also those of electromagnetism.

• *The speed of light in free space has the same value in all inertial frames.* This is a postulate that seems to go against our intuitive concepts of additives of velocities in inertial systems.

Einstein's postulates provided the basis of a consistent picture, which led to the Lorentz transformations mentioned above and other equations for dynamics in inertial systems. We will now examine the outcome of these postulates.

The second postulate of Einstein immediately explains the null result of Michelson and Morley. The constancy of light in all inertial frames immediately explains why there is no phase shift between light traveling parallel and perpendicular to Earth's motion. Since the velocity along the two arms of the interferometer is the same, there is no phase difference between the two light beams.

9.4.1 Relativity of Time

In the previous section, we have discussed how Lorentz introduced his transformations for length and time to account for experimental evidence. Einstein's theory provides a coherent derivation of these transformations. We start with how the time interval between events changes, depending on the relative velocities of two inertial frames.

Let us consider two frames in which observers O and O$'$ make measurements. An experiment to measure time is done in the frame O$'$. The term *proper time* is used for the time interval measured in the frame where the experimental setup is at rest.

In Fig. 9.3a we show a schematic of how a timing device works. The device consists of a laser sending out a signal that reflects from a mirror and is detected by a detector. The total distance traveled is $2L_0$ and the time interval is

$$\Delta t_0 = \frac{2L_0}{c} \tag{9.13}$$

This is the time measured in the system in which the clock is at rest.

Now consider the clock with a timing device in a frame O which is moving at a speed u with respect to the frame O$'$. We will now measure the fine interval between the flash being sent and returning in the frame O. The clock is now moving, while the event is being timed in a frame which is at rest. Observer O now sees a flash emitted at A, reflected at B, and detected at C, as shown in Fig. 9.3b. The clock has moved a distance $u\Delta t$, where Δt is the time interval observed by O. The total distance traveled by the light pulse is $2L$, where

$$L = \sqrt{L_0^2 + (u\Delta t/2)^2}$$

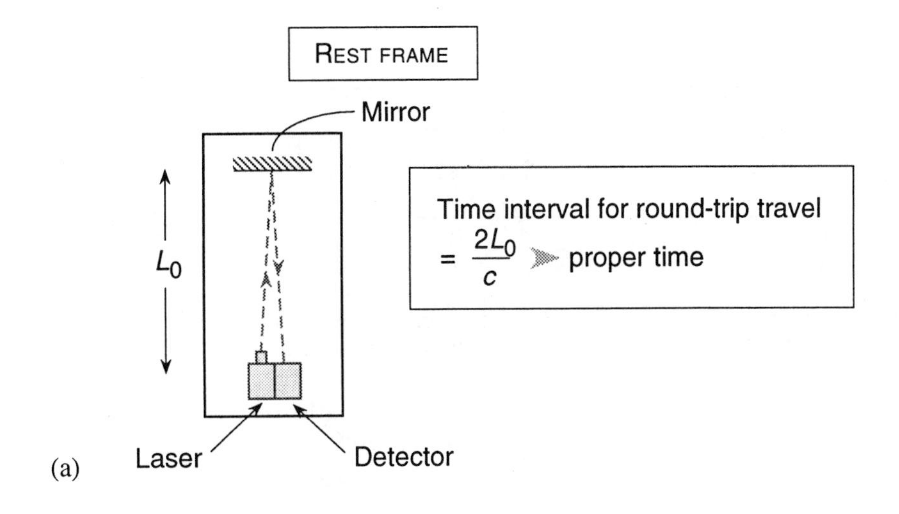

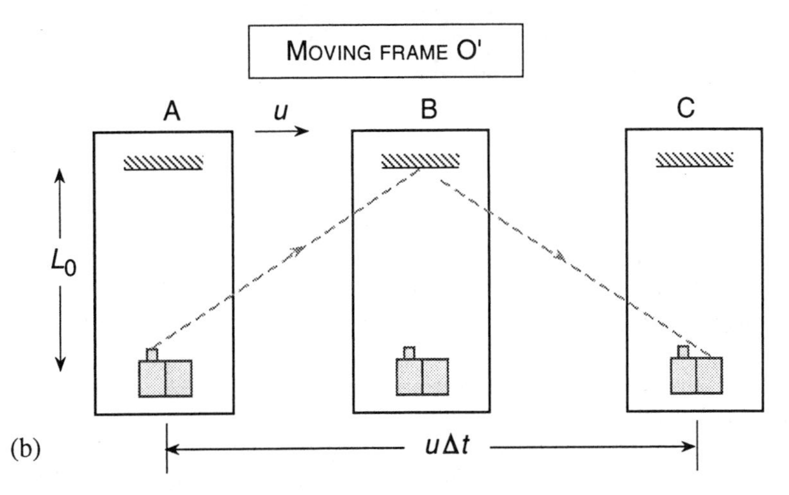

Figure 9.3: (a) Time interval for a timing device measured in a frame where the timing device is at rest; (b) the same event from a frame in which the clock is moving at a speed u.

Since the velocity of light is c in all frames we have

$$\Delta t = \frac{2L}{c} = \frac{2\sqrt{L_0^2 + (u\Delta t/2)^2}}{c}$$

Solving, for Δt, we get

$$\Delta t = \frac{2L_0}{c}\frac{1}{\sqrt{1 - u^2/c^2}} = \frac{\Delta t_0}{\sqrt{1 - u^2/c^2}} \tag{9.14}$$

The increase of time measured in the frame in which the clock is moving is known as time dilation. The equation we have derived is the same as the one given by the Lorentz transformation in the previous section. The slowing down of time for

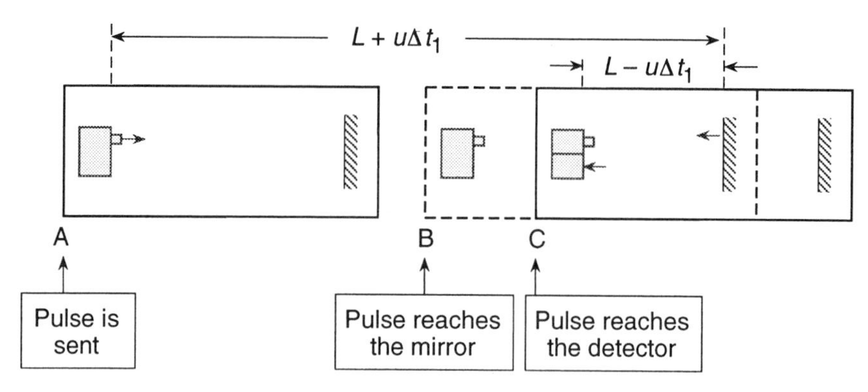

Figure 9.4: An experimental setup to measure the length of an object in two frames of reference moving at a speed u with respect to each other.

moving frames has very real effects that can be observed and carefully measured. The time dilation also predicts the so called "twin paradox," where a person stays young by moving away and then returning in a spaceship. We will discuss this later.

9.4.2 Relativity of Length

Let us see what Einstein's postulates have to say about length as measured in a moving frame and a frame where the object is at rest. To examine this we again consider a device as discussed in the previous subsection. A laser sends a pulse over a distance L and a detector registers the return pulse. In the rest frame, i.e., where the experimental setup is at rest, the distance between the laser and the mirror is L_0.

Let us observe the apparatus from a frame at rest, i.e., the apparatus is in a frame O′ moving at a speed u with respect to our observation frame O. As shown in Fig. 9.4, the flash of light is emitted at A and reaches the mirror, as shown in the time Δt_1. The frame O′ has moved to a point B, as shown. The distance light has to travel is

$$c\Delta t_1 = L + u\Delta t_1 \tag{9.15}$$

or

$$\Delta t_1 = \frac{L}{c-u}$$

The flash, after reaching the mirror, travels back to the detector and reaches the detector at point c in time Δt_2. The distance traveled is now $L - u\Delta t_2$. We have

$$c\Delta t_2 = L - u\Delta t_2$$

or

$$\Delta t_2 = \frac{L}{c+u}$$

The total time is

$$\Delta t = \Delta t_1 + \Delta t_2 = \frac{2L}{c}\frac{1}{1-u^2/c^2} \tag{9.16}$$

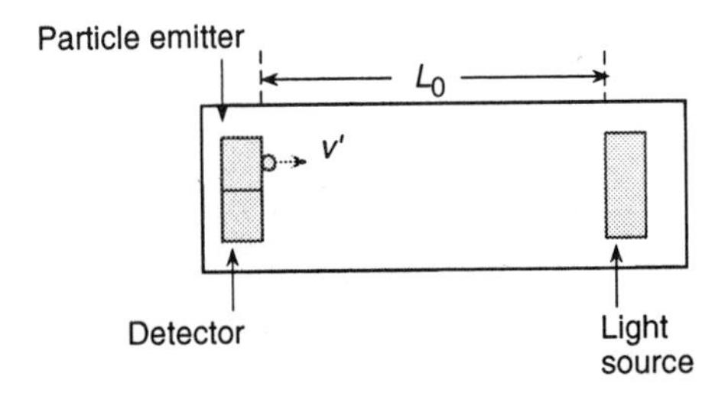

Figure 9.5: An apparatus that emits a particle at velocity v'. When the particle reaches the light source, a pulse is sent to the detector.

We also know that

$$\Delta t = \frac{\Delta t_0}{\sqrt{1-u^2/c^2}} \frac{2t_0}{c} \frac{1}{\sqrt{1-u^2/c^2}} \tag{9.17}$$

Comparing these two equations we have

$$L = L_0\sqrt{1-u^2/c^2} \tag{9.18}$$

This equation is essentially the equation suggested by Lorentz to explain the Michelson-Morley experiment. It summarizes an effect known as length contraction. If an observer in the rest frame measures a length L_0, all other moving observers will measure a shorter length. Note that the contraction only occurs in the direction of motion.

9.4.3 Relativistic Velocity Addition and Doppler Effect

Let us continue our application of Einstein's postulates to see how velocities of moving particles appear in different inertial frames. Let us consider an experimental set up as shown in Fig. 9.5. An experiment is done in the rest frame O$'$ in which a particle is emitted with a velocity V$'$ toward a point F a distance L_0 away. When the particle reaches F, a light pulse is sent back to the detector. The time interval between the particle being emitted and the light signal being received at the detector is

$$\Delta t_0 = \frac{L_0}{v'} + \frac{L_0}{c} \tag{9.19}$$

Let us now consider the frame O, with respect to which O$'$ is moving with a velocity u. In this frame the emitted particle moves with a velocity v. It reaches the point F in a time interval Δt_1 and travels a distance equal to the contracted length L and a distance $u\Delta t_1$. We have

$$v\Delta t_1 = L + u\Delta t_1 \tag{9.20}$$

or

$$\Delta t_2 = \frac{L}{c+u} \tag{9.21}$$

The total time observed by O is

$$\Delta t = \Delta t_1 + \Delta t_2 = L\left(\frac{1}{c+u} + \frac{1}{v-u}\right) \tag{9.22}$$

Writing

$$L = L_0\sqrt{1-u^2/c^2}$$
$$\Delta t = \frac{\Delta t_0}{\sqrt{1-u^2/c^2}}$$

and from Eqn. 9.19

$$L_0 = \frac{\Delta t_0 v' c}{v'+c}$$

We now get

$$\frac{\Delta t_0}{\sqrt{1-u^2/c^2}} = \frac{\Delta t_0 (v'c)}{(v'+c)} \tag{9.23}$$

Solving for v we get

$$v = \frac{v'+u}{1+v'u/c^2} \tag{9.24}$$

This is the relativistic velocity addition law. We have derived this for the velocity component along the direction of the relative motion of the two frames. The equation derived above is to be contracted with the classical relation $v = v' + u$. We see that the new velocity relation ensures that the measured velocity is never greater than c. In fact, if $v' = c$, we see that v is also equal to c.

An important outcome of the special theory of relativity is the relativistic Doppler effect. This has to do with how the frequency of an optical signal depends upon whether the source is moving with respect to an observer. In the classical Doppler effect, say for sound waves, if an observer is moving relative to the source, the frequency that is measured is different from what would be measured if the source were stationary. The frequency v' measured by the observer moving with a velocity v_0 with respect to the medium in which the waves travel is

$$\nu' = \nu\frac{v \pm v_0}{v \mp v_S} \tag{9.25}$$

where v is the speed of the waves in the medium and v_S is the velocity of the source with respect to the medium. The upper sign is to be used when the source S moves toward observer O.

In the classical Doppler effect, the waves propagate whether the source is moving or whether the observer is moving. The frequency does not depend upon the relative speed of the source and the observer, but on their speeds with respect to the medium. In the case of light, there is no medium of travel which defines a special frame of reference. The Doppler shift should only depend on the relative motion of the source and the observer.

To determine the relativistic Doppler shift, let us consider a source that is at rest in the reference frame of observer O. An observer O′ moves relative to the

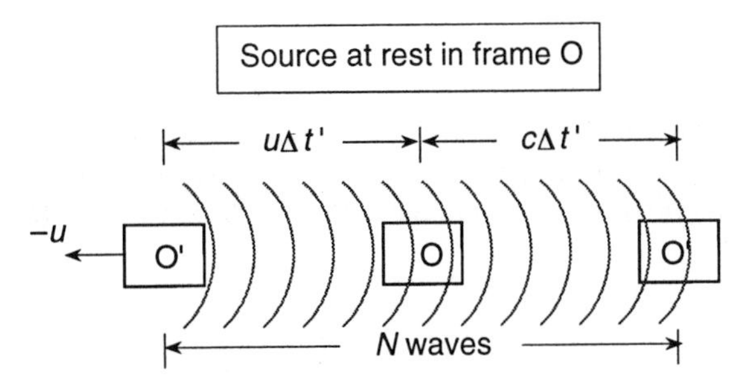

Figure 9.6: A source at rest in frame O is moving at a speed u away from the observer in frame O$'$. Observer O$'$ sees N waves being emitted in time $\Delta t'$.

source at a speed u, as shown in Fig. 9.6. Let us say that observer O finds that the source emits N waves with a frequency ν. The time interval for these waves to be emitted is

$$\Delta t_0 = \frac{N}{\nu} \tag{9.26}$$

The time interval measured in the frame O$'$ is $\Delta t'$, during which O moves a distance $u\Delta t'$. The wavelength measured by O$'$ is the total distance covered by the waves divided by the number of waves. (We assume that the source is moving *away* from the observer)

$$\lambda' = \frac{c\Delta t' + u\Delta t'}{N} = \frac{c\Delta t' + u\Delta t'}{\nu \Delta t_0} \tag{9.27}$$

The frequency is now c/λ', which gives

$$\nu' = \nu \frac{\Delta t_0}{\Delta t'} \frac{1}{1 + u/c}$$

Using the time dilation expression for the relation between $\Delta t'$ and Δt_0, we get

$$\nu' = \nu \sqrt{\frac{1 - u/c}{1 + u/c}} \tag{9.28}$$

If the source is moving toward the observer, u is replaced by $-u$ in this expression. This expression for relativistic Doppler effect is quite different from the expression for the classical effect which is limited to waves that require a medium to travel in. The relativistic effect depends only on the relative velocity between the source and the observer.

9.5 LORENTZ TRANSFORMATION

We have seen that Lorentz suggested some revolutionary ideas about length and time in order to explain the Michelson-Morley experiment. However, Einstein's theory was needed to understand the origin of length contraction and time dilation. Let us express the relationship between the position, time, and velocities measured by two observers in two inertial frames moving at a relative velocity u with respect to each other. Galilean transformations for coordinates, time and velocity are not valid for light or when the relation velocity u is not negligible compared to c.

Using arguments similar to those used in the previous section to derive expressions for length contraction and time dilation, we have the following set of transformations for a frame O′ moving along the x-axis away from frame O at a speed u.

$$\begin{aligned} x' &= \frac{x - ut}{\sqrt{1 - u^2/c^2}} \\ y' &= y \\ z' &= z \\ t' &= \frac{t - (u/c^2)x}{\sqrt{1 - u^2/c^2}} \end{aligned} \tag{9.29}$$

If the frame O moves toward O, u has to be replaced by $-u$.

Lorentz transformations are consistant with length contraction and time dilation. They can also be used to find the relation between velocities measured in two frames. If O observes a particle with velocity $\boldsymbol{v}$ with component v_x, v_y, and v_z, the velocity $\boldsymbol{v}'$ observed by O′ is given by

$$\begin{aligned} v'_x &= \frac{v_x - u}{1 - v_x u/c^2} \\ v'_y &= \frac{v_y\sqrt{1 - u/c^2}}{1 - v_x u/c^2} \\ v'_z &= \frac{v_z\sqrt{1 - u^2/c^2}}{1 - v_x u/c^2} \end{aligned} \tag{9.30}$$

Notice that if $u \ll c$ all of these transformations merge with the classical Galilean transformations.

Simultaneity and Clock Synchronization

An important consequence of the special relativity we have been discussing is the question: How do two observers in different inertial frames synchronize their clocks? The process of clock synchronization requires some sort of signal being sent and this signal can, at most, travel at the speed of light. By synchronization of clocks, we are not referring to the time dilation of times observed by two observers. We are discussing the following issue: Suppose two clocks, separated by some distance, have been synchronized by observer O. Do the clocks also appear synchronized to observer O′ moving at a velocity u with respect to O?

Fig. 9.7 demonstrates a simple mechanism that an observer O could use to synchronize two clocks. O places a light source in the middle of the two clocks that are separated by a distance L. The two clocks start when they receive a pulse of light. Since the light pulse takes the same time to reach the clocks, they will be synchronized.

Let us consider an observer O′ moving with a velocity u along the distance between the clocks. In frame O the coordinate and time of the pulse reaching clock

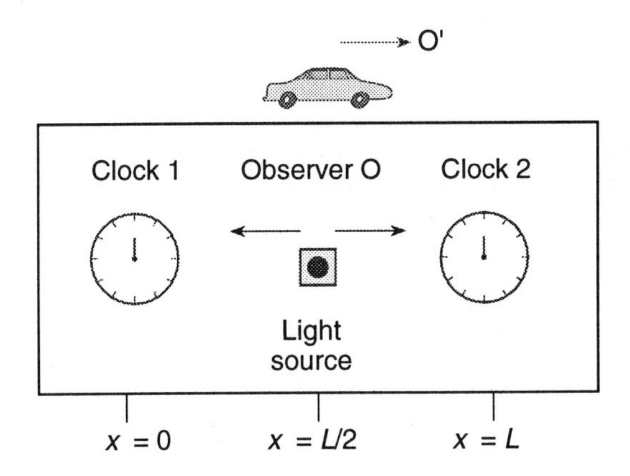

Figure 9.7: Observer O uses a light source placed at the middle of two clocks to synchronize the two clocks. Observer O′ sees that clock 2 starts earlier than clock 1.

1 is

$$\begin{aligned} x_1 &= 0 \\ t_1 &= \frac{L}{2c} \end{aligned} \tag{9.31}$$

The coordinate and time for clock 2 starting is

$$\begin{aligned} x_2 &= 0 \\ t_2 &= \frac{L}{2c} \end{aligned} \tag{9.32}$$

Using the Lorentz transformation, the signal is received at clock 1 at a time

$$t_1' = \frac{t_1(u/c^2)x_1}{\sqrt{1-u^2/c^2}} = \frac{L/2c}{\sqrt{1-u^2/c^2}} \tag{9.33}$$

The time the second clock starts (for observer O′) is

$$t_2' = \frac{t_2-(u/c^2)x_2}{\sqrt{1-u^2/c^2}} = \frac{L/2c-(u/c^2)L}{\sqrt{1-u^2/c^2}} \tag{9.34}$$

We see that t_2' is less than t_1', so that observer O′ sees clock 2 as starting earlier than clock 1. The clocks are thus not synchronized for O′. The time difference is

$$\Delta t' = t_1' - t_2' = \frac{uL/c^2}{\sqrt{1-u^2/c^2}} \tag{9.35}$$

We see from this exercise that two events which occur simultaneously in one frame are not simultaneous in another frame, unless they occur at the same point in space.

EXAMPLE 9.2 Muons are produced in the upper atmosphere when cosmic rays collide with air molecules. These particles are known to decay in 2.2 ns in a rest frame. This time is too small for the muons to cross the atmosphere and make it to the surface of Earth. Yet experiments clearly show that a considerable fraction of muons make it across the atmosphere. Explain this observation in terms of time dilation.

Earth's atmosphere is 100 km wide. For a particle traveling at close to the speed of light, the time taken to cross this distance is

$$\Delta t = \frac{10^5 \text{ m}}{3 \times 10^8 \text{ m/s}} = 333\mu\text{s}$$

In Earth's frame of reference, the muon is traveling at a speed u (close to the speed of light) and the time interval between its creation and decay is related to the proper lifetime by the relation

$$T(\text{Earth}) = \frac{2.2\ \mu\text{s}}{\sqrt{1 - u^2/c^2}}$$

Using $T(\text{Earth})$ as approximately 333 μs, we get

$$u = 0.999978\text{c}$$

EXAMPLE 9.3 Using the results of Example 9.2, calculate the thickness of Earth's atmosphere seen by a muon.

The apparent thickness for a muon is

$$\begin{aligned} L &= L_0\sqrt{1 - u^2/c^2} = 10^5\ \text{m}\sqrt{1 - (0.99978)^2} \\ &= 660\ \text{m} \end{aligned}$$

We can see from this example that the muon sees a much thinner atmosphere as it passes down the atmosphere. But in the muon's frame of reference, its lifetime is also much smaller.

EXAMPLE 9.4 Consider a spacecraft moving away from Earth at a speed of 0.5c. The craft fires two missiles both with a speed of 0.5c relative to the ship. One missile has a velocity parallel to the spaceship and the other in the backward direction. Calculate the speeds of the two missiles as measured by an observer on Earth.

According to the Galilean transformation, the velocities of the two missiles should be 1.0 c and 0.0 c, respectively. Let us see what the special theory of relativity tells us. The velocity of the first missile, as seen by the observer on Earth, is

$$\begin{aligned} V(\text{Earth}) &= \frac{0.5\text{c} + 0.5\text{c}}{1 + (0.5\text{c})(0.5\text{c})/\text{c}^2} \\ &= 0.8\text{c} \end{aligned}$$

The velocity of the second missile is zero.

EXAMPLE 9.5 Emission from hydrogen atoms in a far-away galaxy is recorded to have a wavelength of 4340 Åon Earth. The same emission line from hydrogen atoms at rest on Earth is found to have a wavelength of 6000 Å. Calculate the speed of the galaxy relative to Earth.

Using the equations for relativistic Doppler effect we have ($\nu = c/\lambda$)

$$\begin{aligned} \frac{c}{\lambda'} &= \frac{c}{\lambda}\sqrt{\frac{1 - u/c}{1 + u/c}} \\ \frac{1}{6000} &= \frac{1}{4340}\sqrt{\frac{1 - u/c}{1 + u/c}} \end{aligned}$$

Solving for u/c we find

$$u = 0.31c$$

9.6 RELATIVISTIC DYNAMICS

Einstein's postulates for relativity have forced us to abandon concepts such as "absolute" length and time. What about other concepts, such as mass, kinetic energy, momentum, etc.? And how about conservation laws, such as conservation of energy and momentum? How does the basic law of classical dynamics

$$m\frac{dv}{dt} = F$$

change? According to Einstein's postulates, all the physical laws should have the same form, regardless of the inertial frame in which an experiment is carried out. Thus, for example, if momentum is conserved in one frame, but not in another, this can be used to distinguish one frame from the other. We will see below that as a result of this postulate, our usual definitions of momentum, energy, etc., have to change.

9.6.1 Conservation of Momentum: Relativistic Momentum

Let us consider the implications of the conservation of momentum law on the definition of particle momentum. Fig. 9.8 shows two particles of equal mass m moving with equal and opposite velocities along the x-axis. Let us say they collide at the origin and the experimental conditions are so adjusted that after the collision they move along the y-axis, as shown. If the collision is elastic, no kinetic energy is lost and each particle must move with a velocity v along the y and $-y$ direction. The frame used to define the problem is O and in this frame we have, for the classical components of momentum:

$$\begin{aligned}
\text{Initial :} & \\
p_{xi} &= mv + m(-v) = 0 \\
p_{yi} &= 0 \\
\text{Final :} & \\
p_{xf} &= 0 \\
p_{yf} &= mv + m(-v) = 0
\end{aligned} \tag{9.36}$$

Thus, in this frame momentum is conserved.

Let us now view the same collision from frame O′ moving along the x-axis at a velocity $u = -v$ with respect to frame O. This view is shown in Fig. 9.8b. In frame O′, particle 2 is initially at rest. To find the velocity components of particle 1, we use the Lorentz transformation for velocities. The values are

$$\begin{aligned}
v'_x(1) &= \frac{2v}{1 + v^2/c^2} \\
v'_y(1) &= 0
\end{aligned} \tag{9.37}$$

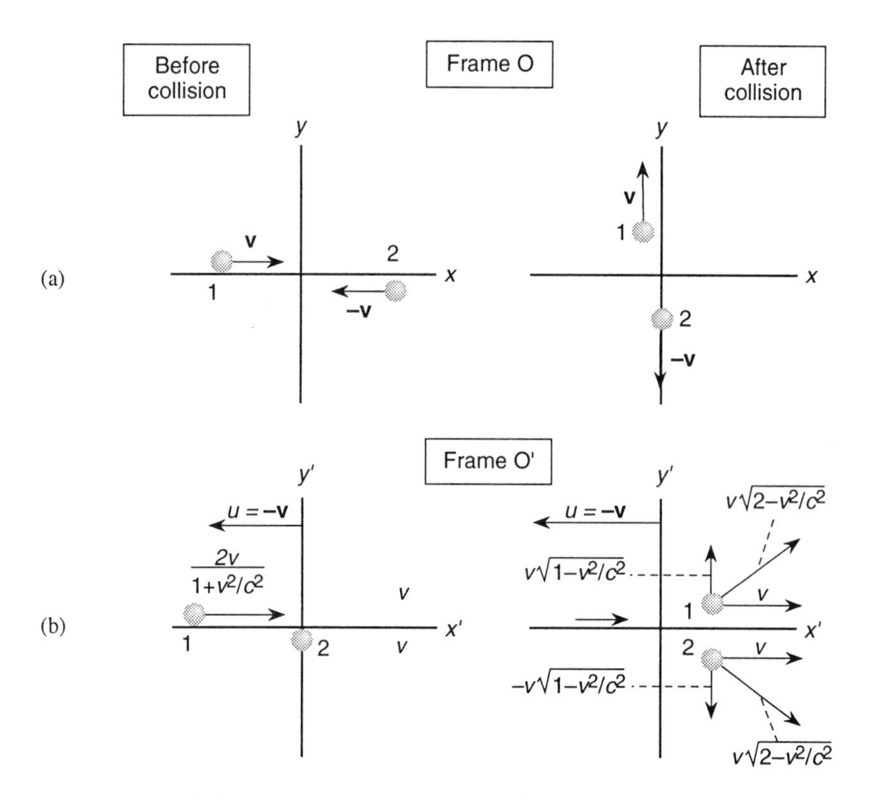

Figure 9.8: (a) A collision between two particles of the same mass viewed in frame O. (b) The same collision viewed in frame O′ moving at a velocity $u = -v$ relative to frame O.

After the collision, we have for the velocity components of particles 1 and 2

$$\begin{aligned} v'_x(1) &= v \\ v'_y(1) &= v\sqrt{1 - v^2/c^2} \\ v'_x(2) &= v \\ v'_y(2) &= -v\sqrt{1 - v^2/c^2} \end{aligned} \tag{9.38}$$

Using the classical definition of momentum $p = mv$, it is easy to see that the x component of momentum is not conserved in the O′ frame. In the O′ frame we have

$$\begin{aligned} \Delta p'_x &= p'_x(\text{initial}) - p'_x(\text{final}) \\ &= 2mv\left(\frac{1}{1 + v^2/c^2} - 1\right) \end{aligned} \tag{9.39}$$

We note that the derivation from conservation of momentum increases as v increases.

We see that with the classical definition of momentum, the conservation law for momentum holds in one frame, but not in another. This is not consistent with Einstein's postulates. How do we reconcile the definition of momentum with Einstein's postulate, which implies that the conservation law of momentum should be obeyed in frames O and O′?

If we define the momentum of a particle not as $m\boldsymbol{v}$, but as

$$p = \frac{mv}{\sqrt{1 - v^2/c^2}} \tag{9.40}$$

where v is the velocity of the particle in frame 2 measurement, it can be seen that the conservation of mass law holds in all inertial frames. Let us reconsider the conservation law in frames O and O$'$. In frame O, the velocities of the two particles are equal and opposite and, as a result, momentum is conserved. In frame O$'$, we have (after some algebra)

Initial state :

$$p'_x = \frac{2mv}{1 - v^2/c^2}; \; p'_y = 0 \tag{9.41}$$

Final state :

$$p'_x = \frac{2mv}{1 - v^2/c^2}; \; p'_y = 0 \tag{9.42}$$

Thus this "new" momentum is conserved in both frames. The relativistic definition of momentum reduces to the classical value when the particle velocity becomes negligible, compared to the velocity of light.

9.6.2 Conservation of Energy: Relativistic Energy

We have seen in the previous subsection that the momentum of a particle has to be redefined to bring consistency between momentum of particles and Einstein's theory. Of course, all of this is forced upon us, not by a clever theory, but by experiment. Let us now consider how the energy of a particle appears when we apply relativistic arguments to the conservation of energy law.

Once again, we examine the collision illustrated in Fig. 9.8. In frame O, the kinetic energy of the system (defined as $1/2mv^2$) is conserved in the collision. However, in frame O$'$, it is not. We have, for the total kinetic energy

Initial state :

$$K'_i = \frac{2mv^2}{1 + v^2/c^2} \tag{9.43}$$

Final state :

$$K'_f = (2 - v^2/c^2) \tag{9.44}$$

We find that the initial and final energies are different. This fact can be used to distinguish one frame from another, something that cannot be done according to the special theory of relativity. This implies that somehow our definition of energy is incorrect and must be modified.

The new definition of the kinetic energy of a particle can be shown to be

$$K = \frac{mc^2}{\sqrt{1 - v^2/c^2}} - mc^2 \tag{9.45}$$

This expression reduces to the classical value if $v \ll c$. We may write Eqn. 9.45 as

$$K = E - E_0 \tag{9.46}$$

where the total relativistic energy is given by

$$E = \frac{mc^2}{\sqrt{1 - v^2/c^2}} \tag{9.47}$$

and the energy of the particle at rest is given by (this is called the rest energy)

$$E_0 = mc^2 \tag{9.48}$$

The rest energy is the energy of the particle in the frame in which it is at rest.

Using Eqns. 9.40 and 9.47 we can write

$$E = \sqrt{(pc)^2 + (mc^2)^2} \tag{9.49}$$

If a particle with mass is traveling with speeds approaching the speed of light, the rest energy becomes negligible, compared to the kinetic energy and we have

$$E \cong pc \tag{9.50}$$

EXAMPLE 9.6 Calculate the momentum of a proton moving at i) 1000 km/s and ii) 0.8c according to Newtonian mechanics and relativistic mechanics.

According to Newtonian mechanics, the momentum of the proton in the two cases is

$$\begin{aligned}
\text{i)} \quad p &= (1.67 \times 10^{-27}\ \text{kg})(10^6\ \text{m/s}) \\
&= 1.67 \times 10^{-21}\ \text{kg.m/s} \\
\text{ii)} \quad p &= (1.67 \times 10^{-27}\ \text{kg})(0.8 \times 3 \times 10^8\ \text{m/s}) \\
&= 4 \times 10^{-19}\ \text{kg.m/s}
\end{aligned}$$

The expressions for relativistic mechanics are

$$\text{i)} \quad p = \frac{p_{\text{classical}}}{\sqrt{1 - v^2/c^2}} \cong 1.67 \times 10^{-21}\ \text{kg.m/s}$$

There is essentially no difference between the classical and relativistic result.

$$\text{ii)} \quad p = \frac{p_{\text{classical}}}{\sqrt{1 - (0.8)^2}} \cong 6.67 \times 10^{-19}\ \text{kg.m/s}$$

We see that the relativistic momentum is much higher than the classical one where the velocity of the particle is close to the velocity of light.

For relativistic particles, it is useful to express momentum in units of energy/c. We write, for the second case,

$$pc = \frac{mcv}{\sqrt{1 - v^2/c^2}} = \frac{mc^2}{\sqrt{1 - v^2/c^2}} \frac{v}{c}$$

Using the value 938 meV for a proton rest mass, we get

$$\begin{aligned} pc &= \frac{938 \text{ MeV}(0.8)}{\sqrt{1-(0.8)^2}} \\ &= 1250.67 \text{ MeV} \end{aligned}$$

or

$$p = 1250.67 \text{ MeV/c}$$

EXAMPLE 9.7 Calculate the kinetic energy and total relativistic energy of a proton moving with a speed of 0.9c. Compare the relativistic kinetic energy with the classical kinetic energy.

The kinetic energy is

$$\begin{aligned} K &= \frac{(1.67 \times 10^{-27} \text{ kg})(3.0 \times 10^8 \text{ m/s})^2}{\sqrt{1-(0.9)^2}} - (1.67 \times 10^{-27} \text{ kg})(3.0 \times 10^8 \text{ m/s})^2 \\ &= 1.945 \times 10^{-10} \text{ J} \\ &= 1215.7 \text{ MeV} \end{aligned}$$

Using the rest energy of 938 MeV, the total energy is 2153.7 MeV.

The classical kinetic energy is

$$\begin{aligned} K &= \frac{1}{2}(1.67 \times 10^{-27} \text{ kg})(0.9 \times 3 \times 10^8 \text{ m/s})^2 \\ &= 380.4 \text{ MeV} \end{aligned}$$

The classical kinetic energy is smaller than the relativistic one.

9.7 APPLICATION EXAMPLES

Fig. 9.9 shows some of the important implications of the special theory of relativity. We will discuss some applications in this section.

9.7.1 Optical Gyroscope

The gyroscope is a very important navigation instrument used to define a fixed direction in space. It can also be used to determine the change in angle or the angular rate of change in the direction of the vehicle on which it is mounted. Gyroscopes (or gyros) are used for guidance, navigation, and stabilization of the carrying vehicle. They are used for heading and orientation of aircraft; to determine the heading of automobiles as they turn on streets; the deviation of flights from set patterns, etc.

Gyros have traditionally relied on the spinning of a mass and the fact that a spinning mass points toward a fixed direction in space unless disturbed by a force. However, in the 1960s, optical gyros were invented. With advances in solid-state technology (lasers, detectors, modulators, etc.), optical gyros have now become highly reliable and compact.

Optical gyros include the ring laser gyros and the fiber optic gyros. These gyros depend upon the principle that the speed of light is c in all frames of reference. The operation of an optical gyro can be understood by examining a schematic of

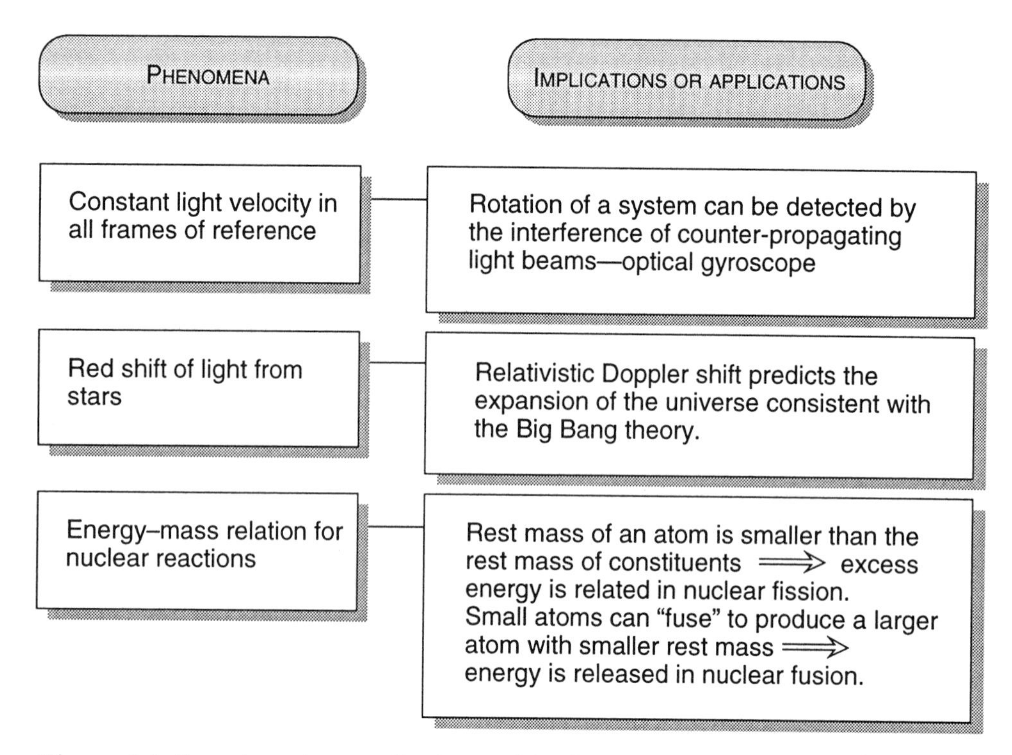

Figure 9.9: Some important applications and implications of the special theory of relativity.

a ring gyro. Fig. 9.10 shows a typical ring gyro configuration. The device consists of a three- or four-sided block which defines a closed optical cavity. The light path is defined by mirrors mounted on corners. Light travels through holes filled in the block. The cavity is filled with a gas, usually a helium-neon mixture which lases when it is excited. Thus there is no need for an external laser. In the fiber gyros, one uses an external laser.

The laser light propagates clockwise and counterclockwise in the cavity. We now have two beams in the cavity with an optical path of $\sim$ 8 cm to 40 cm. A small amount of light can pass through a mirror and the beams can interfere, as shown in Fig. 9.10. A detector can count any changes in the interference pattern.

Optical gyros depend upon the Sagnac effect to detect any rotation. As noted earlier, the speed of light in both beams is c and is independent of the motion of the source. If two identical beams of wavelength λ travel in opposing directions along a closed path undergoing rotation Ω, then the light beam traveling in the same direction as the rotation takes a longer time to travel around the path than the other beam. This results in a change in the interference pattern of the beams. The phase shift ϕ produced is given by

$$\phi = \left(\frac{8\pi AN}{\lambda c}\right)\Omega \tag{9.51}$$

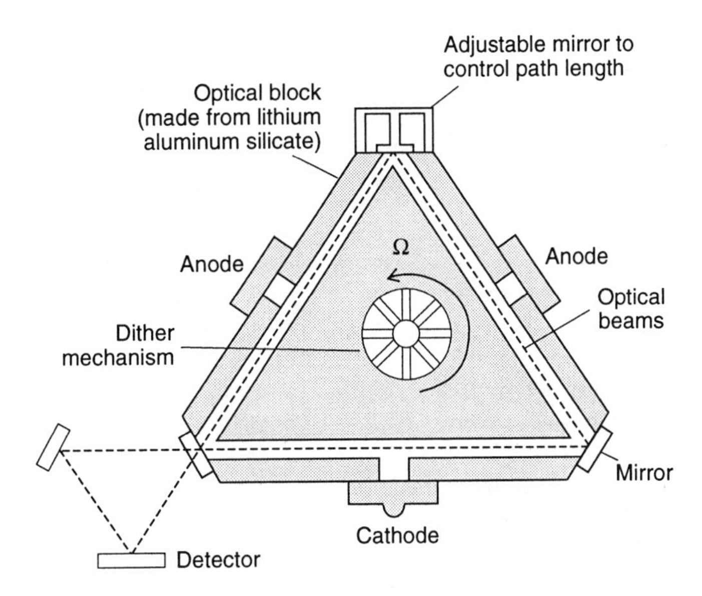

Figure 9.10: A schematic of a ring laser gyroscope. The axis of rotation is normal to the plane of the page.

where A is the area enclosed by the beams path and N is the number of times the beam has gone around the path. The term in the parenthesis is called the scale factor for the gyro. Clearly, the larger the area enclosed by the beams, the better the performance.

In fiber optic gyros, the optical path is defined by an optical fiber coil. The fibers could be as long as 1 km. Since there are no moving parts in optical fibers, they are very reliable. Hence, they have been rapidly replacing mechanical gyros.

9.7.2 Rest Mass and Nuclear Reaction

According to the special theory of relativity, the total energy of a particle is made of the rest energy and the kinetic energy. This concept applies to composite particles such as atoms as well. The total energy of an atom differs from the total energy of its constituents (i.e., when the electrons, neutrons, and protons are all separated) by the binding energy of the atoms. This binding energy is dominated by the nuclear binding energy which is typically in the range of 100 MeV, while the electron–nucleus binding energy is ~10 eV. The difference in total energy means that the mass of a typical atom is smaller than the mass of its constituents. This difference for most atoms is about 0.1 %. When nuclear reactions occur, the rest mass changes produced are manifested as energy.

Nuclear Fission

In the 1930s scientists realized that a uranium atom has about 200 MeV more rest energy than the rest energy of the lighter atoms that it splits into. A typical reaction for uranium fission is

$$^1n +^{235} U \rightarrow^{95} Y +^{139} I + 2^1n \tag{9.52}$$

In this reaction a neutron (^1n) is needed to split the uranium atom. The difference between the rest masses of the constituents on the left-hand side of the reaction and ones on the right-hand side is equivalent to an energy of 172 MeV. This excess energy is released by each uranium atom. The neutrons produced in the reaction can be used for chain reactions, i.e., fission of other uranium atoms.

Nuclear Fusion
Nuclear fusion involves nuclear reactions in which light elements combine to form heavier ones. In these reactions it is found that the rest mass of the lighter atoms is larger than that of the heavier atoms. The difference is released as energy. Such fusion processes are responsible for the energy of the sun and other stars.

Fusion can involve hydrogen, helium, and many other light elements. For example, three helium (^{4}He) atoms can fuse to form one atom of carbon (^{12}C). The change in rest energy is 7.27 MeV. Due to the abundance of helium in the sun this reaction can produce a significant contribution to the sun's energy output.

It certainly seems tempting to have a fusion reactor that would convert the hydrogen in water to helium. A single cup of water could then provide enough energy to power 1,000 homes for a month! However, this possibility has resisted all attempts so far.

Matter–Antimatter
The development of a theory that combines the special theory of relativity and quantum mechanics shows that for every particle there should be an antiparticle. An antiparticle has the same mass as a particle, but has the opposite charge. One can conceive of a universe which is made up of antimatter—all nuclei will be negatively charged with positively charged electrons (or positrons).

The equivalence of mass and energy is demonstrated in experiments where a photon can produce a particle–antiparticle pair. Of course, the photon energy should exceed twice the rest mass energy of the particle. In many radioactive materials photons are released, which can produce an electron–positron pair. When these photons pass through matter, the pair can be produced. It is necessary to have another body present to balance the excess momentum of the photon.

When a particle meets its antiparticle, the pair can be annihilated, producing photons. The conservation of relativistic energy and momentum laws have to be obeyed in this process. If a single photon is present, one must have another body participating in this collision. However, if two photons are produced, there is no need of another body.

9.7.3 Cosmology and Red Shift

The expressions derived on the basis of the special theory of relativity are relevant not only to subatomic particles, but also to massive systems such as stars and galaxies. The Doppler shift expression can be used to examine the speed of recession of a galaxy from Earth. As mentioned in Example 9.5, radiation from elements in distant galaxies are found to be red-shifted by a certain amount. It is possible by independent means to estimate the distance of stars and galaxies from Earth. These estimated distances, in conjunction with the red-shift measurements, indicate that the speed of recession of a galaxy is directly proportional to its distance from Earth.

From this proportionality it has been concluded that the entire universe is expanding. The concept of an expanding universe has led to theories of cosmology that explain numerous details of the known universe. These include the relative abundances of the elements and the presence of low-energy cosmic background radiation that comes uniformly from all directions in space.

EXAMPLE 9.8 The total power radiated by the sun is 10^{26} W. Calculate the rate at which the sun's mass is decreasing.

The decrease in mass is given by

$$\begin{aligned}\frac{dm}{dt} &= \frac{1}{c^2}\frac{dE}{dt} \\ &= \frac{(10^{26}\text{ J/s})}{(3\times 10^8\text{ m/s})^2} \\ &= 1.11\times 10^9\text{ kg/s}\end{aligned}$$

This is an enormous rate of loss, but considering the mass of the sun, it should be around for a while!

9.8 CHAPTER SUMMARY

Table 9.1 summarizes the topics covered in this chapter.

9.9 PROBLEMS

Problem 9.1 Consider a Michelson-Morley experiment done with an interferometer arm of 10 m. Based on classical physics, if a shift of one fringe occurred (i.e., change in round-trip travel time between the two arms is $\sim 2\times 10^{-15}$ s), calculate the velocity through ether.

Problem 9.2 Calculate the speed of an object at which its length appears to be contracted to one-half of its proper length.

Problem 9.3 At what speed would a car have to move to see the usual red color ($\lambda = 6500$ Å) as a green color ($\lambda = 5500$ Å)?

Problem 9.4 An astronaut must journey to a distant planet, which is 100 light-years from Earth. What speed will be necessary if the astronaut wishes to age only 10 years during the round-trip?

Problem 9.5 High-energy particles are observed in laboratories by photographing the tracks they leave in certain detectors; the length of the track depends on the speed of the particle and its lifetime. A particle moving at 0.995c leaves a track 1.25 mm long. What is the proper lifetime of the particle?

Problem 9.6 One of the strongest emission lines observed from distant galaxies comes from hydrogen and has a wavelength of 122 nm (in the ultraviolet region). How fast must a galaxy be moving away from us in order for that line to be observed in the visible region at 366 nm?

Problem 9.7 For what range of velocities of a particle of mass m can we use the classical expression for kinetic energy $\frac{1}{2}mv^2$ to within an accuracy of 1 percent?

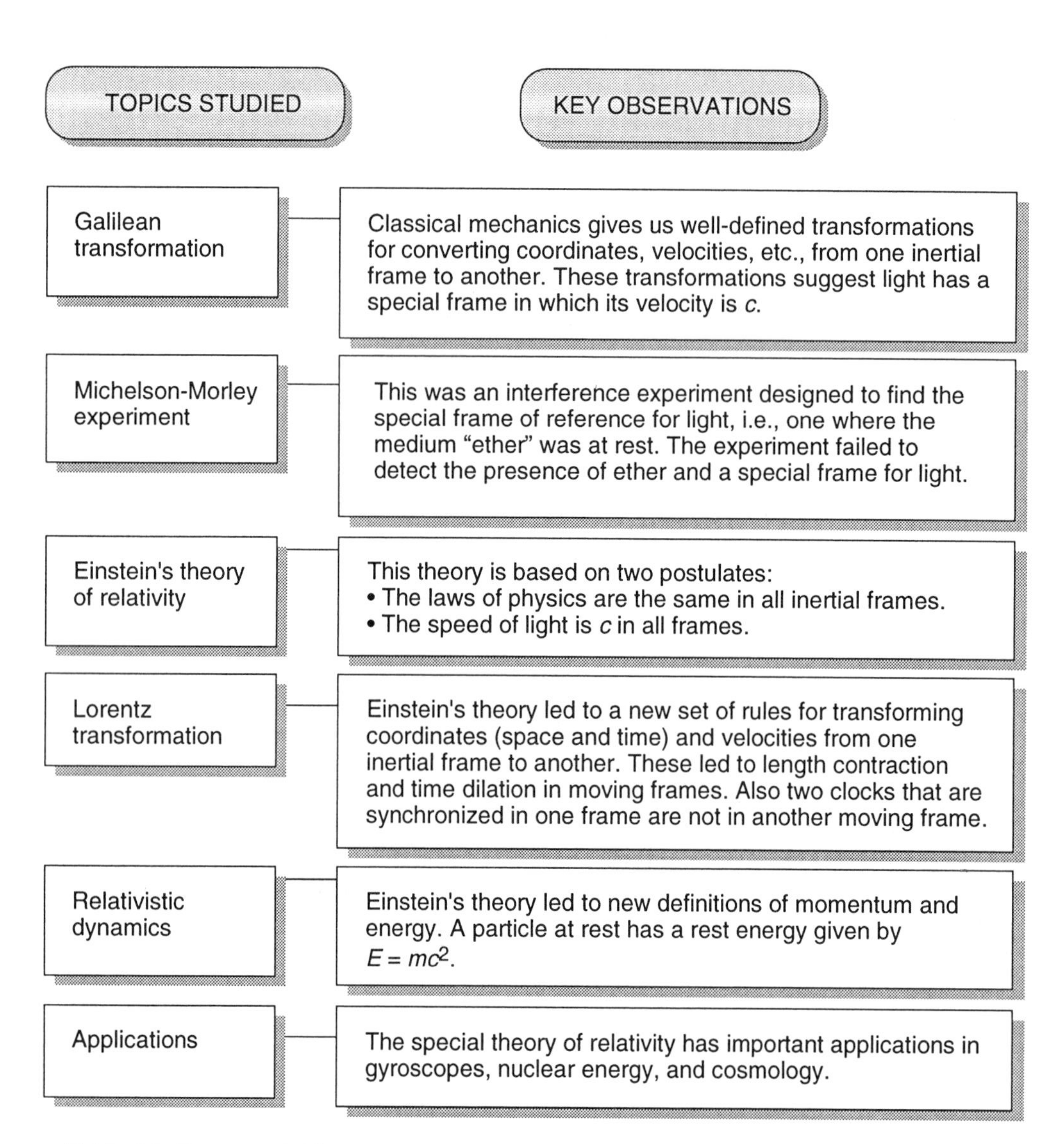

TOPICS STUDIED	KEY OBSERVATIONS
Galilean transformation	Classical mechanics gives us well-defined transformations for converting coordinates, velocities, etc., from one inertial frame to another. These transformations suggest light has a special frame in which its velocity is c.
Michelson-Morley experiment	This was an interference experiment designed to find the special frame of reference for light, i.e., one where the medium "ether" was at rest. The experiment failed to detect the presence of ether and a special frame for light.
Einstein's theory of relativity	This theory is based on two postulates: • The laws of physics are the same in all inertial frames. • The speed of light is c in all frames.
Lorentz transformation	Einstein's theory led to a new set of rules for transforming coordinates (space and time) and velocities from one inertial frame to another. These led to length contraction and time dilation in moving frames. Also two clocks that are synchronized in one frame are not in another moving frame.
Relativistic dynamics	Einstein's theory led to new definitions of momentum and energy. A particle at rest has a rest energy given by $E = mc^2$.
Applications	The special theory of relativity has important applications in gyroscopes, nuclear energy, and cosmology.

Table 9.1: Summary table.

Problem 9.8 Find the kinetic energy of an electron moving at a speed of a) $v = 1.00 \times 10^{-4}$c; b) $v = 1.00 \times 10^{-2}$c; c) $v = 0.300$c; and d) $v = 0;999$c.

Problem 9.9 Electrons are accelerated to high speeds by a two-stage machine. The first stage accelerates the electrons from rest to $v = 0.99$c. The second stage accelerates the electrons from 0.99c to 0.999c. a) How much energy does the first stage add to the electrons? b) How much energy does the second stage add?

Problem 9.10 An electron and a proton are each accelerated through a potential difference of 20 MV. Calculate the momentum and kinetic energy of each particle using relativistic and classical results.

Problem 9.11 Consider a nuclear reaction where a uranium atom releases 200 MeV during fission. Calculate the change in mass that occurs when 1 kg of uranium undergoes fission.

Problem 9.12 An electron and a positron make a head-on collision. Each particle has a speed of 0.99999c. After the collision it is observed that two muons are produced and move off in opposite directions. If $mc^2 = 105.7$ MeV for a muon, calculate the kinetic energy of each muon.

Problem 9.13 A cosmic-ray μ-meson is moving vertically through the atmosphere with a speed of 0.99c. Its mean life expectancy for radioactive decay into an electron and two neutrinos is 2.22 μs (measured in its own rest frame). What is the lifetime measured by an observer on Earth?

Problem 9.14 An airplane 20 m long is flying at 350 ms^{-1}. How much shorter does this aircraft appear to an observer on the ground?

Problem 9.15 A star is moving away from Earth at 50 km/s. Calculate the shift in the $H\alpha$ line ($\lambda = 6563$ Å).

Problem 9.16 An observer sees two particles traveling in opposite directions, each at a speed of 0.99c. Calculate the speed of one particle with respect to the other.

Problem 9.17 Calculate the kinetic energy of a proton with momentum 900 MeV/c.

Problem 9.18 Consider a proton with energy 10^{13} MeV. This represents about the highest energy of cosmic ray particles. The proton travels the entire galaxy (10^5 light-years in diameter). How long does it take the proton in its own reference frame?

Problem 9.19 The "classical radius" r_0 of an electron is defined through the relation

$$E = m_0 c^2 = \frac{e^2}{4\pi\epsilon_0 r_0}$$

Calculate the classical radius of an electron.

Problem 9.20 An electron–positron pair can be produced by a photon-striking stationary electron according to the reaction

$$\hbar\omega + e^- \to e^- + (e^+ + e^-)$$

Show that the threshold energy for the photon for this reaction is $4m_0c^2$.

Problem 9.21 A collision in which a photon strikes an electron and is scattered by the electron is called Compton scattering. If the photon is deflected by an angle θ during the collision, show that the change in the photon wavelength is

$$\Delta\lambda = \frac{h}{m_0 c}(1 - \cos\theta)$$

Problem 9.22 Calculate the maximum change in the wavelength of a photon when it suffers Compton scattering.

APPENDIX A

FERMI GOLDEN RULE

In Chapter 8 we have seen that scattering of electrons from one state to another plays a key role in almost all physical properties of materials. Optical and transport phenomena are linked to scattering processes. We have used the Fermi golden rule to evaluate several important scattering processes in materials. In this appendix we will give a derivation of this important equation. The general Hamiltonian of interest is of the form

$$H = H_0 + H' \tag{A.1}$$

where H_0 is a simple Hamiltonian with known solutions

$$H_0 u_k = E_k u_k \tag{A.2}$$

and E_k, u_k are known. In the absence of $H'(t)$, if a particle is placed in a state u_k, it remains there forever. The effect of H' is to cause time-dependent transitions between the states u_k. The time-dependent Schrödinger equation is

$$i\hbar \frac{\partial \psi}{\partial t} = H\psi \tag{A.3}$$

The approximation will involve expressing ψ as an expansion of the eigenfunctions $u_n \exp(-iE_n t/\hbar)$ of the unperturbed time-dependent functions

$$\psi = \sum_n a_n(t) u_n e^{-iE_n t/\hbar} \tag{A.4}$$

The time-dependent problem is solved when the coefficients $a_n(t)$ are known. In the spirit of the perturbation approach, these coefficients are determined to different orders. Hopefully, the first- or second-order terms would suffice and higher-order terms would be negligible.

Substituting for ψ (given by Eqn. A.4) in Eqn. A.3, using Eqn. A.2, we get

$$\sum_n i\hbar\dot{a}_n(t)u_n e^{-iE_nt/\hbar} + \sum_n a_n E_n u_n e^{-iE_nt/\hbar} = \sum_n a_n(H_0 + H^{'})u_n e^{-iE_nt/\hbar} \quad (A.5)$$

Multiplying by u_k^* and integrating over space, we get (using orthogonality of u_k's)

$$i\hbar\dot{a}_k e^{-iE_kt/\hbar} = \sum_n a_n e^{-iE_nt/\hbar}\langle k|H^{'}|n\rangle \quad (A.6)$$

Note that we have used the Dirac notation for u_k and u_n here. Writing

$$\omega_{kn} = \frac{E_k - E_n}{\hbar} \quad (A.7)$$

$$\dot{a}_k = \frac{1}{i\hbar}\sum_n \langle k|H^{'}|n\rangle \; a_n \; e^{i\omega_{kn}t} \quad (A.8)$$

To find the corrections to various orders in $H^{'}$, we can write the perturbation as $\lambda H^{'}$, where λ is a parameter that goes from 0 (no perturbation) to 1:

$$\begin{aligned} H^{'} &\rightarrow \lambda H^{'} \\ a_n &= a_n^{(0)} + \lambda a_n^{(1)} + \lambda^2 a_n^{(2)} + \cdots \end{aligned} \quad (A.9)$$

Here $a_n^{(0)}, a_n^{(1)}, \ldots$ are the different orders of the expansion coefficients of the wavefunction. Substituting this expansion in Eqn. A.8 and comparing the coefficients of the same powers of λ, we get

$$\begin{aligned} \dot{a}_k^{(0)} &= 0 \\ \dot{a}_k^{(s+1)} &= \frac{1}{i\hbar}\sum_n \langle k|H^{'}|n\rangle a_n^{(s)} e^{i\omega_{kn}t} \end{aligned} \quad (A.10)$$

In principle, these equations can be integrated to any order to obtain the desired solution. To study the time evolution of the problem, we assume that the perturbation is absent at time $t < 0$ and is turned on at $t = 0$. With this assumption, the system is in a time-independent state up to $t = 0$. From Eqn. A.10, we see that the zeroth-order coefficients $a_n^{(0)}$ are constant time and are simply given by the initial conditions of the problem, before the perturbation is applied. We assume that initially the system is in a single, well-defined state $|m\rangle$:

$$\begin{aligned} a_m^{(0)} &= 1 \\ a_k^{(0)} &= 0 \text{ if } k \neq m \end{aligned} \quad (A.11)$$

Integration of the first-order term in Eqn. A.10 gives

$$a_k^{(1)}(t) = \frac{1}{i\hbar}\int_{-\infty}^{t} \langle k|H^{'}(t^{'})|m\rangle e^{i\omega_{km}t^{'}} \, dt^{'} \quad (A.12)$$

We choose the constant of integration to be zero since $a_k^{(1)}$ is zero at time $t \longrightarrow -\infty$ when the perturbation is not present.

We see from Eqn. A.12 that if the perturbation is of finite duration, the amplitude of finding the system in a state $|k\rangle$ different from the initial state $|m\rangle$ is proportional to the Fourier component of the matrix element of the perturbation between the two states.

Harmonic Perturbation

A number of important problems in quantum mechanics involve a perturbation which has time dependence with a harmonic form. Examples include interaction of electrons with electromagnetic radiation (or photons), electrons in crystals interacting with lattice vibrations (or phonons), etc. For such perturbations, time-dependent perturbation theory gives some simple results that have been widely applied in understanding and designing experiments. Consider the case where the perturbation is harmonic, except that it is turned on at $t = 0$ and turned off at $t = t_0$. Let us assume that the time dependence is given by

$$\langle k|H(t^{'})|m\rangle = 2\langle k|H^{'}(0)|m\rangle \sin \omega t^{'} \tag{A.13}$$

Carrying out the integration until time $t \geq t_0$ in Eqn. A.12, we get

$$a_k^{(1)}(t \geq t_0) = -\frac{\langle k|H^{'}(0)|m\rangle}{i\hbar} \left(\frac{\exp[i(\omega_{km} + \omega)t_0] - 1}{\omega_{km} + \omega} - \frac{\exp[i(\omega_{km} - \omega)t_0] - 1}{\omega_{km} - \omega} \right) \tag{A.14}$$

The structure of this equation tells us that the amplitude is appreciable only if the denominator of one term or the other is close to zero. The first term is important if $\omega_{km} \approx -\omega$ or $E_k \approx E_m - \hbar\omega$. The second term is important if $\omega_{km} \approx \omega$ or $E_k \approx E_m + \hbar\omega$. Thus in the first-order, the effect of a harmonic perturbation is to transfer, or to receive from the system, the quantum of energy $\hbar\omega$.

If we focus on a system where $|m\rangle$ is a discrete state, $|k\rangle$ is one of the continuous states, and $E_k > E_m$, so that only the second term of Eqn. A.14 is important, the first-order probability of finding the system in the state k after the perturbation is removed is

$$\left| a_k^{(1)}(t \geq t_0) \right|^2 = 4|\langle k|H^{'}(0)|m\rangle|^2 \frac{\sin^2 \left[\frac{1}{2}(\omega_{km} - \omega)t_0\right]}{\hbar^2(\omega_{km} - \omega)^2} \tag{A.15}$$

The probability function has an oscillating behavior, as shown in Fig. A.1. The probability is maximum when $\omega_{km} = \omega$ and the peak is proportional to t_0^2. However, the uncertainty in frequency $\Delta\omega = \omega_{km} - \omega$ is non-zero if the time t_0 over which the perturbation is applied is finite. This uncertainty is in accordance with the Heisenberg uncertainty principle

$$\Delta\omega \; \Delta t = \Delta\omega \; t_0 \sim 1 \tag{A.16}$$

If the perturbation extends over a long time, the function plotted in Fig. A.2 approaches the Dirac δ-function and the probability is non-zero only for $E_k = E_m + \hbar\omega$.

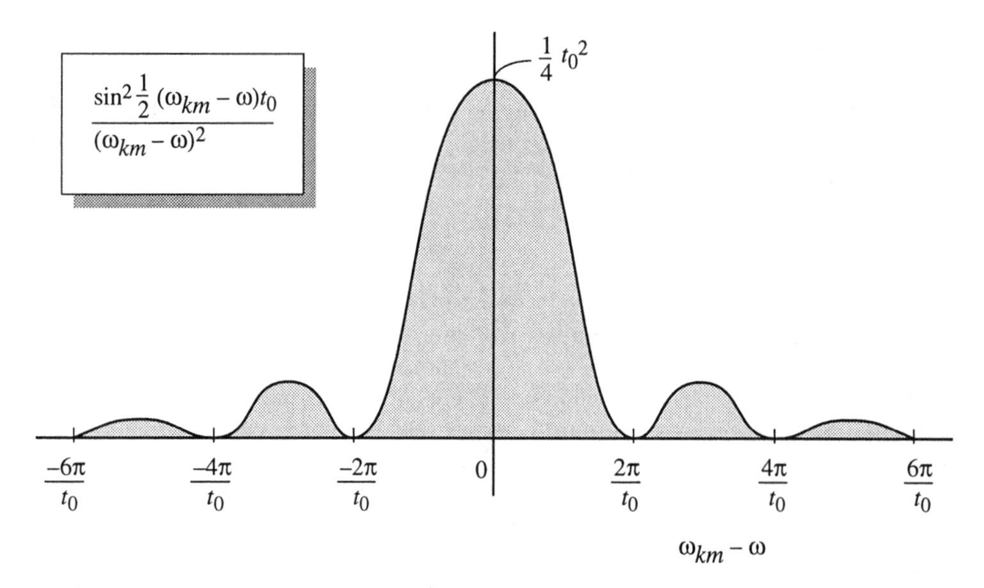

Figure A.1: The ordinate is proportional to the probability of finding the system in a state k after the perturbation has been applied to time t_0.

Transition Probability

An important class of problems falls in the category where either the perturbing potential or the unperturbed states are described by continuous spectra. For example, as shown in Fig. A.2a, the perturbation may have a spread of frequencies as in the case of electromagnetic radiation with a finite frequency spread. Or the states $|k\rangle$ or $|m\rangle$ may be in a continuum. In either case, this leads to a spread in the allowed values of $(\omega_{km} - \omega)$. In such cases, it is possible to define a scattering rate per unit time. We can see from Fig. A.2b that the probability of finding the system in a state $|k\rangle$ has a shape where the peak is proportional to t_0^2, while the width of the main peak decreases inversely as t_0. Thus the area under the curve is proportional to t_0. Thus, if we were to define the probability of finding the system anywhere in a spread of states covering the width of the main peak, the total probability will be proportional to t_0. This would allow us to define a transition rate, i.e., transition probability per unit time. The total rate per unit time for scattering into any final state, is given by

$$W_m = \frac{1}{t_0} \sum_{\text{final states}} \left| a_k^{(1)}(t \geq t_0) \right|^2$$

If t_0 is large, the sum over the final states includes only the final states where $\omega_{km} - \hbar\omega \cong 0$.

In summing over the final states, we can use the concept of density of states, which gives us the number of states per unit volume per unit energy:

$$W_m = \frac{1}{t_0} \int | a_k^{(1)}(t \geq t_0) |^2 \rho(k)\, dE_k \tag{A.17}$$

where $\rho(k)$ is the density of states near the final state.

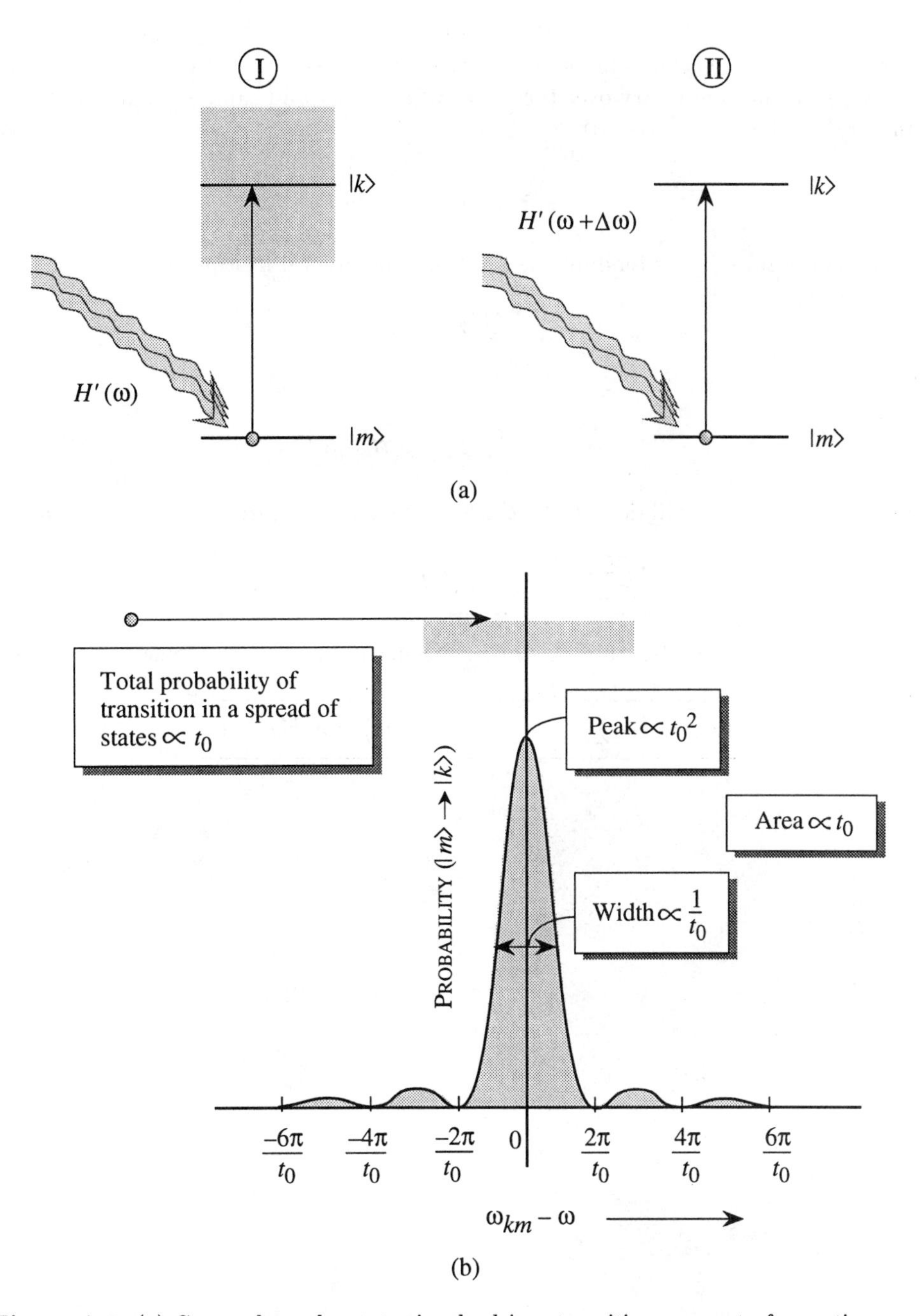

Figure A.2: (a) Cases where the states involved in a transition are part of a continuum. Case I shows the case where the unperturbed state is in a continuum. In case II, the perturbation has a continuum of frequencies. (b) A schematic of transitions in continuous spectra. The total transition probability in the continuum is proportional to the area under the curve.

From Fig. A.2b we see that the peak of the probability function becomes very narrow as t_0 becomes large. As a result, we can assume that the matrix element $\langle k|H'(0)|m\rangle$ does not vary over the width of the peak and can be taken outside the integral in Eqn. A.17. We write

$$x = \frac{1}{2}(\omega_{km} - \omega)\, t_0$$

and use the integral (extending the limits of the integral in Eqn. A.17 to $\pm\infty$)

$$\int_{-\infty}^{\infty} x^{-2} \sin^2 x = \pi \tag{A.18}$$

to get

$$W_m = \frac{2\pi}{\hbar}\rho(k) \mid \langle k|H^{'}|m\rangle \mid^2$$

or equivalently (we will denote the time-independent amplitude $H^{'}(0)$ by $H^{'}$)

$$\boxed{W_m = \frac{2\pi}{\hbar} \sum_{\text{final states}} \mid \langle k|H^{'}|m\rangle \mid^2 \delta\left(\hbar\omega_{km} - \hbar\omega\right)} \tag{A.19}$$

This is the Fermi golden rule. A similar calculation for the case where $\omega_{km} = -\omega$ gives

$$\boxed{W_m = \frac{2\pi}{\hbar} \sum_{\text{final states}} \mid \langle k|H^{'}|m\rangle \mid^2 \delta\left(\hbar\omega_{km} + \hbar\omega\right)} \tag{A.20}$$

APPENDIX B

BOLTZMANN TRANSPORT THEORY

Transport of electrons in solids is the basis of many modern technologies. The Boltzmann transport theory allows us to develop a microscopic model for macroscopic quantities such as mobility, diffusion coefficient, and conductivity. This theory has been used in Chapter 8 to study transport of electrons and holes in materials. In this appendix we will present a derivation of this theory.

B.1 BOLTZMANN TRANSPORT EQUATION

In order to describe the transport properties of an electron gas, we need to know the distribution function of the electron gas. The distribution would tell us how electrons are distributed in momentum space or k-space (and energy-space) and from this information all of the transport properties can be evaluated. We know that at equilibrium the distribution function is simply the Fermi-Dirac function

$$f(E) = \frac{1}{\exp\left(\frac{E - E_F}{k_B T}\right) + 1} \tag{B.1}$$

This distribution function describes the equilibrium electron gas and is *independent* of any collisions that may be present. While the collisions will continuously remove electrons from one $\boldsymbol{k}$-state to another, the net distribution of electrons is always given by the Fermi-Dirac function as long as there are no external influences to disturb the equilibrium.

To describe the distribution function in the presence of external forces, we develop the Boltzmann transport equation. Let us denote by $f_{\mathbf{k}}(\boldsymbol{r})$ the local concentration of the electrons in state $\boldsymbol{k}$ in the neighborhood of $\boldsymbol{r}$. The Boltzmann approach begins with an attempt to determine how $f_{\mathbf{k}}(\boldsymbol{r})$ changes with time. Three possible reasons account for the change in the electron distribution in k-space and r-space:

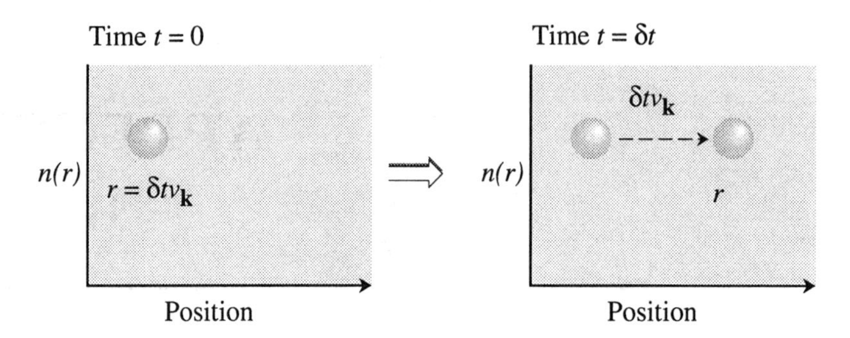

Figure B.1: At time $t = 0$ particles at position $r - \delta t v_{\mathbf{k}}$ reach the position r at a later time δt. This simple concept is important in establishing the Boltzmann transport equation.

1. Due to the motion of the electrons (diffusion), carriers will be moving into and out of any volume element around $\boldsymbol{r}$.
2. Due to the influence of external forces, electrons will be changing their momentum (or $\boldsymbol{k}$-value) according to $\hbar\, d\boldsymbol{k}/dt = \boldsymbol{F}_{ext}$.
3. Due to scattering processes, electrons will move from one $\boldsymbol{k}$-state to another.

We will now calculate these three individual changes by evaluating the partial time derivative of the function $f_{\mathbf{k}}(\boldsymbol{r})$ due to each source.

B.1.1 Diffusion-Induced Evolution of $f_{\mathbf{k}}(\boldsymbol{r})$

If $\boldsymbol{v}_{\mathbf{k}}$ is the velocity of a carrier in the state $\boldsymbol{k}$, in a time interval t, the electron moves a distance $t\ \boldsymbol{v}_{\mathbf{k}}$. Thus the number of electrons in the neighborhood of $\boldsymbol{r}$ at time δt is equal to the number of carriers in the neighborhood of $\boldsymbol{r} - \delta t\ \boldsymbol{v}_{\mathbf{k}}$ at time 0, as shown in Fig. B.1

We can thus define the following equality due to the diffusion

$$f_{\mathbf{k}}(\boldsymbol{r}, \delta t) = f_{\mathbf{k}}(\boldsymbol{r} - \delta t\ \boldsymbol{v}_{\mathbf{k}}, 0) \tag{B.2}$$

or

$$\begin{aligned} f_{\mathbf{k}}(\boldsymbol{r}, 0) + \frac{\partial f_{\mathbf{k}}}{\partial t} \cdot \delta t &= f_{\mathbf{k}}(\boldsymbol{r}, 0) - \frac{\partial f_{\mathbf{k}}}{\partial \boldsymbol{r}} \cdot \delta t\ \boldsymbol{v}_{\mathbf{k}} \\ \left.\frac{\partial f_{\mathbf{k}}}{\partial t}\right|_{\text{diff}} &= -\frac{\partial f_{\mathbf{k}}}{\partial \boldsymbol{r}} \cdot \boldsymbol{v}_{\mathbf{k}} \end{aligned} \tag{B.3}$$

B.1.2 External Field-Induced Evolution of $f_{\mathbf{k}}(\boldsymbol{r})$

The crystal momentum $\boldsymbol{k}$ of the electron evolves under the action of external forces according to Newton's equation of motion. For an electric and magnetic field ($\boldsymbol{E}$ and $\boldsymbol{B}$), the rate of change of $\boldsymbol{k}$ is given by

$$\dot{\boldsymbol{k}} = \frac{e}{\hbar}\left[\boldsymbol{E} + \boldsymbol{v}_{\mathbf{k}} \times \boldsymbol{B}\right] \tag{B.4}$$

In analogy to the diffusion-induced changes, we can argue that particles at time $t = 0$ with momentum $\boldsymbol{k} - \dot{\boldsymbol{k}}\,\delta t$ will have momentum $\boldsymbol{k}$ at time δt and

$$f_{\mathbf{k}}(\boldsymbol{r}, \delta t) = f_{\mathbf{k}-\dot{\mathbf{k}}\delta t}(\boldsymbol{r}, 0) \tag{B.5}$$

which leads to the equation

$$\begin{aligned} \left.\frac{\partial f_{\mathbf{k}}}{\partial t}\right|_{\text{ext. forces}} &= -\dot{\boldsymbol{k}}\frac{\partial f_{\mathbf{k}}}{\partial \boldsymbol{k}} \\ &= \frac{-e}{\hbar}\left[\boldsymbol{E} + \frac{\boldsymbol{v} \times \boldsymbol{B}}{c}\right] \cdot \frac{\partial f_{\mathbf{k}}}{\partial \boldsymbol{k}} \end{aligned} \tag{B.6}$$

B.1.3 Scattering-Induced Evolution of $f_{\mathbf{k}}(\boldsymbol{r})$

We will assume that the scattering processes are *local* and *instantaneous* and change the state of the electron from $\boldsymbol{k}$ to $\boldsymbol{k}'$. Let $W(\boldsymbol{k}, \boldsymbol{k}')$ define the rate of scattering from the state $\boldsymbol{k}$ to $\boldsymbol{k}'$ if the state $\boldsymbol{k}$ is occupied and $\boldsymbol{k}'$ is empty. The rate of change of the distribution function $f_{\mathbf{k}}(\boldsymbol{r})$ due to scattering is

$$\left.\frac{\partial f_{\mathbf{k}}}{\partial t}\right)_{\text{scattering}} = \int \left[f_{\mathbf{k}'}\left(1 - f_{\mathbf{k}}\right) W(\boldsymbol{k}', \boldsymbol{k}) - f_{\mathbf{k}}\left(1 - f_{\mathbf{k}'}\right) W(\boldsymbol{k}, \boldsymbol{k}')\right] \frac{d^3k'}{(2\pi)^3} \tag{B.7}$$

The $(2\pi)^3$ in the denominator comes from the number of states allowed in a k-space volume d^3k'. The first term in the integral represents the rate at which electrons are coming from an occupied $\boldsymbol{k}'$ state (hence the factor $f_{\mathbf{k}'}$) to an unoccupied $\boldsymbol{k}$-state (hence the factor $(1 - f_{\mathbf{k}})$). The second term represents the loss term.

Under steady-state conditions, there will be no net change in the distribution function and the total sum of the partial derivative terms calculated above will be zero.

$$\left.\frac{\partial f_{\mathbf{k}}}{\partial t}\right)_{\text{scattering}} + \left.\frac{\partial f_{\mathbf{k}}}{\partial t}\right)_{\text{fields}} + \left.\frac{\partial f_{\mathbf{k}}}{\partial t}\right)_{\text{diffusion}} = 0 \tag{B.8}$$

Let us define

$$g_{\mathbf{k}} = f_{\mathbf{k}} - f_{\mathbf{k}}^0 \tag{B.9}$$

where $f_{\mathbf{k}}^0$ is the equilibrium distribution.

We will attempt to calculate $g_{\mathbf{k}}$, which represents the deviation of the distribution function from the equilibrium case.

Substituting for the partial time derivatives due to diffusion and external fields we get

$$-\boldsymbol{v}_{\mathbf{k}} \cdot \nabla_r f_{\mathbf{k}} - \frac{e}{\hbar}\left(\boldsymbol{E} + \frac{\boldsymbol{v}_{\mathbf{k}} \times \boldsymbol{B}}{c}\right) \cdot \nabla_k f_{\mathbf{k}} = \left.\frac{-\partial f_{\mathbf{k}}}{\partial t}\right)_{\text{scattering}} \tag{B.10}$$

Substituting $f_{\mathbf{k}} = f_{\mathbf{k}}^0 + g_{\mathbf{k}}$

$$\begin{aligned} &-\boldsymbol{v}_{\mathbf{k}} \cdot \nabla_r f_{\mathbf{k}}^0 - \tfrac{e}{\hbar}\left(\boldsymbol{E} + \boldsymbol{v}_{\mathbf{k}} \times \boldsymbol{B}\right) \nabla_k f_{\mathbf{k}}^0 \\ &= -\left.\tfrac{\partial f_{\mathbf{k}}}{\partial t}\right)_{\text{scattering}} + \boldsymbol{v}_{\mathbf{k}} \cdot \nabla_r g_{\mathbf{k}} + \tfrac{e}{\hbar}\left(\boldsymbol{E} + \boldsymbol{v}_{\mathbf{k}} \times \boldsymbol{B}\right) \cdot \nabla_k g_{\mathbf{k}} \end{aligned} \tag{B.11}$$

We note that the magnetic force term on the left-hand side of Eqn. B.11 is proportional to

$$\boldsymbol{v}_{\mathbf{k}} \cdot \frac{e}{\hbar} (\boldsymbol{v}_{\mathbf{k}} \times \boldsymbol{B})$$

and is thus zero. We remind ourselves that (the reader should be careful not to confuse $E_{\mathbf{k}}$, the particle energy and $\boldsymbol{E}$, the electric field)

$$\boldsymbol{v}_{\mathbf{k}} = \frac{1}{\hbar} \frac{\partial E_k}{\partial \boldsymbol{k}} \tag{B.12}$$

and (in semiconductor physics, we often denote μ by E_F)

$$f_{\mathbf{k}}^0 = \frac{1}{\exp\left[\frac{E_{\mathbf{k}} - \mu}{k_B T}\right] + 1} \tag{B.13}$$

Thus

$$\begin{aligned}
\nabla_r f^0 &= \frac{-\left[\exp\left(\frac{E_{\mathbf{k}} - \mu}{k_B T}\right)\right]}{\left[\exp\left(\frac{E_{\mathbf{k}} - \mu}{k_B T}\right) + 1\right]^2} \nabla_r \left(\frac{E_{\mathbf{k}} - \mu(\boldsymbol{r})}{k_B T(r)}\right) \\
&= k_B T \cdot \frac{\partial f^0}{\partial E_{\mathbf{k}}} \left[-\frac{\nabla \mu}{k_B T} - \frac{(E_{\mathbf{k}} - \mu)}{k_B T^2} \nabla T\right]
\end{aligned}$$

$$\nabla_r f^0 = \frac{\partial f^0}{\partial E_{\mathbf{k}}} \left[-\nabla \mu - \frac{(E_{\mathbf{k}} - \mu)}{T} \nabla T\right] \tag{B.14}$$

Also

$$\begin{aligned}
\nabla_k f^0 &= \frac{\partial f^0}{\partial E_{\mathbf{k}}} \cdot \nabla_k E_k \\
&= \hbar \boldsymbol{v}_{\mathbf{k}} \frac{\partial f^0}{\partial E_{\mathbf{k}}}
\end{aligned} \tag{B.15}$$

Substituting these terms and retaining terms only to second-order in electric field (i.e., ignoring terms involving products $g_{\mathbf{k}} \cdot \boldsymbol{E}$), we get, from Eqn. B.11,

$$\begin{aligned}
&-\frac{\partial f^0}{\partial E_{\mathbf{k}}} \cdot \boldsymbol{v}_{\mathbf{k}} \cdot \left[-\frac{(E_{\mathbf{k}} - \mu)}{T} \nabla T + e\boldsymbol{E} - \nabla \mu\right] \\
&= -\left.\frac{\partial f}{\partial t}\right)_{\text{scattering}} + \boldsymbol{v}_{\mathbf{k}} \cdot \nabla_r g_{\mathbf{k}} + \frac{e}{\hbar} (\boldsymbol{v}_{\mathbf{k}} \times \boldsymbol{B}) \cdot \nabla_k g_{\mathbf{k}}.
\end{aligned} \tag{B.16}$$

The equation derived above is the Boltzmann transport equation.

We will now apply the Boltzmann equation to derive some simple expressions for conductivity, mobility, etc., in semiconductors. We will attempt to relate the microscopic scattering events to the measurable macroscopic transport properties. Let us consider the case where we have a uniform electric field $\boldsymbol{E}$ in an infinite system maintained at a uniform temperature.

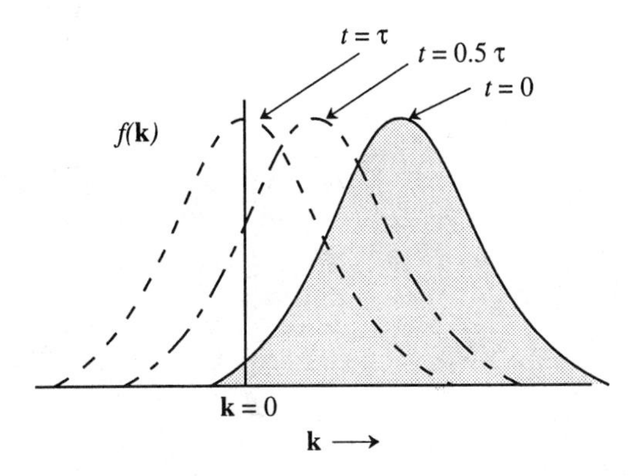

Figure B.2: This figure shows that at time $t = 0$, the distribution function is distorted by some external means. If the external force is removed, the electrons recover to the equilibrium distribution by collisions.

The Boltzmann equation becomes

$$-\frac{\partial f^0}{\partial E_{\mathbf{k}}} \boldsymbol{v}_{\mathbf{k}} \cdot e\boldsymbol{E} = -\left.\frac{\partial f_{\mathbf{k}}}{\partial t}\right)_{\text{scattering}} \tag{B.17}$$

Note that only the deviation $g_{\mathbf{k}}$ from the equilibrium distribution function above contributes to the scattering integral.

As mentioned earlier, this equation, although it looks simple, is a very complex equation which can only be solved analytically under fairly simplifying assumptions. We make an assumption that the scattering induced change in the distribution function is given by

$$-\left.\frac{\partial f_{\mathbf{k}}}{\partial t}\right)_{\text{scattering}} = \frac{g_{\mathbf{k}}}{\tau} \tag{B.18}$$

We have introduced a time constant τ whose physical interpretation can be understood when we consider what happens when the external forces have been removed. In this case the perturbation in the distribution function will decay according to the equation

$$\frac{-\partial g_{\mathbf{k}}}{\partial t} = \frac{g_{\mathbf{k}}}{\tau}$$

or

$$g_{\mathbf{k}}(t) = g_{\mathbf{k}}(0)e^{-t/\tau} \tag{B.19}$$

The time τ thus represents the time constant for relaxation of the perturbation as shown schematically in Fig. B.2. The approximation which allows us to write such a simple relation is called the relaxation time approximation (RTA).

According to this approximation

$$\begin{aligned} g_{\mathbf{k}} &= -\left.\frac{\partial f_{\mathbf{k}}}{\partial t}\right)_{\text{scattering}} \cdot \tau \\ &= \frac{-\partial f^0}{\partial E_{\mathbf{k}}} \tau \boldsymbol{v}_{\mathbf{k}} \cdot e\boldsymbol{E} \end{aligned} \tag{B.20}$$

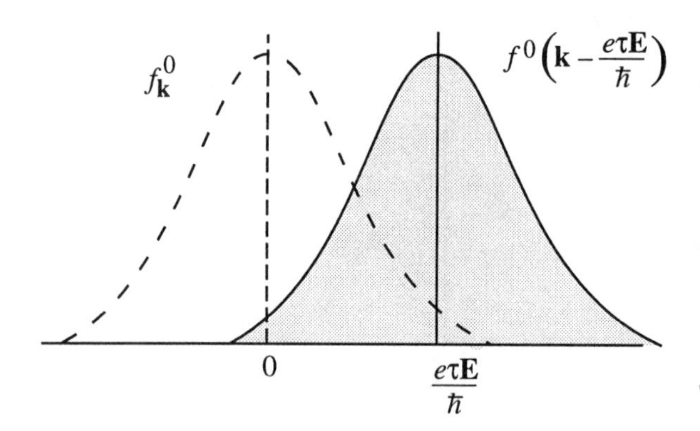

Figure B.3: The displaced distribution function shows the effect of an applied electric field.

Note that we have not defined how τ is to be calculated. We have merely introduced a simpler unknown that still needs to be determined. The k-space distribution function may be written as

$$\begin{aligned} f_{\mathbf{k}} &= f_{\mathbf{k}}^0 - \left(\frac{\partial f_{\mathbf{k}}^0}{\partial E_{\mathbf{k}}}\right) e\tau \boldsymbol{v}_{\mathbf{k}} \cdot \boldsymbol{E} \\ &= f_{\mathbf{k}}^0 - \left(\nabla_k f_{\mathbf{k}}^0\right) \cdot \frac{\partial \boldsymbol{k}}{\partial E_{\mathbf{k}}} \cdot e\tau \boldsymbol{v}_{\mathbf{k}} \cdot \boldsymbol{E} \end{aligned}$$

Using the relation

$$\hbar \frac{\partial \boldsymbol{k}}{\partial E_{\mathbf{k}}} \cdot \boldsymbol{v}_{\mathbf{k}} = 1$$

We have

$$\begin{aligned} f_{\mathbf{k}} &= f_{\mathbf{k}}^0 - \left(\nabla_{\mathbf{k}} f_{\mathbf{k}}^0\right) \cdot \frac{e\tau \boldsymbol{E}}{\hbar} \\ &= f_{\mathbf{k}}^0 \left(\boldsymbol{k} - \frac{e\tau \boldsymbol{E}}{\hbar}\right) \end{aligned} \tag{B.21}$$

This is a very useful result which allows us to calculate the non-equilibrium function $f_{\mathbf{k}}$ in terms of the equilibrium function f^0. The recipe is very simple—shift the original distribution function for $\boldsymbol{k}$ values parallel to the electric field by $e\tau\boldsymbol{E}/\hbar$. If the field is along the z-direction, only the distribution for k_z will shift. This is shown schematically in Fig. B.3. Note that for the equilibrium distribution function, there is an exact cancellation between positive velocities and negative velocities. When the field is applied, there is a net shift in the electron momenta and velocities given by

$$\begin{aligned} \delta \boldsymbol{p} &= \hbar \delta \boldsymbol{k} = -e\tau \boldsymbol{E} \\ \delta \boldsymbol{v} &= -\frac{e\tau \boldsymbol{E}}{m^*} \end{aligned} \tag{B.22}$$

This gives, for the mobility,

$$\mu = \frac{e\tau}{m^*} \tag{B.23}$$

If the electron concentration is n, the current density is

$$\begin{aligned} \boldsymbol{J} &= ne\delta \boldsymbol{v} \\ &= \frac{ne^2\tau \boldsymbol{E}}{m^*} \end{aligned}$$

or the conductivity of the system is

$$\sigma = \frac{ne^2\tau}{m^*} \tag{B.24}$$

This equation relates a microscopic quantity τ to a macroscopic quantity σ.

So far we have introduced the relaxation time τ, but not described how it is to be calculated. We will now relate it to the scattering rate $W(\boldsymbol{k}, \boldsymbol{k}^{'})$, which can be calculated by using the Fermi golden rule. We have, for the scattering integral,

$$\left.\frac{\partial f}{\partial t}\right)_{\text{scattering}} = \int \left[f(\boldsymbol{k}^{'})(1 - f(\boldsymbol{k}))W(\boldsymbol{k}^{'}, \boldsymbol{k}) - f(\boldsymbol{k})(1 - f(\boldsymbol{k}^{'}))W(\boldsymbol{k}, \boldsymbol{k}^{'})\right] \frac{d^3\boldsymbol{k}^{'}}{(2\pi)^3}$$

Let us examine some simple cases where the integral on the right-hand side becomes simplified.

Elastic Collisions

Elastic collisions represent scattering events in which the energy of the electrons remains unchanged after the collision. Impurity scattering and alloy scattering discussed in Chapter 8 fall into this category. In the case of elastic scattering the principle of microscopic reversibility ensures that

$$W(\boldsymbol{k}, \boldsymbol{k}^{'}) = W(\boldsymbol{k}^{'}, \boldsymbol{k}) \tag{B.25}$$

i.e., the scattering rate from an initial state $\boldsymbol{k}$ to a final state $\boldsymbol{k}^{'}$ is the same as that for the reverse process. The collision integral is now simplified as

$$\begin{aligned} \left.\frac{\partial f}{\partial t}\right)_{\text{scattering}} &= \int \left[f(\boldsymbol{k}^{'}) - f(\boldsymbol{k})\right] W(\boldsymbol{k}, \boldsymbol{k}^{'}) \frac{d^3\boldsymbol{k}^{'}}{(2\pi)^3} \\ &= \int \left[g(\boldsymbol{k}^{'}) - g(\boldsymbol{k})\right] W(\boldsymbol{k}, \boldsymbol{k}^{'}) \frac{d^3\boldsymbol{k}^{'}}{(2\pi)^3} \end{aligned} \tag{B.26}$$

The simple form of the Boltzmann equation is (from Eqn. B.17)

$$\begin{aligned} \frac{-\partial f^0}{\partial E_{\mathbf{k}}} \boldsymbol{v}_{\mathbf{k}} \cdot e\boldsymbol{E} &= \int \left(g_{\mathbf{k}} - g_{\mathbf{k}'}\right) \, W(\boldsymbol{k}, \boldsymbol{k}^{'}) d^3k^{'} \\ &= \left.\frac{-\partial f}{\partial t}\right)_{\text{scattering}} \end{aligned} \tag{B.27}$$

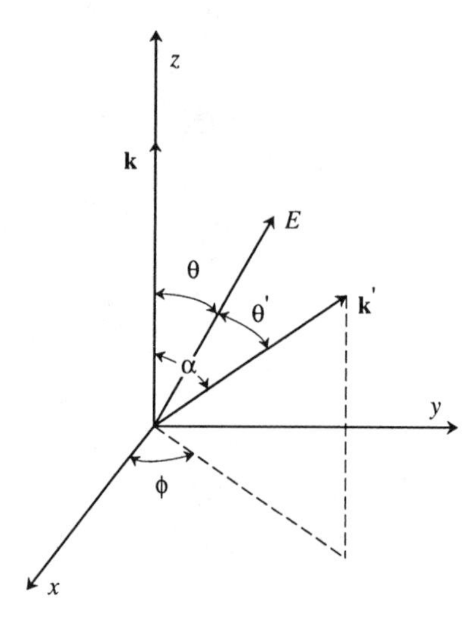

Figure B.4: Coordinate system illustrating a scattering event.

The relaxation time was defined through

$$\begin{aligned} g_{\mathbf{k}} &= \left(\frac{-\partial f^0}{\partial E}\right) e\boldsymbol{E} \cdot \boldsymbol{v}_{\mathbf{k}} \cdot \tau \\ &= \left.\frac{-\partial f}{\partial t}\right)_{\text{scattering}} \cdot \tau \end{aligned} \tag{B.28}$$

Substituting this value in the integral on the right-hand side, we get

$$\frac{-\partial f^0}{\partial E_{\mathbf{k}}} \boldsymbol{v}_{\mathbf{k}} \cdot e\boldsymbol{E} = \frac{-\partial f^0}{\partial E_{\mathbf{k}}} e\tau \boldsymbol{E} \cdot \int (\boldsymbol{v}_{\mathbf{k}} - \boldsymbol{v}_{\mathbf{k}'})\, W(\boldsymbol{k}, \boldsymbol{k}')\, d^3k' \tag{B.29}$$

or

$$\boldsymbol{v}_{\mathbf{k}} \cdot \boldsymbol{E} = \tau \int (\boldsymbol{v}_{\mathbf{k}} - \boldsymbol{v}_{\mathbf{k}'})\, W(\boldsymbol{k}, \boldsymbol{k}')\, d^3k' \cdot \boldsymbol{E} \tag{B.30}$$

and

$$\frac{1}{\tau} = \int W(\boldsymbol{k}, \boldsymbol{k}') \left[1 - \frac{\boldsymbol{v}_{\mathbf{k}'} \cdot \boldsymbol{E}}{\boldsymbol{v}_{\mathbf{k}} \cdot \boldsymbol{E}}\right] d^3k' \tag{B.31}$$

In general, this is a rather complex integral to solve. However, it becomes considerably simplified for certain simple cases. Consider, for example, the case of isotropic parabolic bands and elastic scattering. In Fig. B.4 we show a geometry for the scattering process. We choose a coordinate axis where the initial momentum is along the z-axis and the applied electric field is in the y-z plane. The wavevector after scattering is given by $\boldsymbol{k}'$ represented by the angles α and ϕ. Assuming that the energy bands of the material is isotropic, $|\boldsymbol{v}_{\mathbf{k}}| = |\boldsymbol{v}_{\mathbf{k}'}|$. We thus get

$$\frac{\boldsymbol{v}_{\mathbf{k}'} \cdot \boldsymbol{E}}{\boldsymbol{v}_{\mathbf{k}} \cdot \boldsymbol{E}} = \frac{\cos\theta'}{\cos\theta} \tag{B.32}$$

We can easily see from Fig. B.4 that

$$\cos\theta' = \sin\theta\sin\alpha\sin\phi + \cos\theta\cos\alpha$$

or

$$\frac{\cos\theta'}{\cos\theta} = \tan\theta\sin\alpha\sin\phi + \cos\alpha$$

When this term is integrated over ϕ to evaluate τ, the term involving $\sin\phi$ will integrate to zero for isotropic bands since $W(\boldsymbol{k},\boldsymbol{k}')$ does not have a ϕ dependence, only an α dependence. Thus

$$\frac{1}{\tau} = \int W(\boldsymbol{k},\boldsymbol{k}')\,(1-\cos\alpha)\,d^3k' \tag{B.33}$$

This weighting factor $(1-\cos\alpha)$ confirms the intuitively apparent fact that large-angle scatterings are much more important in determining transport properties than small-angle scatterings. Forward-angle scatterings ($\alpha = 0$), in particular, have no detrimental effect on σ or μ for the case of elastic scattering.

Inelastic Collisions

In the case of inelastic scattering processes, we cannot assume that $W(\boldsymbol{k},\boldsymbol{k}') = W(\boldsymbol{k}',\boldsymbol{k})$. As a result, the collision integral cannot be simplified to give an analytic result for the relaxation time. If, however, the system is non-degenerate, i.e., $f(E)$ is small, we can ignore second-order terms in f and we have

$$\left.\frac{\partial f}{\partial t}\right|_{\text{scattering}} = \int \left[g_{\mathbf{k}'}W(\boldsymbol{k}',\boldsymbol{k}) - g_{\mathbf{k}}W(\boldsymbol{k},\boldsymbol{k}')\right]\frac{d^3\boldsymbol{k}'}{(2\pi)^3} \tag{B.34}$$

Under equilibrium we have

$$f^0_{\mathbf{k}'}W(\boldsymbol{k}',\boldsymbol{k}) = f^0_{\mathbf{k}}W(\boldsymbol{k},\boldsymbol{k}') \tag{B.35}$$

or

$$W(\boldsymbol{k}',\boldsymbol{k}) = \frac{f^0_{\mathbf{k}}}{f^0_{\mathbf{k}'}}W(\boldsymbol{k},\boldsymbol{k}') \tag{B.36}$$

Assuming that this relation holds for scattering rates in the presence of the applied field, we have

$$\left.\frac{\partial f}{\partial t}\right|_{\text{scattering}} = \int W(\boldsymbol{k},\boldsymbol{k}')\left[g_{\mathbf{k}'}\frac{f^0_{\mathbf{k}}}{f^0_{\mathbf{k}'}} - g_{\mathbf{k}}\right]\frac{d^3\boldsymbol{k}'}{(2\pi)^3} \tag{B.37}$$

The relaxation time then becomes

$$\frac{1}{\tau} = \int W(\boldsymbol{k},\boldsymbol{k}')\left[1 - \frac{g_{\mathbf{k}'}}{g_{\mathbf{k}}}\frac{f^0_{\mathbf{k}}}{f^0_{\mathbf{k}'}}\right]\frac{d^3\boldsymbol{k}'}{(2\pi)^3} \tag{B.38}$$

The Boltzmann is usually solved iteratively using numerical techniques.

B.2 AVERAGING PROCEDURES

We have so far assumed that the incident electron is on a well-defined state. In a realistic system the electron gas will have an energy distribution and τ, in general, will depend upon the energy of the electron. Thus it is important to address the appropriate averaging procedure for τ. We will now do so under the assumptions that the drift velocity due to the electric field is much smaller than the average thermal speeds so that the energy of the electron gas is still given by $3k_BT/2$.

Let us evaluate the average current in the system.

$$\boldsymbol{J} = \int e\, \boldsymbol{v}_{\mathbf{k}}\, g_{\mathbf{k}}\, \frac{d^3k}{(2\pi)^3} \tag{B.39}$$

The perturbation in the distribution function is

$$\begin{aligned} g_{\mathbf{k}} &= \frac{-\partial f^0}{\partial E_{\mathbf{k}}} \tau \boldsymbol{v}_{\mathbf{k}} \cdot e\boldsymbol{E} \\ &\approx \frac{f^0}{k_BT}\, \boldsymbol{v}_{\mathbf{k}} \cdot e\boldsymbol{E} \end{aligned} \tag{B.40}$$

If we consider a field in the x-direction, the average current in the x-direction is from Eqns. B.39 and B.40

$$\langle J_x \rangle = \frac{e^2}{k_BT} \int \tau\, v_x^2\, f^0\, \frac{d^3k}{(2\pi)^3}\, E_x \tag{B.41}$$

The assumption made on the drift velocity ensures that $v_x^2 = v^2/3$, where $\boldsymbol{v}$ is the total velocity of the electron. Thus we get

$$\langle J_x \rangle = \frac{e^2}{3k_BT} \int \tau\, v^2\, f^0(\boldsymbol{k})\, \frac{d^3k}{(2\pi)^3}\, E_x \tag{B.42}$$

Now we note that

$$\begin{aligned} \frac{1}{2} m^* \langle v^2 \rangle &= \frac{3}{2} k_BT \\ \Rightarrow k_BT &= m^* \langle v^2 \rangle / 3 \end{aligned}$$

also

$$\begin{aligned} \langle v^2\, \tau \rangle &= \frac{\int v^2\, \tau\, f^0(\boldsymbol{k})\, d^3k/(2\pi)^3}{\int f^0(\boldsymbol{k})\, d^3k/(2\pi)^3} \\ &= \frac{\int v^2\, \tau\, f^0(\boldsymbol{k})\, d^3k/(2\pi)^3}{n} \end{aligned} \tag{B.43}$$

Substituting in the right-hand side of Eqn. B.42, we get (using $3k_BT = m\langle v^2 \rangle$)

$$\begin{aligned} \langle J_x \rangle &= \frac{ne^2}{m^*} \frac{\langle v^2 \tau \rangle}{\langle v^2 \rangle} E_x \\ &= \frac{ne^2}{m^*} \frac{\langle E\tau \rangle}{\langle E \rangle} E_x \end{aligned} \tag{B.44}$$

Thus, for the purpose of transport, the proper averaging for the relaxation time is

$$\langle\langle \tau \rangle\rangle = \frac{\langle E\tau \rangle}{\langle E \rangle} \tag{B.45}$$

Here the double brackets represent an averaging with respect to the perturbed distribution function while the single brackets represent averaging with the equilibrium distribution function.

For calculations of low-field transport where the condition $v_x^2 = v^2/3$ is valid, one has to use the averaging procedure given by Eqn. B.45 to calculate mobility or conductivity of the semiconductors. For most scattering processes, one finds that it is possible to express the energy dependence of the relaxation time in the form

$$\tau(E) = \tau_0 (E/k_B T)^s \tag{B.46}$$

where τ_0 is a constant and s is an exponent which is characteristic of the scattering process. We will be calculating this energy dependence for various scattering processes in the next two chapters. When this form is used in the averaging of Eqn. B.45, we get, using a Boltzmann distribution for $f^0(\boldsymbol{k})$

$$\langle\langle \tau \rangle\rangle = \tau_0 \frac{\int_0^\infty [p^2/(2m^* k_B T)]^s \, \exp[-p^2/(2m^* k_B T)] \, p^4 \, dp}{\int_0^\infty \exp[-p^2/(2m^* k_B T)] \, p^4 \, dp} \tag{B.47}$$

where $\boldsymbol{p} = \hbar \boldsymbol{k}$ is the momentum of the electron.

Substituting $y = p^2/(2m^* k_B T)$, we get

$$\langle\langle \tau \rangle\rangle = \tau_0 \frac{\int_0^\infty y^{s+(3/2)} e^{-y} \, dy}{\int_0^\infty y^{3/2} e^{-y} \, dy} \tag{B.48}$$

To evaluate this integral, we use Γ-functions which have the properties

$$\begin{aligned} \Gamma(n) &= (n-1)! \\ \Gamma(1/2) &= \sqrt{\pi} \\ \Gamma(n+1) &= n \, \Gamma(n) \end{aligned} \tag{B.49}$$

and have the integral value

$$\Gamma(a) = \int_0^\infty y^{a-1} e^{-y} \, dy \tag{B.50}$$

In terms of the Γ-functions we can then write

$$\langle\langle \tau \rangle\rangle = \tau_0 \frac{\Gamma(s+5/2)}{\Gamma(5/2)} \tag{B.51}$$

If a number of different scattering processes are participating in transport, the following approximate rule (Mathiesen's rule) may be used to calculate mobility:

$$\frac{1}{\tau_{tot}} = \sum_i \frac{1}{\tau_i} \tag{B.52}$$

$$\frac{1}{\mu_{tot}} = \sum_i \frac{1}{\mu_i} \tag{B.53}$$

where the sum is over all different scattering processes.

APPENDIX C

QUANTUM INTERFERENCE DEVICES

C.1 INTRODUCTION

In quantum mechanics we see that classical particles behave as waves. Is it possible for these particle-waves to display phenomena that "normal waves" such as light waves display? For example, can particle waves interfere and diffract? The answer is yes and in Chapter 4, Section 4.4.8 we briefly discussed how particle waves can be used for determining crystal structures. In this appendix we will address the following two questions: i) under what conditions can electrons in solids display interference behavior? ii) How can we exploit particle wave interference to design information-processing devices?

C.2 PARTICLE-WAVE COHERENCE

In Chapter 4 we have seen that in a perfectly periodic potential the electron wavefunction has the form

$$\psi_k(r) = u_k(r)e^{ik \cdot r} \tag{C.1}$$

In this perfect structure the electron maintains its phase coherence as it propagates in the structure. However, in a real material, we see from our discussions of Chapter 8 that electrons scatter from a variety of sources. As a result, the particle wave's coherence is lost after a distance of a mean free path. If v is the average speed of the electron and τ the time interval between scattering, the particle loses its phase coherence in a distance

$$\lambda \sim v\tau \tag{C.2}$$

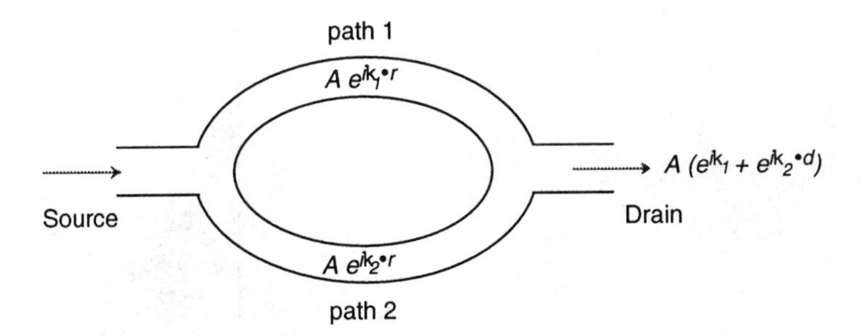

Figure C.1: A schematic of a coherent electron beam traveling along two paths and interfering.

In the absence of phase coherence, effects such as interference are not observable unless the dimensions of the material (or device) are smaller than the mean free path.

In high-quality semiconductors (the material of choice for most information-processing devices) the mean free path is $\sim$ 100 Å at room temperature and $\sim$ 1000 Å at liquid helium. It is possible to see quantum interference effects at very low temperatures in semiconductor devices. These effects can be exploited to design digital devices and switches operating at very low power levels. The general principle of operation is shown in Fig. C.1. Electron waves travel from a source to a drain via two paths. At the output the intensity of the electron wave is

$$I(d) = \mid \psi_1(d) + \psi_2(d)^2 \tag{C.3}$$

If the waves are described by

$$\begin{aligned} \psi_1(x) &= Ae^{ik_1x} \\ \psi_2(x) &= Ae^{ik_2x} \end{aligned} \tag{C.4}$$

we have

$$I(d) = 2A^2[1 - \cos(k_1 - k_2)d] \tag{C.5}$$

If we can now somehow alter the wavelengths of the electron (i.e., the value of $(k_1 - k_2)$) we can modulate the signal at the drain. This modulation can be done by i) using an electric bias to alter the kinetic energy of the electrons in one arm or ii) by introducing a magnetic field within the loop formed by the electron flow path and varying the magnetic field.

In quantum interference transistors, a gate bias is sued to alter the potential energy seen by the electrons. The electron k-vector is given by (E_c is the bandedge)

$$E = E_c + \frac{\hbar^2 k^2}{2m^*} \tag{C.6}$$

By changing the position of E_c, one can later the k-value. Thus one can develop quantum interference transistors. Unfortunately, these effects are only observable

Element	T_c(K)
Al	1.196
Cd	0.56
Ga	1.091
In	3.4
Pb	7.19
Tc	7.77

Table C.1: Values of the critical temperature T_c for several superconducting elements.

at very low temperatures. Material systems in which phase coherence can be maintained over long distances are superconductors. We have discussed in Chapter 4 how in these materials at low temperatures electrons pair up to form bosons. These pairs can move with perfect phase coherence and this feature is exploited in superconducting devices.

Many elements when cooled below a critical temperature T_c display superconductivity. In Table C.1 we show the critical temperature of several metals. We see that these "traditional" superconductors have a very low T_c value ranging from mK to $\sim$ 10 K. This low temperature makes the use of these superconductors quite expensive. As a result, one uses these materials only in applications where the benefits outweigh the cooling costs.

In the 1980s a new class of superconductors was discovered. These materials are called high T_c superconductors and many of them have T_c values well above liquid N_2 temperature. This makes them very attractive for novel device applications. Materials such as yttrium-barium-cuprous oxide have even been developed in thin-film form so that they can be used in ultra-small devices.

To understand the applications of superconducting devices in technology (particularly information-processing technology) it is important to examine how electron waves behave in magnetic fields.

C.3 FREE ELECTRONS IN MAGNETIC FIELDS

In Chapter 7 we have discussed how electrons in atomic systems are influenced by magnetic fields. Here we will examine electrons which are in perfect crystals or in free space and have a wavefunction given by (in the absence of a magnetic field)

$$\psi_k^0(r) = A\ e^{ik\cdot r} \tag{C.7}$$

For electrons in crystals $A = u_k(r)$, the cell periodic part as required by Bloch's theorem. We have discussed, in Chapter 4, that for the outside world, $\hbar\mathbf{k}$ acts as an effective momentum.

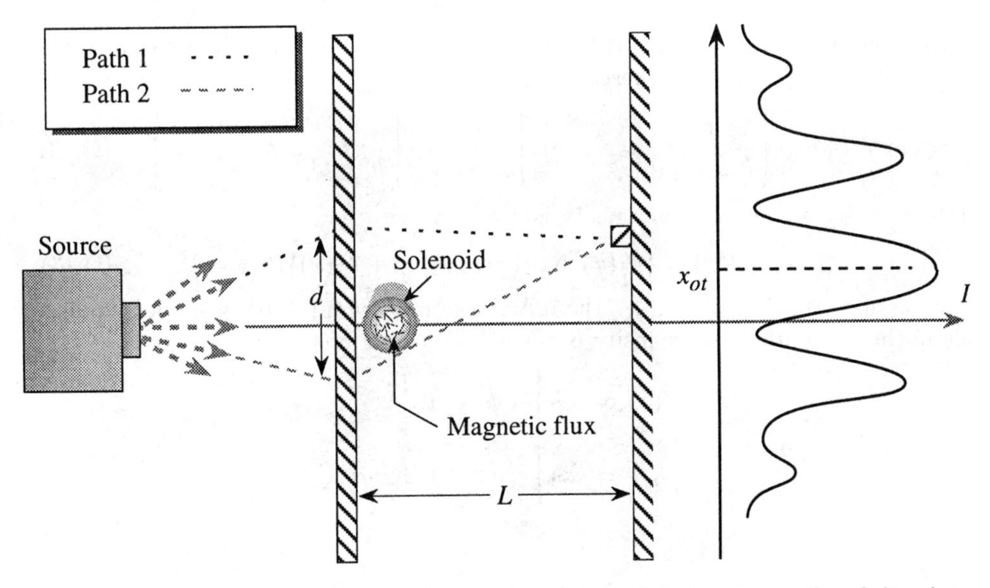

Figure C.2: A magnetic field can influence the motion of electrons even though it exists only in regions where there is an arbitrarily small probability of finding the electrons. The interference pattern of the electrons can be shifted by altering the magnetic field.

Consider the electron in the presence of a magnetic field $\boldsymbol{B}$ described by a vector potential $\boldsymbol{A}$

$$\boldsymbol{B} = \nabla \times \boldsymbol{A} \tag{C.8}$$

We know from classical physics that the vector potential causes the electron momentum to alter via the transformation

$$\boldsymbol{p} \rightarrow \boldsymbol{p} - e\boldsymbol{A} \tag{C.9}$$

For the electron represented in quantum mechanics this corresponds to

$$\boldsymbol{k} \rightarrow \boldsymbol{k} - \frac{e\boldsymbol{A}}{\hbar} \tag{C.10}$$

The electron wavefunction in the presence of a magnetic field is now given by

$$\psi(x) = \psi^0(x) \exp\left[\frac{ie}{\hbar}\int^{S(x)} \mathbf{A}(x^{'}) \cdot d\boldsymbol{s}^{'}\right] \tag{C.11}$$

where $\psi^0(x)$ satisfies the Schrödinger equation with $\mathbf{A}(x) = 0$. The line integral in Eqn. C.11 can be along any path as long as the end point $S(x)$ is the point x and $\nabla \times \mathbf{A}$ is zero along the integral.

Let us now consider the problem described by Fig. C.2. Here a beam of coherent electrons is separated into two parts and made to recombine at an interference region. This is similar to the double-slit experiment from optics, except now we have a region of magnetic field enclosed by the electron paths as shown. The wavefunction of the electrons at the point where the two beams interfere is given by (we assume that phase coherence is maintained)

$$\psi(x) = \psi_1^0 \exp\left[\frac{ie}{\hbar}\int_{\text{path 1}}^{S(x)} \mathbf{A}(x')\cdot ds'\right] + \psi_2^0 \exp\left[\frac{ie}{\hbar}\int_{\text{path 2}}^{S(x)} \mathbf{A}(x')\cdot ds'\right] \quad \text{(C.12)}$$

The intensity or the electron density is given by

$$I(x) = \{\psi_1(x) + \psi_2(x)\}\{\psi_1(x) + \psi_2(x)\}^* \quad \text{(C.13)}$$

If we assume that $\psi_1^0 = \psi_2^0$, i.e., the initial electron beam has been divided equally along the two paths, the intensity produced after interference is

$$\begin{aligned} I(x) &\propto \cos\left[\frac{e}{\hbar}\oint \mathbf{A}\cdot ds\right] \\ &= \cos\left[\frac{e}{\hbar}\int_{\text{area}} \mathbf{B}\cdot\mathbf{n}\, da\right] \\ &= \cos\frac{e\Phi}{\hbar} \end{aligned} \quad \text{(C.14)}$$

where we have converted the line integral over the path enclosed by the electrons to a surface integral and used $\mathbf{B} = \nabla \times \mathbf{A}$. The quantity Φ is the magnetic flux enclosed by the two electron paths. It is interesting to note that, even though the electrons never pass through the $\mathbf{B} \neq 0$ region, they are still influenced by the magnetic field. From Eqn. C.14 it is clear that if the magnetic field is changed the electron density will undergo modulation. This phenomenon has been observed in semiconductor structures as well as in metallic structures.

Let us now examine the implications of our results in a superconductor. As noted earlier, in superconductors, the electrons form pairs. These pairs, called *Cooper pairs*, do not suffer collisions because of the existence of an energy gap between their energy and the energies of state into which they could scatter. We use $2e$ instead of e to describe the wavefunction of the Cooper pairs:

$$\psi(x) = \psi^0 \exp\left[\frac{2ie}{\hbar}\int^{S(x)} \mathbf{A}(x')\cdot ds'\right] \quad \text{(C.15)}$$

If we consider a superconducting ring as shown in Fig. C.3 enclosing a magnetic field region, the fact that the electron wavefunction should not be multivalued if we go around the ring gives us the condition

$$\frac{2e}{\hbar}\oint \mathbf{A}\cdot ds = 2n\pi \quad \text{(C.16)}$$

or

$$\frac{2e\Phi}{\hbar} = 2n\pi \quad \text{(C.17)}$$

The flux enclosed by the superconducting ring is thus quantized

$$\boxed{\Phi = \frac{n\pi\hbar}{e}} \quad \text{(C.18)}$$

This effect was used to confirm that the current in superconductors is carried by a pair of electrons rather than individual electrons.

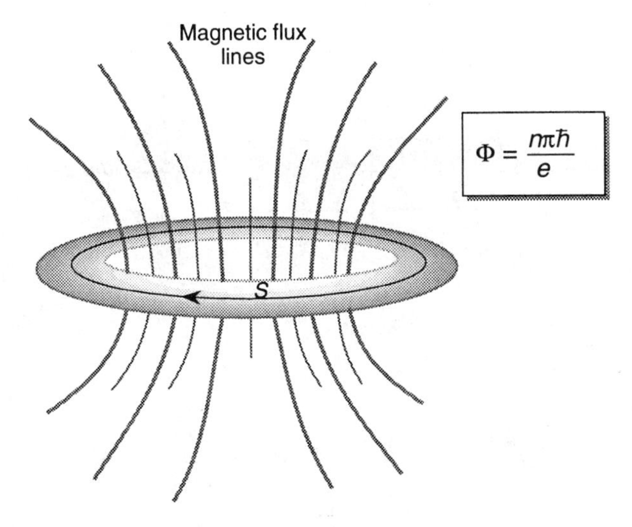

Figure C.3: Flux quantization in superconducting rings.

C.4 SUPERCONDUCTING DEVICES

In Chapter 6, Section 6.6 we discussed the current flow through a Josephson junction. The junction consists of a thin insulator sandwiched between two superconductors. A remarkable feature of the current–voltage relation of such a junction, as shown in Fig. C.4a, is that the current is given by

$$J = J_0 \sin \delta_0 \tag{C.19}$$

where J_0 is dependent upon the tunneling probability of the electron pairs through the junction. A number of very interesting devices can be designed with a superconductor loop in which two arms are formed by Josephson junctions, as shown in Fig. C.4b. These devices are based on the fact that the phase difference around a closed superconducting loop which encompasses a total magnetic flux Φ is an integral product of $2e\Phi/\hbar$.

Let us consider two Josephson junctions in parallel as shown in Fig. C.4b. The junctions enclose a magnetic flux, as shown. The total current through the loop can be shown to be

$$J_{Total} = 2J_0 \sin \delta_0 \cos \frac{e\Phi}{\hbar} \tag{C.20}$$

The current can be seen to vary with Φ and has a maxima when

$$\frac{e\Phi}{\hbar} = n\pi \tag{C.21}$$

where n is an integer.

The control of the current through a Josephson loop by a magnetic field is the basis of many important superconducting devices. It is important to note that the magnetic flux needed to alter the current through a loop is very small.

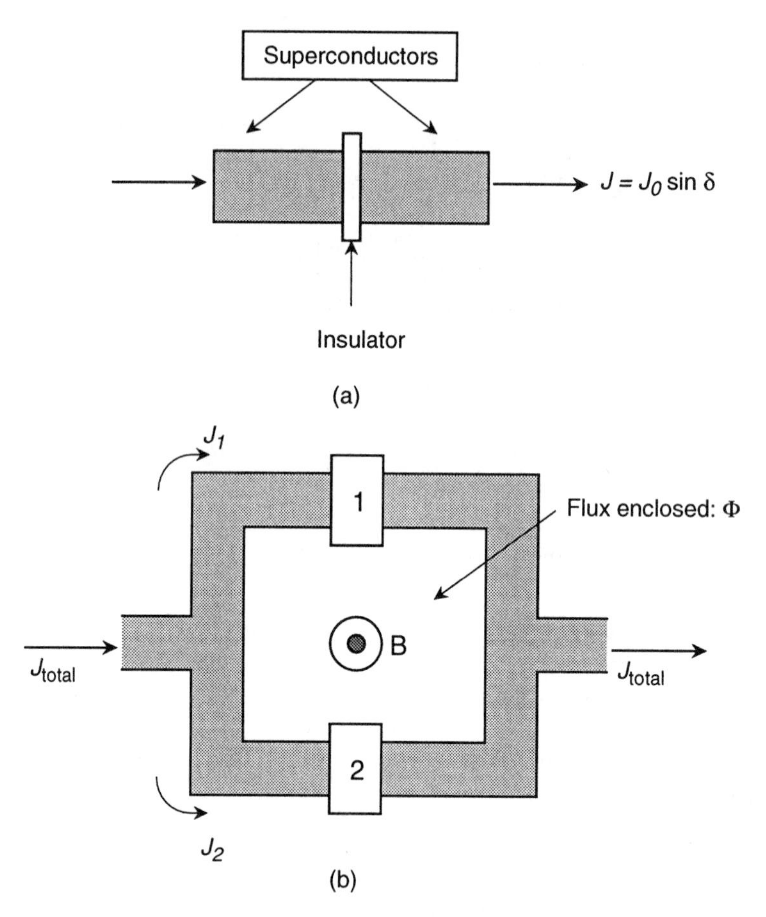

Figure C.4: (a) A schematic of a Josephson junction. (b) Quantum interference in a loop formed by two Josephson junctions.

Superconducting loops can be used in the fabrication of magnetometers, digital logic devices, signal processing devices, detectors, and for measurement standards. We will briefly discuss applications in magnetometers and digital logic devices.

SQUID Magnetometers

The superconducting quantum interference device (SQUID) is an extremely sensitive device for measuring changes in magnetic flux. Clever coupling of the device with other circuits also allows it to be used as a very sensitive voltmeter to measure tiny voltages in, say, Hall effect measurements, or as a gradiometer to measure field gradients.

The SQUID is used in either a dc or ac configuration as shown in Fig. C.5. In the dc configuration shown in Fig. C.5a, the Josephson loop encloses the flux Φ to be detected. The operation depends upon the fact that the maximum dc supercurrent as well as the I–V relations depend upon the flux Φ. This is shown

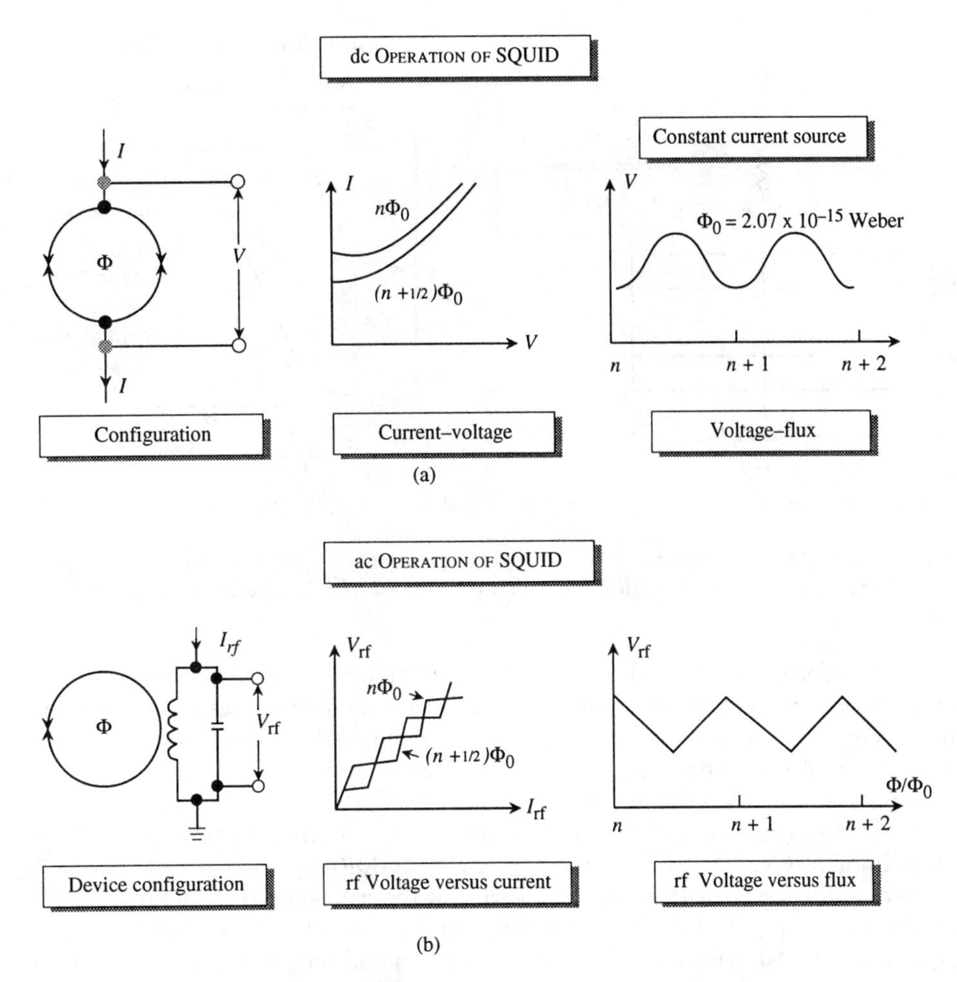

Figure C.5: A schematic of the operation of (a) dc and (b) ac SQUIDs.

schematically in Fig. C.5a. The dc device uses a constant current source in which case the voltage across the device oscillates with changes in the flux through the loop.

In the rf SQUID, the device consists of a single Josephson junction incorporated into a superconducting loop and circuit operates with an rf bias. The SQUID is coupled to the inductor of an LC circuit excited at its resonant frequency. The rf voltage across the circuit versus the rf current is shown in Fig. C.5b and oscillates with the applied flux.

Due to the extreme sensitivity of SQUID, the device (in various configurations) finds use in biomagnetism, geophysical exploration, gravitational experiments, Hall effect, magnetic monopole detection, relativity, and many other fields.

Digital Devices

An important application of Josephson tunneling junctions is in the area of micro-

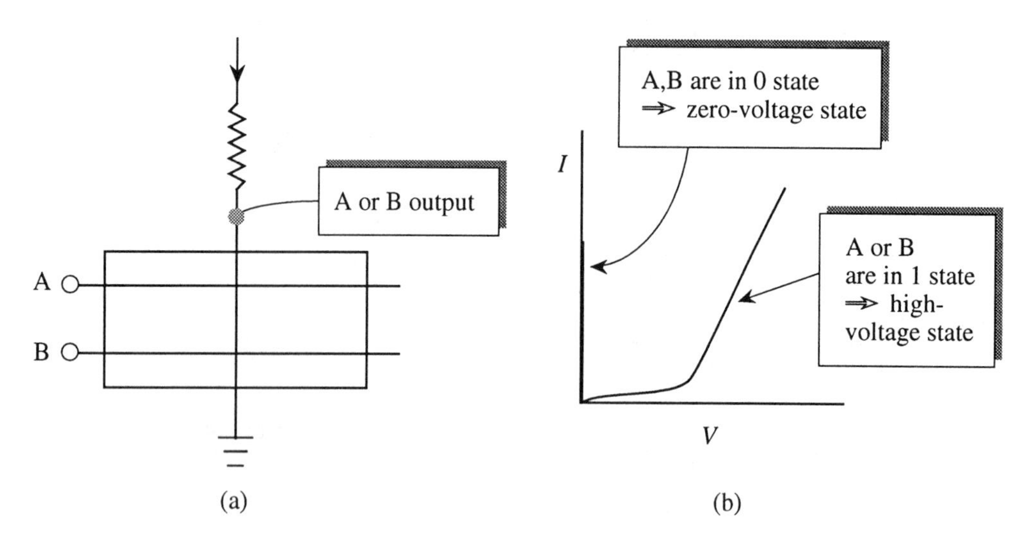

Figure C.6: Use of Josephson devices in logic implementation. (a) A schematic of a circuit to perform A or B operation; and (b) the states of the I–V characteristics of the circuit.

electronics for computers. Although it is not clear whether Josephson devices will ever compete with semiconductor devices due to manufacturing and cost concerns, in principle, the superconducting devices offer high-speed and low-power operation. The devices depend upon the use of magnetic flux (induced by a current flowing in an input) to alter the critical current that can flow in a superconducting loop.

The general operation of superconducting logic circuits can be appreciated by examining Fig. C.6a, which shows an OR gate. Initially, the device is biased in the zero voltage state with a current level below the critical current. A current pulse in either of the inputs A or B couples magnetic flux into the loop so that the critical current is reduced. The device thus switches into a non-zero voltage state, as shown in Fig. C.6b.

Index